NDE OF MICROSTRUCTURE FOR PROCESS CONTROL

Proceedings of a symposium sponsored by the Materials Testing and Quality Control Division, Nondestructive Testing Committee of the American Society for Metals at the ASM Metals Congress, Detroit, Michigan, September 18-19, 1984

Edited by

H.N.G. Wadley
National Bureau of Standards
Washington, D.C.

A Publication of

The American Society For Metals

**Cover photograph courtesy of AISI*

Library of Congress Catalog Card Number: 85-073216

ISBN: 0-87170-251-7

SAN: 204-7586

Printed in the United States of America

FOREWORD

Recently an increased awareness of the importance of advanced processing of materials has begun to emerge. On the one hand it has been recognized that advanced materials require the most stringent process control if the microstructures that bestow superior properties to laboratory samples are to be reliably attained during commercial production. On the other, it has also become clear that as scientists and engineers grapple with the design of tomorrow's factories, automation of the processing of basic materials is vital if competitiveness in the international marketplace is to be maintained or enhanced. Automation of the processing of both basic and advanced materials leads to more uniformity of properties, optimal productivity and better quality.

Advanced processing of materials can in general be viewed as three separate but interlinked problems: Measurement of key process and microstructural variables (sensors), development of predictive process models, and the application of automated control techniques. Process modeling and innovative control strategies (some featuring artificial intelligence/expert systems) have already been or are planned, to be extensively reviewed and assessed. The area of sensor development has received much less focused attention and is the area of emphasis in this symposium proceedings.

The sensors to which we refer in materials processing are not the relatively straightforward devices that measure say air temperatures or pressures: rather they are novel in-process applications of traditional NDE techniques coupled with new analysis methods for the in-process measurement of key microstructural and process variables. For example, ultrasonic techniques are being developed to locate and characterize the solid-liquid interface and to map internal temperature fields during continuous casting, and to measure grain size distributions and characterize the evolution of microporosity during sintering in powder metals/ceramics. The non-invasive measurement of a material's response to nondestructive probes, and the analysis of this to facilitate process control, are major research and development challenges often requiring multidisciplinary approaches. Since sensor needs are very numerous it would be most helpful if the generic methodologies developed for one materials processing sensor were more generally applicable to another. One objective of the symposium then was to foster the search for generic issues in nondestructive characterization for process control.

Those who have been involved with traditional NDE technologies have a unique contribution to offer to the emerging field of materials processing sensors. However, some remain unaware of this opportunity while others are, at this time, unable to identify specific sensor needs. Thus, a second objective of the meeting was to encourage a dialogue between the NDE community and those people in industry able to identify their industries respective sensor needs.

This book contains twenty-three papers presented at a symposium on Nondestructive Microstructure Characterization and Process Control. It was held as part of the 1984 ASM Metals Congress in Detroit. Three sections are contained. The first is a Metals Processing Overview where the intent was to identify the emerging Nondestructive Characterization needs of principally the metals processing community.

Section II is concerned with current research directed at non-destructive microstructure characterization. Here, work of high scientific excellence has been in progress for several years for the characterization of laboratory samples, but not for process control per se. This, however, must be considered the necessary first step in the ultimate development of in-process monitoring capabilities.

The final section is concerned with in-process application of NDE to process control problems such as reheat furnace control, laser drilling, machining, inclusion detection in molten metals etc. They are amongst the first tentative efforts to apply the resevoir of NDE expertise to automated materials processing.

In organizing this meeting, I received much help. Dr. Harry Berger, Chairman of the ASM Committee on NDE, was very supportive in our efforts to develop a symposium in this new area. The session chairmen served joint roles: those of the program committee and the running of their respective sessions. To Drs. B. Droney, M. Buckley, H. Vanderveldt, J. Danko, D. C. Rogers and H. T. Yolken, I express my gratitude for their enlightened contributions and support. The American Society for Metals organized the meeting and have promoted the publication of the proceedings, and thus, to Robert Uhl, Shari Gerstenberger and Joan Bellian, I express my sincere thanks. Finally, I would like to thank the authors, both for their clear verbal presentations at the symposium and their timely manuscripts. Their contributions have laid a new groundwork for these emerging technologies.

Haydn Wadley
National Bureau of Standards
Gaithersburg, Maryland

TABLE OF CONTENTS

III: PROCESS CONTROL

NEEDS FOR PROCESS CONTROL IN ADVANCED PROCESSING OF MATERIALS

R. Mehrabian
University of California
College of Engineering
Santa Barbara, CA 93106

Haydn N. G. Wadley
U.S. Department of Commerce
National Bureau of Standards
Washington, D.C. 20234

ABSTRACT

Advanced materials processing is currently limited by our inability to properly monitor and control the processes. "Intelligent" computer-aided process control systems are needed for advanced processing of materials to ensure reproducibility and reliability of sophisticated microstructures. Components of such systems are predictive process models relating process variables to microstructure/properties, sensors (both direct and indirect) to characterize microstructure/-physical properties during processing, and "intelligent" feedback control systems based on artifical intelligence, expert systems and control theory. Rapid solidification, continuous casting, welding, and ceramic consolidation processing are used as prototypes.

THE TRADITIONAL ROLE of nondestructive evaluation (NDE) has been the detection and sizing of flaws in completed engineering structures. Numerous techniques have evolved for this including x-ray radiography, ultrasonics, acoustic emission and eddy-currents. The application of these techniques often contends with annoying problems associated with heterogeneous phase distributions and the polycrystalline nature of engineering materials. Theories are being developed to understand these phenomena and ingenious methods are evolving to overcome their deleterious effects during the important task of flaw detection and sizing.

During this same period the materials processing community has been grappling with the problems of advanced processing of materials with optimal microstructure/-properties. New laboratory and commercial processes, e.g., rapid solidification processing (RSP), composite processing, directed high energy beam welding, superplastic forming, advanced ceramics and polymer blends processing, have evolved. Concurrently, a deeper understanding of process-microstructure-property relationship has emerged in both the new processes and established technologies such as continuous casting, rolling and other forming processes. These developments have created the possibility to produce optimal materials with more reproducible and reliable quality by cost competitive methods. On the other hand, full exploitation of both emerging and existing technologies is hampered by the complexity of controlling the processes and the resulting microstructures. The trend is toward the utilization of automated techniques using on-line measurements of process variables and evolving microstructures.

While some strides have been made in process modeling, the measurement methods and their associated analysis techniques (sensors in the nomenclature of process control) have definitely lagged behind, and for many important processes are the single obstacle to the implementation of automated materials processing. These sensors must not disturb processing and so are true nondestructive evaluation (NDE) techniques. They often rely upon precisely the same phenomena that plague the use of NDE techniques for detecting flaws. For example, the scattering of ultrasound by the grains of metals, which precludes the detection of small flaws in welds, provides a physical basis for the potential in-process measurement of grain size (a key microstructure parameter) during materials processing. It should also be noted that the lack of in-process monitoring tools has reduced the incentive for work on computer-aided process control systems specifically designed for advanced processing of materials.

The economic gains for those industries choosing to exploit this new approach to materials processing are potentially great because of improved relative productivity and quality, key factors in controlling the market place. There thus exists a strong driving force within the materials processing community to couple the fields of NDE and materials processing and develop sensors for on-line microstructure characterization and process control [1]. The knowledge that will be learned about these NDE methods, may in turn, also improve their application to more traditional inspection and in-service monitoring of finished products.

ADVANCED PROCESSING OF MATERIALS

The trend today is toward what we refer to as "intelligent" materials processing systems, Figure 1. These systems, for convenience, can be broken into three components: a process model, sensors and control. Sensors are to be the emphasis of this paper, but in order to understand their role fully it is necessary to consider briefly the two other components.

PROCESS MODELS - Process models relate process variables (temperature, pressure, time, compositon, etc.) to microstructure/-properties. Numerous models have been developed in recent years. To illustrate information they convey and the role they serve, consider a modern model used for alloy solidification, Figure 2. [2]. During solidification, the morphology of growth (i.e. plane-front, cellular, or dendritic) and the compositional changes at the solid/-liquid interface (i.e. with equilibrium, partial or no partitioning of solute atoms between the liquid and solid phases) determine the microstructural consequences of the solidification process and, hence, the process-structure-property relationships. During conventional (slow) solidification processes, compositional equilibrium (partitioning) is maintained at the moving growth front and the interfacial morphology is a function of thermal gradient (G_L) in the melt at the interface and the interfacial velocity or growth rate (R). At high values of the ratio G_L/R, left side of Figure 2, the interface advances on a planar (or smooth) front and there will be no microsegregation of solute atoms transverse to the solidification front. In constrast, at lower values of G_L/R the interface becomes morphologically unstable and leads to cellular and dendritic growth morphologies, with solute-enriched microsegregation building up in the remaining liquid between the cells or dendritic arms.

During rapid solidification, high interfacial velocities, the regime of morphological instability is capped off, as illustrated in Figure 2. Morphological stability is reestablished and plane-front ensues, because potential pertubations in the growth front are now much finer in scale and surface tension imposes a dominant stabilizing influence. The equilibrium-partitioning assumption is not realistic at the high growth rates in question and various models are available for the dependence of partitioning on interfacial velocity. In the limit, when the growth front moves at a rate in excess of the diffusive velocity of the solute atoms in the liquid no partitioning takes place and a completely homogeneous solid is produced. This range of velocities is shown as a cross-hatched bar in Figure 2. In general, as the solute content in a given alloy system is increased, the dome-shaped region in Figure 2 expands outward and planar solidification becomes more difficult.

Figure 3 shows a comparison of experiment and theory in the Ag-Cu system [3,4]. The structures were obtained by electron beam surface melting and resolidification of Ag-Cu alloys at various scanning velocities - growth rates. Here the data are plotted as a function of electron beam scan velocity and alloy composition. For dilute alloys, with less than 10 wt% Cu, they correspond to the vertical line on the right side of the dome-shaped curve in Figure 2. The calculated curve in Figure 3, denoted as the absolute stability line, is within a factor of 2 of the experimental observations. The agreement between theory and experiment is acceptable considering the uncertainty of the physical data used in the calculation. Theory for higher solute concentrations is still qualitative and requires further development. Nevertheless, present understanding of solidification is sufficiently advanced that when combined with thermodynamics (phase diagrams), kinetics and heuristic knowledge should permit formulation of conditions necessary for microsegregation - free solidification.

"INTELLIGENT CONTROL - The trend in processing is toward "intelligent" computer-aided control based on artificial intelligence (AI), expert systems and control theory. Expert systems are applied when there is a human expert and when the knowledge-base is primarily heuristic. Model-based systems are applied when rules are mathematical, as on the left side of Figure 3. In general, coupling of heuristic and model-based systems is the anticipated initial approach for "intelligent" materials process control systems. Artificial learning systems can be applied to help identify trends and to formulate predictive process models.

As an example of how a materials processing control system might be implemented, consider atomizing methods for the production of rapidly solidified metal powders. All atomization processes involve the break-up of a liquid metal stream either by direct impingement by a gas jet or the use of a mechanical device such as a rotating disc used in centrifugal atomization. Regardless of the process used, it is now clear that powder microstructure and properties of consolidated products depend on the cooling and solidification rates which are profoundly influenced by powder particle size. In addition, powder particle size distribution influences consolidation processes in important ways. It is thus desirable to monitor and control particle size and distribution during atomization using an on-line sensor, a predictive model relating process variables (melt temperature, gas flow rate, pressure, etc.) to particle size distribution and a system for automated feed back control. Figure 4 shows a hypothetical process control chart for atomization where a predetermined ideal particle size distribution is used as the desired product. The two loops are for sensing and feedback control. On the left side of the diagram a particle size sensor establishes actual size distribution for comparison with the desired distribution shown in the center box. On the right side a feedback control loop utilizes a process model and control laws to make the appropriate changes in the process variables to affect the particle size distribution. Thus, the system functions in a self correcting mode and the desired (ideal) distribution of powders is consistently produced. Finally, present atomization processes are essentially batch operated; modifications will probably be desirable to permit continuous operation when automated monitoring and control systems are utilized.

SENSORS - Sensors characterize microstructure/physical properties and measure critical process variables during processing. Few exist today but current NDE research is leading to methodologies (measurement methods, imaging, and interpretation) to characterize microstructural features such as grain size, composition variations, inclusion size and content, second phase distributions, state of cure, etc. together with process variables such as internal temperature. Sensors are also under development for in-process metrology and for detecting defects such as porosity. Sensors currently are viewed as the weak link in the "intelligent" materials processing chain.

The development of sensors is often complicated by a lack of knowledge base; the relationships between sensing mechanism/measurement and microstructure are yet to fully emerge. Furthermore, the hostile environment in which sensors are used, the limited time available for measurements, and the need to avoid interference with the processing itself all introduce constraints upon practical sensor systems that ultimately result in less than ideal data. Thus, there is a need to develop better models and algorithms of sensor-microstructure interaction, both for the analysis of collected data and to reduce the data into useful form for particular processes (5,6). It will be necessary to apply emerging high speed imaging and associated data aquisition, reconstruction, manipulation and display techniques to represent the necessary information on-line in a readily interpretable form.

Presently, sensor-microstructure interactions are very much a topic of research and have not generally reached the stage where models may be reliably used. Therefore, we need to also explore the use of what we term "indirect sensors" (sensors which cannot be used in-process) to 1) calibrate direct sensors and 2) aid the development of predictive models.

DIRECT SENSORS FOR MATERIALS PROCESSING

Without better sensor systems, there can be no real advancement in process control, and without the advancements in process control, there can be no significant advancement in the processes themselves. Numerous needs for sensors during the processing of metals, ceramics and polymers exist. Each material and process imposes its own particular set of contraints upon measurement methodology. However, important principles common to many applications exist, and these can be most easily illustrated by considering some examples from current research.

CONTINUOUS METAL SOLIDIFICATION - A schematic illustration of a continuous metal caster is shown in Figure 5. For aluminum alloys the direct chill casting process produces ingots of several meters at a time. In iron-base alloys, strand casters can operate continuously with the solidified product being fed continuously through a support roller system as shown in Figure 5. In either process, the metal is poured from a ladle into a distributor or a tundish, from which it enters a water-cooled mold where surface solidification begins. By the time the metal leaves the mold a solidified "skin" has formed, typically of the order of several centimeters; sufficient to physically constrain the molten material within. Further direct water cooling removes heat through the "skin" and facilitates complete solidification.

Numerous sensor needs exist for continuous casting. Figure 6 identifies some for continuous casting of steel. These include sensors to assure melt cleanliness, measure oxygen content, melt temperature, location of liquidus and solidus isotherms during solidification, post solidification temperature distribution etc. For the purpose of this discussion we will focus on sensor needs during the solidification process.

Solidification in engineering alloys occurs over a range of temperatures, hence finite distances. In the schematic of Figure 5 it is confined in the space between the liquidus and solidus isotherms. This region of liquid plus solid co-existence is termed the "mushy" zone and its shape, depth, and extent control both segregation of alloy elements on the small dendritic scale (micro-segregation) and on a large scale (macro-segregation) approaching the dimensions of the ingot itself. The latter type of segregation is most troublesome and can lead to significant surface-to-centerline composition variations which can not be eliminated by subsequent thermomechanical processing. In general, the depth of the liquid metal pool and the "mushy" zone is much shallower in aluminum alloys than in steels.

The productivity of a continuous caster is improved by increasing its casting speed. However, as the speed increases, the extent of the solidified "skin" is reduced and the depth of the metal pool and the "mushy" zone are increased. This can cause ruptures (breakouts) to occur requiring complete shutdown of the system or increased macro-segregation and hot tearing in the ingot. Thus, continuous casters are operated under conservative conditions to eliminate breakouts and reduce macrosegregation to an acceptable level.

A critical balance between productivity (casting speed) and quality (macrosegregation) can be achieved if an "intelligent" control system is adopted in these processes. The solidification/segregation models relating heat and fluid flow to composition variations in the metal pool and the "mushy" zone are quite well developed [7]. Combined with good sensory data on the size and shape of the "mushy" zone they could provide the basis of an "intelligent" control system. Further, a sensor capable of measuring the strength of the solidified "skin" (i.e. its thickness and temperature) could enable safe speed-up of the casting process.

Ultrasonics is currently being explored for several of the sensor needs identified in Figure 6, particularly for the detection of inclusions in the melt, location/characterization of the "mushy" zone, temperature distribution in the solidified ingot, and detection of internal discontinuities (hot tears and porosity). Consider as an example the application of ultrasonics to "mushy" zone location/shape characterization.

Both the velocity of sound and the density of alloys vary with temperature. For example, in 304 stainless steel the velocity of ultrasound varies with temperature in the manner shown in Figure 7 [8,9]. The difference in density and velocity of liquid and solid metal results in a finite reflection coefficient, about 0.1 for many metals, Figure 8. Thus, for planar solidification a medium strength ultrasonic echo, about 10% of the incident signal, is anticipated to be reflected. In conventionally solidified alloys where the temperature gradients are low, dendritic solidification is more common, and the region of reflection is distributed throughout the "mushy" zone resulting in a much less coherent reflected signal of reduced amplitude.

The detection of the reflected signal affords an opportunity to locate and perhaps characterize liquid-solid interfaces. The work of R.L.Parker [10] has demonstrated the feasibility of locating a single interface in several metals and alloys including iron, Figure 9. Little however appears to have been reported on the characterization of the "mushy" zone by detailed analysis of the reflected echo. In addition to continuous casting, the development of this sensor will lead to many other applications in solidification technology (e.g. the determination of location/shape of "mushy" zone during growth of single crystal turbine blades).

A second application of this variation of ultrasonic velocity with temperature is that of actually mapping the internal temperature distribution of solid metals, Figure 10. If the relation between velocity and temperature is known for a particular grade of steel (it varies from grade to grade) then tomographic reconstruction schemes can be devised for the recovery of the spatial variation of velocity and hence temperature. Work in progress, for example, attempts to determine average temperature within concentric rings (annular pixels) in cylinders by measuring time of flight values for an appropriate set of ultrasonic rays [11]. Initial results, obtained using noncontact laser generated ultrasound have shown encouraging success, Figure 11, although to date only a few pixels have been recovered per sample. In the future, the two evolving techniques for determination of location/shape of the "mushy" zone, and temperature distribution in the solidified shell of a casting need to be combined to maximize useful information for the control of solidification processes. In addition, changes in velocity due to temperature gradients in the solidified shell must be accounted for in interface location measurements.

WELDING AND SURFACE MODIFICATION - The melting/resolidification of a metal surface is basic to both welding and many surface modification processes employing directed high energy beams such as lasers and electron beams. Two of the critically needed sensors for these processes are on-line measurement of the liquid-solid interface position/velocity and detection of defects. These processes are much more rapid than continuous metal solidification, and so less time is available to gather information. Even so, measurement techniques are beginning to emerge for these important technologies.

Utilizing techniques similar to those of Parker, Lott et al [12] have explored the use of ultrasonics as a sensor for the closed-loop control of arc welding, Figure 12. Controlled experiments were done to determine the feasibility of measuring the instantaneous weld pool depth. It was necessary to not only detect and characterize the interface echos (from the liquid/solid, solid/air, and liquid/air interfaces), but to make two significant corrections to the time-of-flight values due to high temperature gradients. The temperature gradients changed the time-of-flight because a) the average velocity was shifted, and b) the ray paths were bent (refracted). Using a heat flow model to calculate the thermal field, and relatively simple codes to compute correction factors to time-of-flight values, Lott et al were able to deduce a linear relation between weld pool depth and modified ultrasonic transit time that demonstrated feasibility, Figure 12. Considerable problems remain before the technique may be applied in practice to moving weld pools. However, the initial studies provide encouraging results that a sensor for weld process control may ultimately be developed.

It would be a major advance if similar techniques could be applied to rapid surface solidification where surface melts, typically less than 1 mm in depth, solidify in a small fraction of a second. Under these conditions the properties of the modified layer are critically controlled by solidification interface velocity and temperature gradient. For this process, acoustic emission has been explored. While it could not characterize either variable it has shown some promise of detecting flaw growth during solidification [13]. Using electron beam pulses 8 ms in duration, surface melting and resolidification of aluminum alloys has been achieved. Typically, acoustic emission signals are observed for 10-20 ms during cooling, by which time the melt has cooled to 200°C, Figure 13. The intensity of the emission in commercial aluminum alloys is markedly enhanced when hot tearing occurs, Figure 14. This work promises the opportunity to detect defective processing enabling the area to be retreated immediately. Acoustic emission techniques are also being investigated for other (slower) solidification processes, for example, fusion welding [14].

POWDER PROCESSING - Complex metal alloy and fine ceramic powders are increasingly finding, or being identified for, applications for the manufacture of high temperature/high performance components. While metal powders are presently used in the production of engine discs in advanced jet aircraft, their full exploitation is limited by lack of process control, including assurance of powder cleanliness. In high performance ceramics the overriding problem is again the fact that components with desired microstructures/properties can not be reliably and reproducibly manufactured. Important new advances have recently been made in the production of both types of powders. In metals new processes such as centrifugal atomization permit production of more rapidly solidified powders with tighter size distribution. Direct production of unagglomerated ceramic powders are realized through precipitation from liquids, condensation from vapor reactions and crystallization of aerosols. Many sensor needs are recognized for control of powder production, rheological behavior of particulate assemblies, consolidation and microstructure evolution. The need to monitor powder particle size distribution has already been mentioned. For metals, with coarser particles in the size range of 5 to 200 µm or more, direct light scattering or optical imaging techniques are available that can be adapted to particular processes. For very fine, less than 0.1 µm, and intermediate, 0.1 to a few micrometers, sizes of ceramic particles both dynamic and direct light scattering techniques may have to be investigated.

In both metals and ceramics, the consolidation process is particularly important to monitor. On the one hand, it is important to sufficiently sinter to eliminate porosity and attain full density, yet on the other hand, it is desirable to still retain small grain and second phase size distributions. Several research groups have been exploring the use of ultrasonics to characterize pore size distribution, though not, to date, for in-process characterization purposes [15, 16].

Tittman et al [16] have initiated studies of ultrasonic characterization of micropores in IN100 powder alloy used for engine discs. Using ultrasonic backscattering techniques they have examined materials with different metallographically determined pore densities and size distributions, Figure 15. The backscattering spectra indicate much lower scattering at all frequencies for their sample C, the material with the fewest pores, Figure 16. Furthermore, theories based on scattering and diffraction are successful in predicting the backscattering behavior, thus providing the potential to ultimately deduce size distribution from measured scattering spectra.

INDIRECT SENSORS

The development of analysis methods to deduce quantitative information about microstructure, such as pore size distribution or precipitate distribution, from the scattering of ultrasound etc., could be simplified and calibrated by the use of what we term indirect sensors. These interrogate the bulk structure and give statistical estimates of the distribution of microstructure constituents. Examples include small angle neutron scattering (SANS) and synchrotron radiation. Both are finding increased application in metals, ceramics and polymers.

Neutrons are an excellent nondestructive probe of microstructure. Since thermal neutron energies are low and they interact only with the atomic nucleus, they are not significantly absorbed and penetrate deeply into most materials without disturbing them. Scattering is controlled by properties of the nucleus and varies non-systematically with atomic number so that microstructure constituents rich in elements with similar atomic number to the matrix may be readily observed (e.g., magnesium and aluminum or iron and manganese).

Hardman-Rhyne and co-workers [17] have reviewed numerous recent applications of SANS to ceramics. One of generic interest to both metals and ceramics is the characterization of the sintering process. Using the beam broadening regime they studied the densification process associated with the sintering of $MgAl_2O_4$ (spinel) as a function of temperature. Figure 17 shows the intensity of scattered neutrons as a function of position (scattering vector) on a neutron detector. Although the spinel powders used for the study had been heated at 1300°C for 12 hours, little sintering had actually occurred and considerable scattering from porosity in the 1-10 μm size was observed. Sintering at higher temperatures (e.g., 1500°C) reduced the porosity and the neutron scattering. Even the material treated at 1500°C is not fully sintered in that its full width at half maximum is significantly larger than that of the fully dense blank.

Historically, the development of SANS theory has been limited to single particle diffraction. The theory is presently being expanded to include neutron scattering dominated by multiple scattering. Further advancement of this theory coupled with the availability a cold source (such as the one planned for the National Bureau of Standards reactor in 1985) should permit quantitative analysis of pore size distribution and volume fraction during in-situ sintering experiments using large neutron wavelengths (1.5 μm).

This research highlights the opportunity for the calibration of direct nondestructive sensors such as those stemming from the work of Tittman et al [16] on porosity measurement. Related work offers the opportunity of calibrating other direct sensors, e.g., those to characterize precipitates, inclusions, microcracks, etc., whose size range is 1 nm-50 μm.

SUMMARY

Recent advances in the synthesis of new materials with complex microstructures coupled with improved understanding of process/microstucture/property relationships has created a new challenge for NDE - the redirection of a technology originally conceived for detection/characterization of flaws to the nondestructive measurement of process and microstructure variables during materials processing. This review has shown how NDE techniques could play the role of sensors in automated process control. The techniques originally developed for detecting cracks show merit as a means of monitoring solidification; other ultrasonic techniques show promise of characterizing temperature distributions and porosity. The few examples considered highlight the promise of NDE process control sensors. Problems of course remain: the need for inverse modeling, calibration, high speed data acquisition, reconstruction, and display and hardware able to survive the harsh processing environment are but a few. Nevertheless, the future looks bright for this new challenge to NDE.

For the materials processing specialist, a tremendous opportunity exists to integrate improved process understanding with advances in sensor and intelligent feedback control technologies to produce high quality products reliably and economically.

REFERENCES

1. R. Mehrabian, R.E. Whiteley, E.C. van Reuth, and H.N.G. Wadley, "Process Control Sensors for the Steel Industry," NBSIR 82-2618.
2. M. Cohen and R. Mehrabian, "Some Fundamental Aspects of Rapid Solidification," in Proceedings of Rapid Solidification Processing: Principals and Technologies III, R. Mehrabian, Ed., Washington, D.C., 1983, p. 1.
3. W.J. Boettinger, D. Shechtman, R.J. Schaefer and F.S. Biancaniello, "The Effect of Rapid Solidification Velocity on the Microstructure of Ag-Cu Alloys," Met. Trans. A, Vol. 15A, 1984, p. 55.
4. W.J. Boettinger and S.R. Coriell, "Mechanisms of Microsegregation-free Solidification," Materials Science and Engineering, Vol. 65, 1984, p. 27.
5. S.J. Norton and H.N.G. Wadley, "Tomographic Reconstruction of Internal Temperature," Review of Progress in Quantitative NDE, La Jolla, California, July 1984.
6. H.N.G. Wadley and R. Mehrabian, "Acoustic Emission Sensors for Materials Processing: A Review," Materials Science and Engineering, 65, p. 245-263.
7. S.D. Ridder, R. Mehrabian, and S. Kou, "A Review of Our Present Understanding of Macrosegregation in Axi-Symmetric Ingots," in Proceedings of Modelling of Casting and Welding Processes, D. Apelian and H. Brody, Eds., The

Metallurgical Society of AIME, Warrendale, PA, 1981, p. 261.
8. W. Kurtz, B. Lux, Huttenmannische Monatshefte 114(5), 1969, p. 123.
9. F. Mauer, C. Turner, K. Sandstorm, and H.N.G. Wadley, NBS, unpublished work.
10. R.L. Parker, in Physics in the Steel Industry, F.C. Schwerer, Ed., American Institute of Physics, New York, AIP Conference Proceedings, No. 84, 1982, p. 255.
11. H.N.G. Wadley, S.J. Norton, F.S. Biancaniello, and R. Mehrabian, "Ultrasonic Measurement of Internal Temperature Distribution," Proceedings of the Symposium on Application and Development of NDE for Use in Materials Processing, American Society for Metals, Philadelphia, PA, October 1983.
12. L.A. Lott, J.A. Johnson, and H.B. Smartt, "Real-Time Ultrasonic Sensing of Arc Welding Processes," Proceeding 1983 TMS-AIME Conference.
13. R.B. Clough, H.N.G. Wadley, and R. Mehrabian, "Experimental Studies of the Dynamics of Rapid Solidification Using Acoustic Emission," Applications of Lasers and Materials Processing, E.A. Metzbower and S.M. Copley, eds.,The American Society for Metals, 1983.
14. P.G. Bentley, D.G. Dawson, and D. Prine, RNPDE Report ND-R-767(R), 1982.
15. B.R. Patterson, K.L. Miljus, and W.V. Knopp, in Nondestructive Evaluation: Application to Materials Processing, Otto Buck and Stanley M. Wolf, Eds., American Society for Metals, 1984.
16. B.R. Tittman, L.A. Ahlberg, and K. Fertig, Review of Progress in Quantitative NDE, 3A, D.O. Thompson and D.E. Chimenti, Eds., Plenum Press, 1984, p. 57.
17. K. Hardman-Rhyne, N.G. Berk, and E.R. Fuller, Journal of Research of NBS, 89(1), 1984.

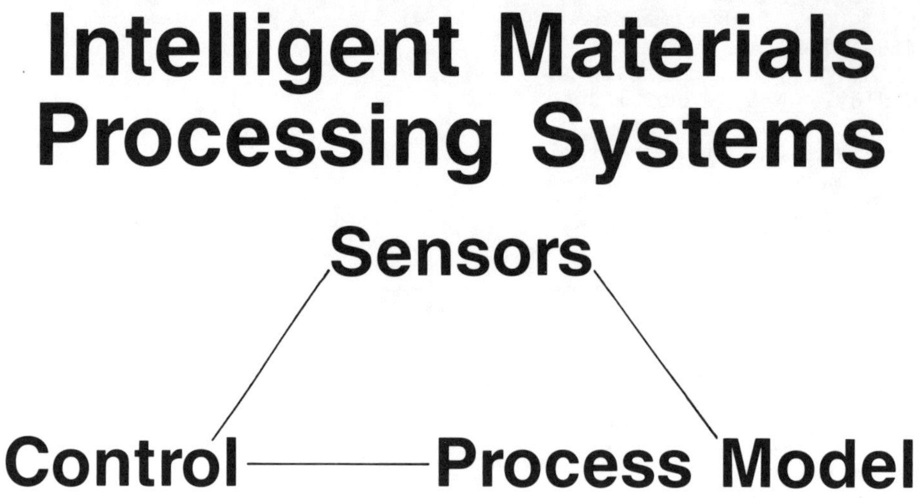

Figure 1 - The three components of an automated materials process control system.

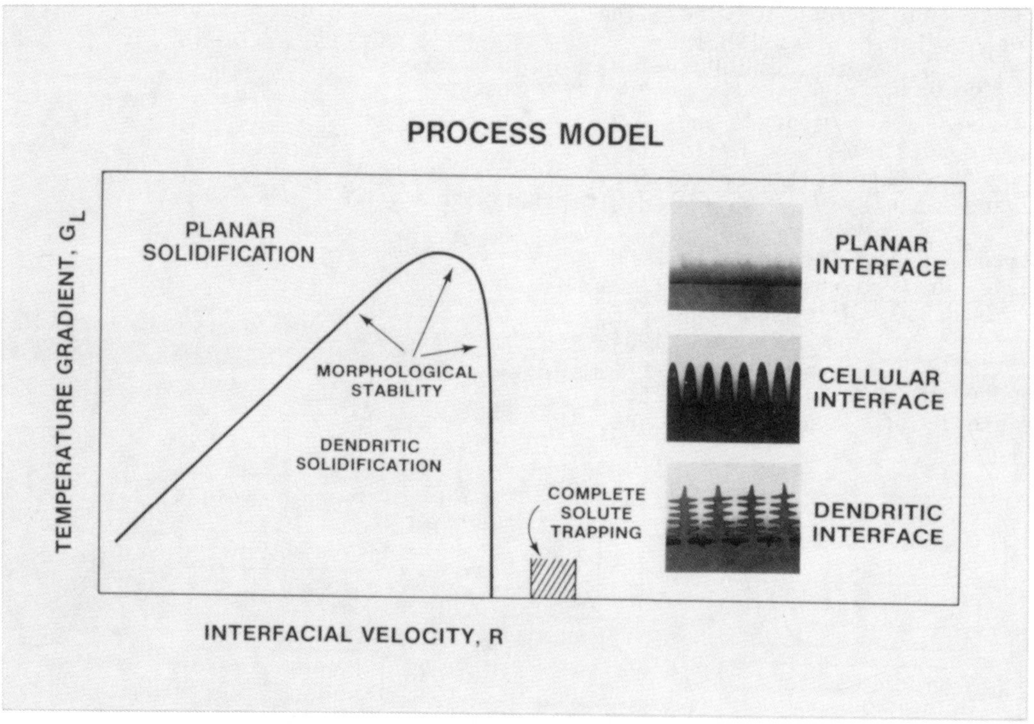

Figure 2 - Regimes of liquid temperature gradient (G_L) and interfacial velocity (R) for plane-front growth (morphologically stable interface) and for cellular/dendritic growth. The schematics on the right show the three modes of solidification normally observed. The cross-hatched band indicates interfacial velocity which is fast enough to achieve partitionless solidification.

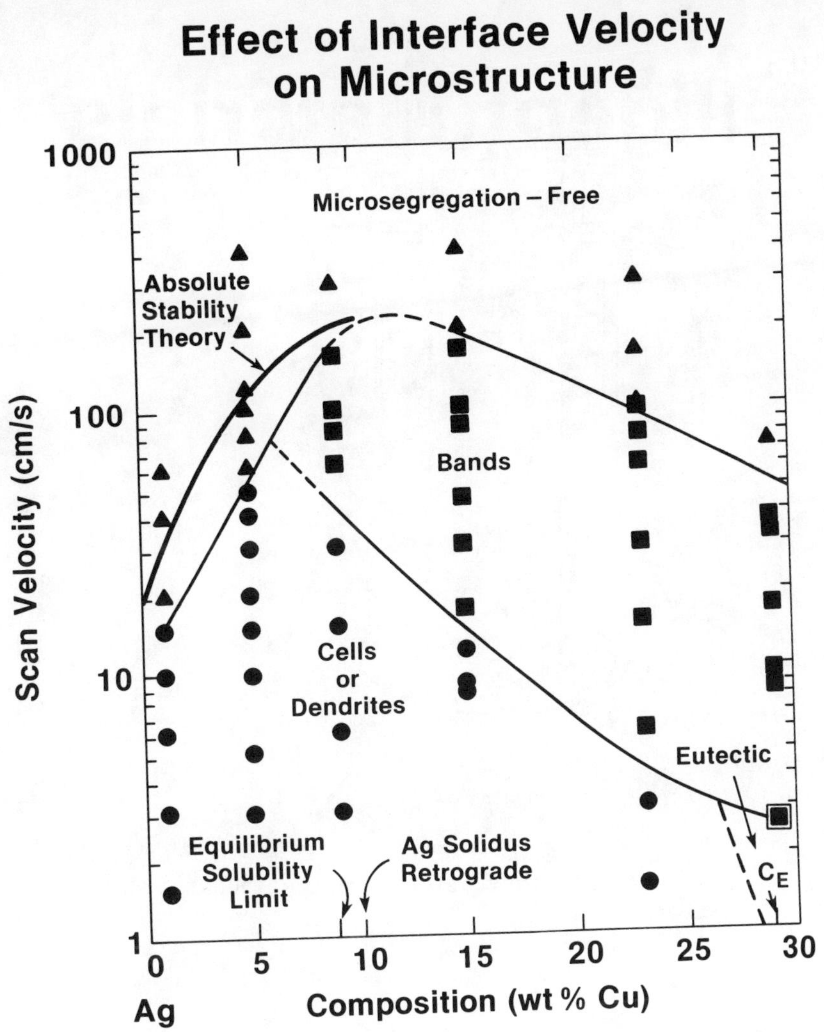

Figure 3 - A comparison of solidification theory and experiment in the Ag-Cu alloy system [3,4].

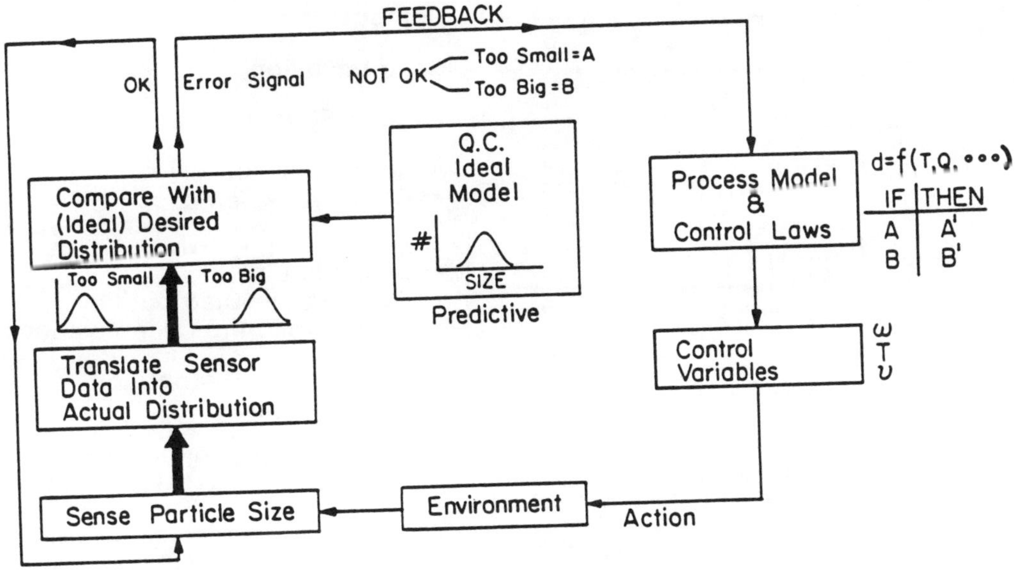

Figure 4 - Process control chart for atomization of metal powders. A predetermined particle size distribution is used as the desired product.

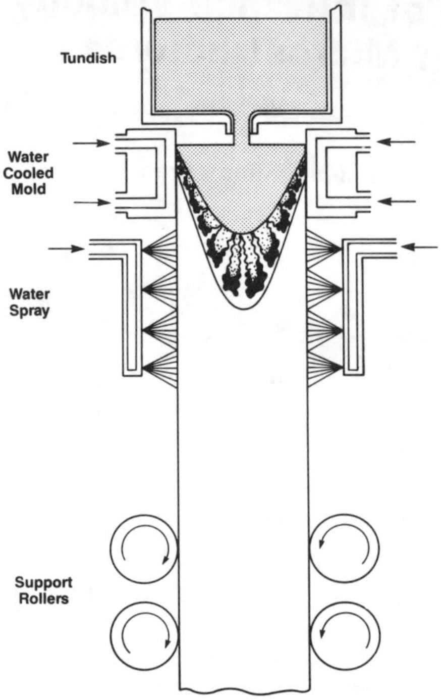

Figure 5 - A schematic illustration of a continuous metal caster.

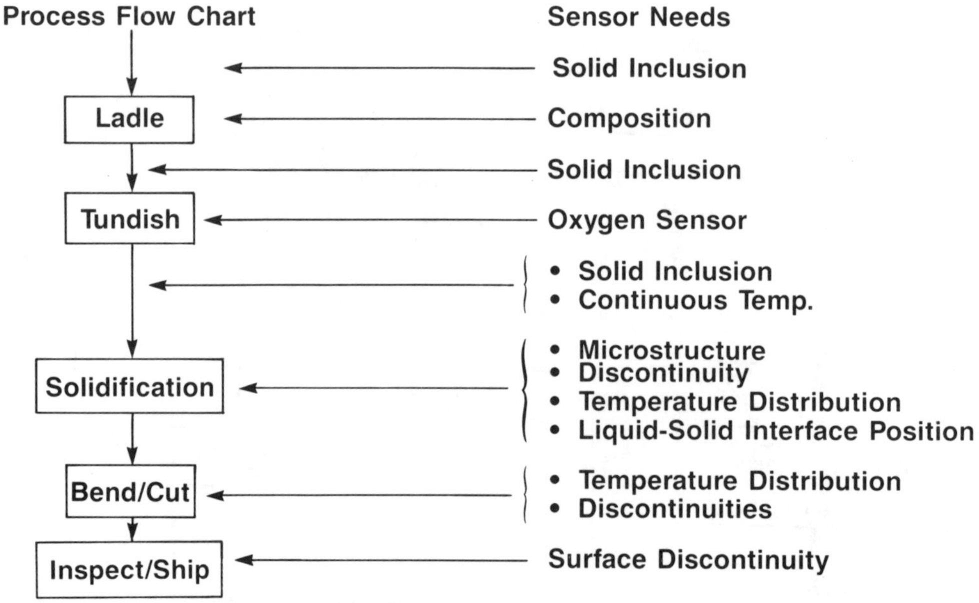

Figure 6 - Sensor needs for automated process control of continuous steel casting.

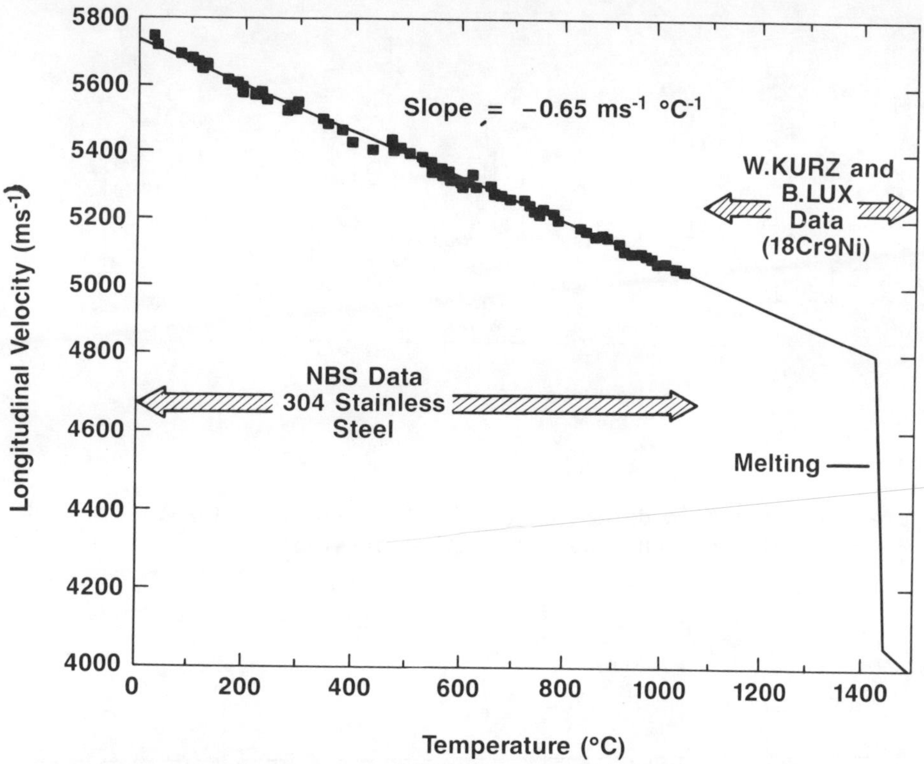

Figure 7 - The variation of longitudinal ultrasonic velocity with temperature for type 304 stainless steel.

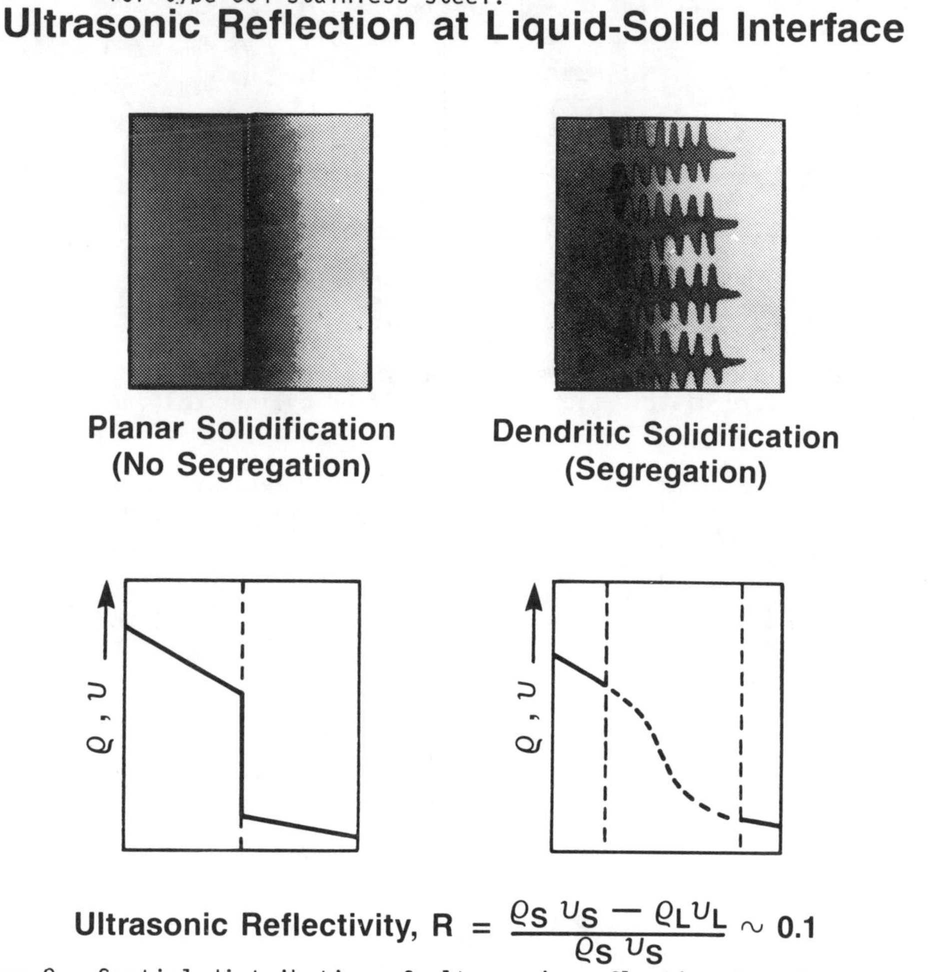

Figure 8 - Spatial distribution of ultrasonic reflection for planar and dendritic solidification.

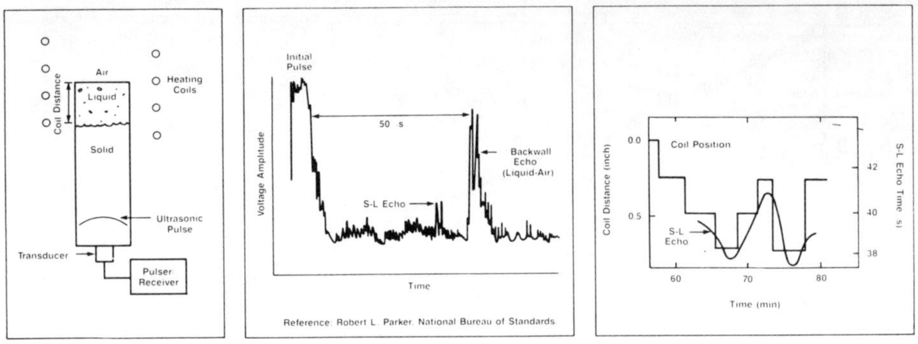

Figure 9 - Ultrasonic detection of liquid-solid interface during unidirectional solidification of steel.

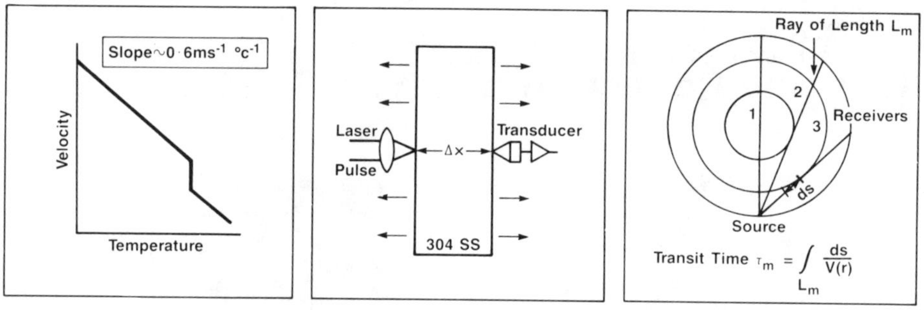

Figure 10 - Tomographic reconstruction of ultrasonic velocity/temperature distribution in a solidified cylindrical ingot.

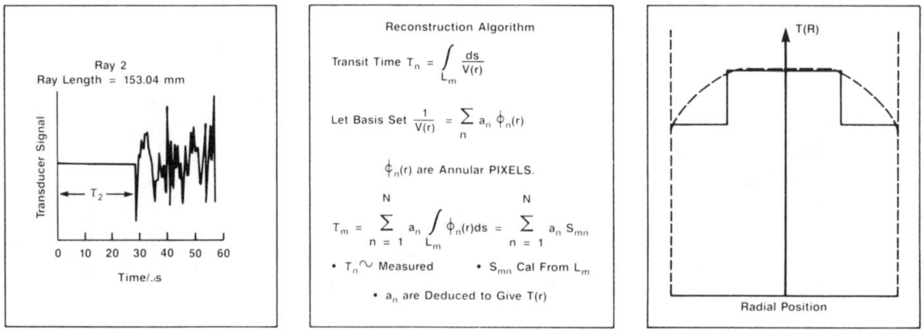

Figure 11 - Example of temperature reconstruction in a solidified 304 stainless steel ingot with an actual centerline temperature of 350°C.

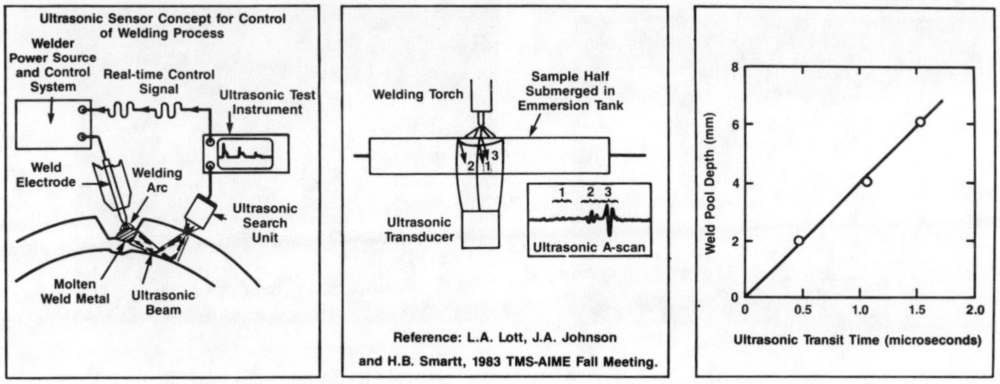

Figure 12 - Weld process monitoring and control using ultrasonic sensors. Arc-welding concept is shown on the left side; middle and right side of figure shows laboratory bead-on-plate weld configuration and correlation of weld-pool depth with time difference between echoes (12).

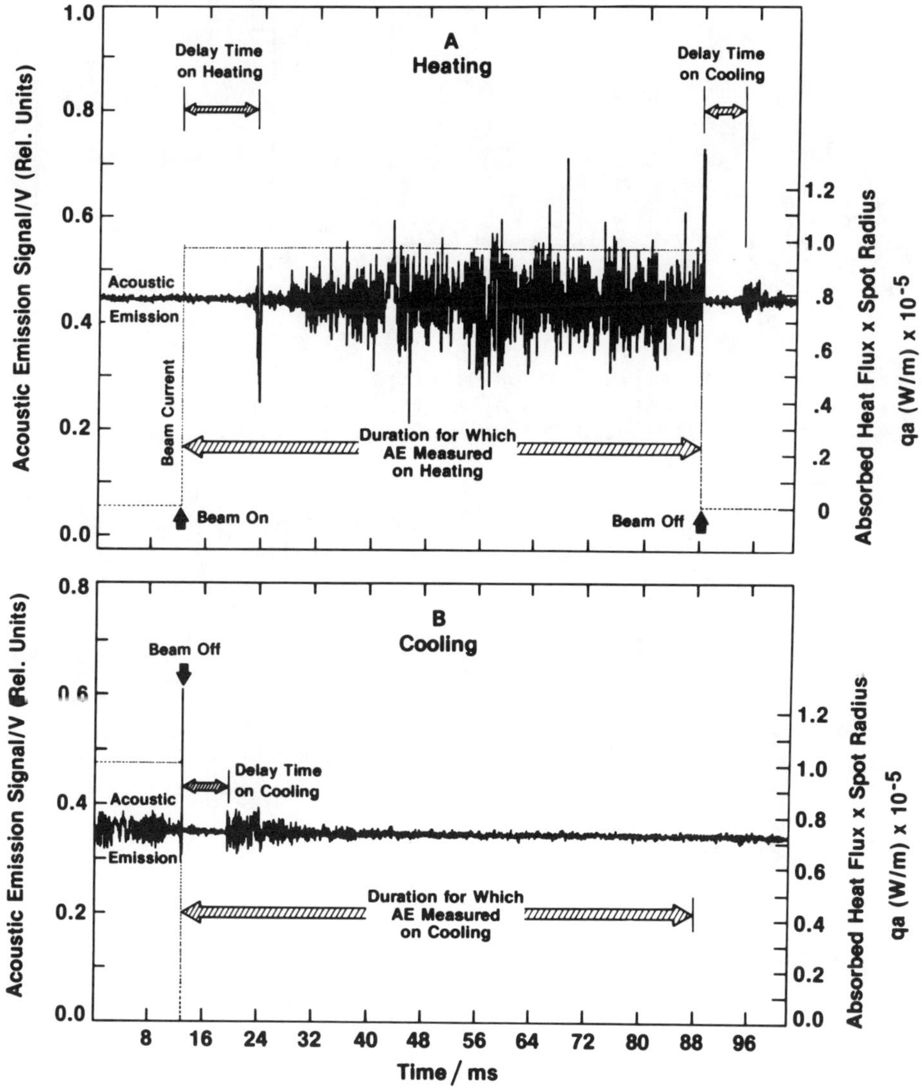

Figure 13 - Acoustic emission as a function of time for the heating and cooling phases of electron beam surface melting.

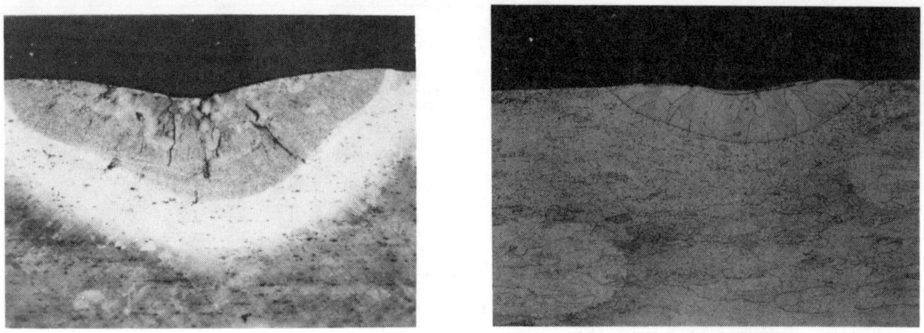

Figure 14 - Cross-sections of electron beam surface melted regions; hot tearing occurs in an Al-Cu alloy (left), no tearing occurs in pure aluminum (right).

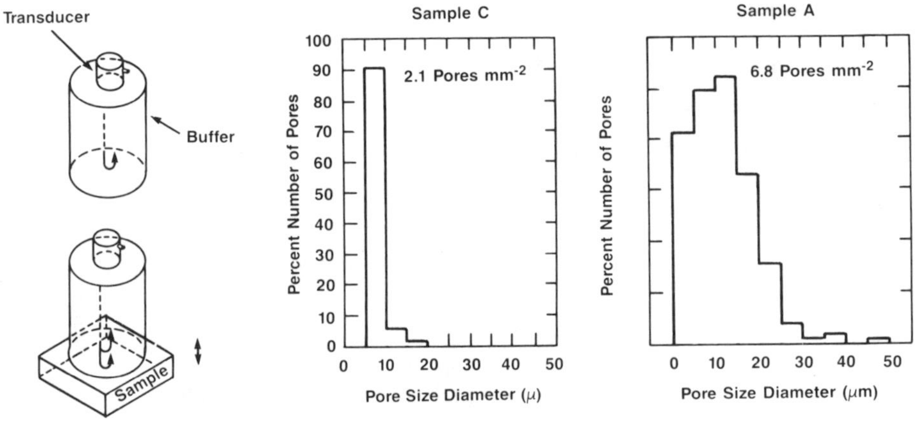

Figure 15 - Schematic illustration of ultrasonic sensor/sample configuration for characterization of sintered metal powder compacts. Pore size distributions for IN100 samples (16).

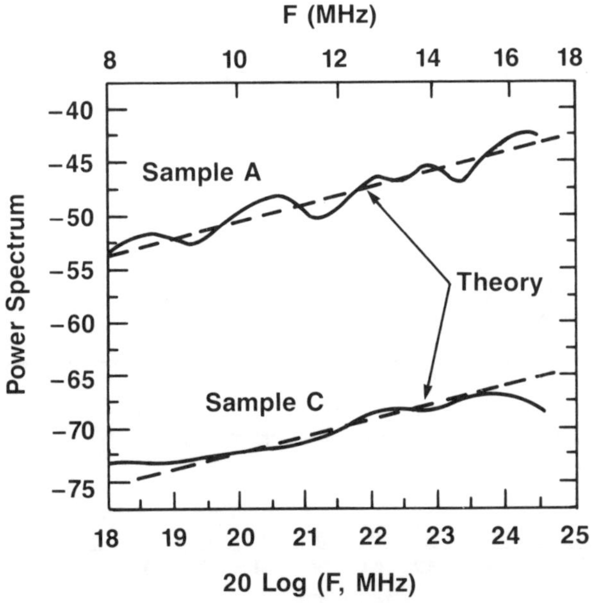

Figure 16 - Ultrasonic backscattering power spectrum. Sample C containing fewer pores scatters less ultrasound at all frequencies (16).

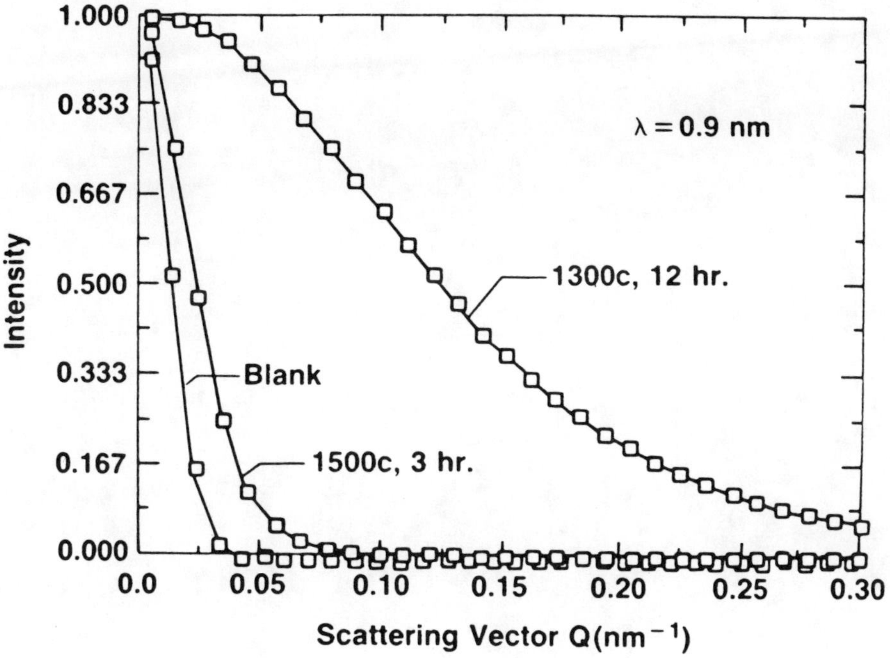

Figure 17 - Scattered neutron intensity versus scattering vector (row number on a position sensitive dector) for spinel components after various sintering treatments (17).

PROCESS CONTROL NEEDS FOR FUTURE STEELMAKING

Keith K. Kappmeyer
U.S. Steel Technical Center
Monroeville, Pennsylvania

Waldo Rall
U.S. Steel Technical Center
Monroeville, Pennsylvania

ABSTRACT

Future steelmaking will entail more continuous and higher speed operations, higher purity and more consistent quality requirements, and prompter response to demand. It will likely involve reducing and melting the ore in a high-temperature, coal-based chemical reactor, adjusting composition and achieving purity requirements in a ladle, converting the liquid steel to near net shape by high-rate continuous casting, and achieving the final shape and properties through heat-treating and forming. The earlier steps will require compositional, rheological, temperature, and pressure information; the intermediate steps, both compositional and structural information; and the final steps, structural and surface information.

IN THIS PRESENTATION I will discuss our philosophy of process control and quality management, and then I will take you through a complete future steelmaking process from iron ore to automotive sheets. I will then describe some of the specific processing steps in terms of our quality management philosophy, identifying each of the inputs to the process and outputs from the process and the measurements, both on-stream and off-line, that must be made for control of the process. I will conclude by singling out a few key nondestructive characterization needs with which to challenge you. Along the way I will have a few asides on processes that must be controlled that are less amenable to measurement and control than steelmaking.

QUALITY MANAGEMENT

Our philosophy of quality management is not unique. It goes back to the classic work and implementation programs of Deming and Juran. If your inputs and your process control are within the statistical limits you have established for the system to perform properly, then your outputs will also be within the limits you have established. These in turn will be inputs to the next processing step. So, if your internal and external inputs to the next step are within the limits prescribed, and the process is controlled within the limits established for the system, the outputs from this process will also be under control. To do this you must establish measures, have the means for measuring (that's where you come in), and establish frequencies of measuring. I want to emphasize that the system for which the statistical limits have been established includes both the equipment and the personnel operating the equipment. Thus we must continually assure that process outputs meet the inprocess product standards, that external inputs meet the material quality standards, and that the process functions according to the standard operating procedures and meets the standard quality practices. And all of these must be continually updated based on feedback from the customer and his changing needs.

I'm sure you are all familiar with this approach. You would not be developing or applying nondestructive characterization techniques if you did not recognize how essential they are to the proper control of any manufacturing process. Parenthetically, I might comment that we in manufacturing have it easy. We have tangible quantities to measure and nondestructive ways to measure them. The service industries have their quality problems, too, and the equivalent of our manufacturing processes. They can establish quality ratings for the service that is their product, but the process steps that go into their end product are tougher to measure. They deal even more with people, their attitudes and their interactions. I suppose, instead of nondestructive characterization, you would have to call the measurement needs nonintrusive or noninvasive characterization.

FUTURE STEELMAKING

Back to steelmaking. The selection of the winner among the myriad iron and steelmaking systems that have been proposed and evaluated requires a much better futurist than I. The 1971 edition of the classic MAKING, SHAPING AND TREATING OF STEEL describes 49 direct-reduction iron and steelmaking processes and a 1982 study prepared by Davis, McFarlin and Pratt at the U. S. Steel Research Laboratory, under the sponsorship of the Department of Energy, provided economic and energy analyses of 20 alternative steelmaking processes. Fortunately, for the purposes of this conference on nondestructive characterization needs, there is not a great penalty associated with picking an also-ran instead of the winner. Few characterization needs will be overlooked, if any, and probably none identified will be superfluous.

Therefore, we'll be resolute and somewhat arbitrary; we'll pick a future steelmaking process. Because of our large domestic coal reserves, we will assume coal to be the reductant and major source of energy. Because we have an intuitive belief in the greater efficiency of high temperature processes, we will select a high temperature process. And because we want to cast liquid steel close to finished shape for efficiency and lower capital cost, we will select a process that yields liquid iron. The liquid iron can then be purified and converted to liquid steel.

The KR[1)] or coal-reduction process that has been developed by Korf Engineering of Germany and VOEST Alpine of Austria and is operating on a pilot scale in Kehl, West Germany, is one of the many processes now under development that nearly fits the description just presented, Figure 1. It uses rough ground coal as the fuel, it operates at high temperature, and it produces liquid iron. The process takes place in two stages: in a reduction shaft furnace for producing direct reduced hot sponge iron, and in a pressurized melter gasifier where the iron is melted by the excess heat generated by the combustion of coal. The coal is fractured by the heat and partially combusted and gasified in an oxygen-fed fluidized bed where it gives off its heat to the iron and produces a reducing gas that is about 25% H_2, 70% CO and 5% CO_2, N_2, water vapor and impurities. This hot reducing gas, perhaps after desulfurization, is used to reduce the iron ore or pellets in the shaft furnace. The top gas or off-gas from the shaft furnace contains both sensible heat and a fuel value of about 190 Btu/scf so it becomes a valuable resource. Incidentally, the energy requirements are about the same as for the blast furnace route if you give full credit for the fuel gas produced. Such fuel credit assumes you have a good use for this low Btu gas.

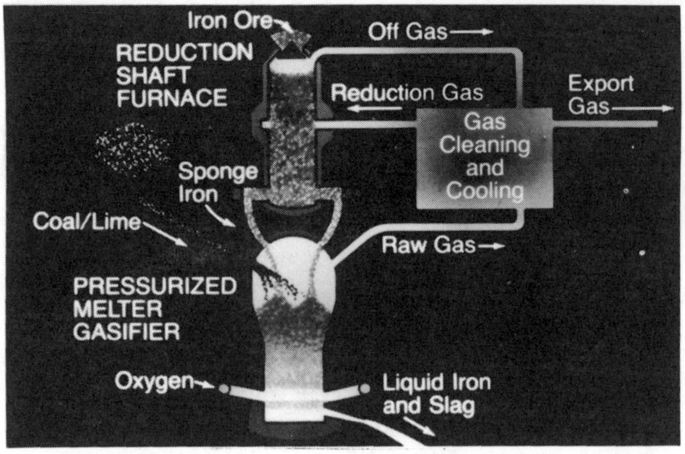

Fig. 1 – Schematic of KR process

The hot metal from this process is similar to blast furnace hot metal except that it is higher in sulfur (dependent on the degree of hot gas desulfurization) and at 1480° to 1540° Celsius (2700 to 2800° Fahrenheit) is similar in temperature to blast furnace hot metal.

The next process step is to desulfurize and dephosphorize the hot metal, Figure 2. But in order to do this we must first desiliconize the hot metal. We can remove the silicon by injecting mill scale or iron oxide that reacts exothermically with the silicon to produce molten iron and silicon dioxide. The reaction is enhanced by nitrogen stirring. We then inject lime, or another powdered flux reagent, and stir with nitrogen to remove the phosphorus and sulfur. We must then skim off the slag to prevent downstream reversion of phosphorus and sulfur from the slag.

Next we proceed to a vessel where we decarburize the low sulfur, low phosphorus hot metal to the aim carbon content by blowing with oxygen and stirring with an inert gas such as argon. This high purity liquid steel is now transferred to an argon-shrouded vessel where the steel is deoxidized and final alloy additions such as manganese and grain refiners such as aluminum are added. The shrouding is necessary to prevent reoxidation of our carefully compounded and purified liquid steel. If necessary, the liquid steel temperature is adjusted by electric heating to obtain the proper temperature for the continuous casting of that grade and composition. We may employ vacuum degassing at this point if we want a liquid steel with low concentrations of dissolved gases.

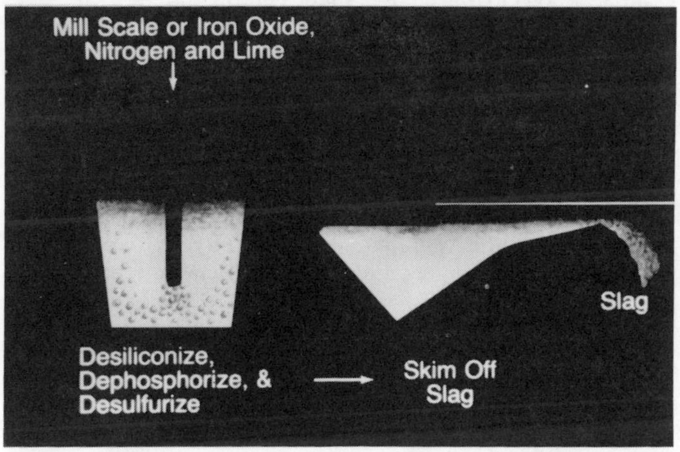

Fig. 2 – Ladle for desiliconizing, desulfurizing and dephosphorizing

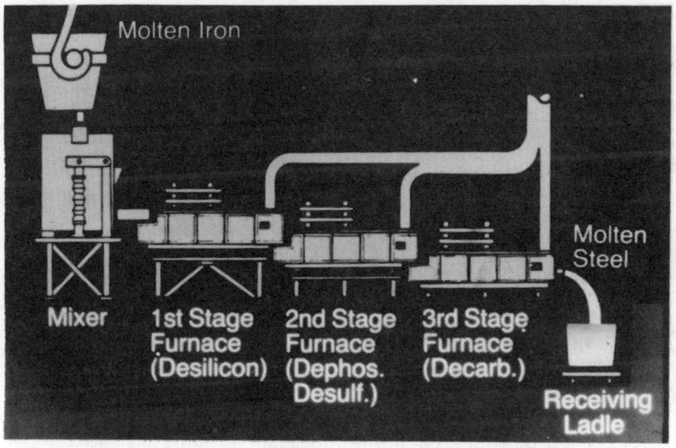

Fig. 3 – Schematic of continuous-stream cascade steelmaking

If we are clever enough in the future, these refining steps in our process might be carried out as a continuous process on a stream that passes through closely coupled vessels, Figure 3, rather than in batches, poured from one vessel to the next, as is done in our current steelmaking practices.

We now want to cast the steel continuously to as near final shape as possible. With our objective of making automotive sheet steel, we bypass the old ingot-casting, slabbing mill, hot rolling mill steps, and even the current slab casting, hot-strip mill steps in our future steelmaking process by casting full-width sections about 1 inch thick and direct rolling them to hot band in a hot-strip finishing mill or coiling them and insulating them to maintain homogeneous temperature for transfer to a hot-strip finishing mill. (Parenthetically, I will add that, in conjunction with Bethlehem Steel, Hazelett Strip Casting, Harbison-Walker Refractories and Westinghouse, we are beginning a 5-year cost sharing program with the Department of Energy to develop such thin-section casting technology.) To accomplish this thin-section casting will require precise metering of the high-purity, essentially inclusion-free, steel into a moving-mold caster, Figure 4. Although we are dealing with 100 to 200 tons per hour of liquid steel at 1600°C (2900°F), we must meter just the right amount in terms of cross section and feed rate into an aperture that may be 25 mm (1 inch) high and 1500 mm (60 inches) wide going into a moving belt casting machine at a rate of 15 meters per minute (50 feet per minute). Such a caster must operate like a precision instrument in its dimensional tolerances and rigidity requirements despite the high-cooling-rate water jets that flood the interior to provide the rapid heat extraction needed.

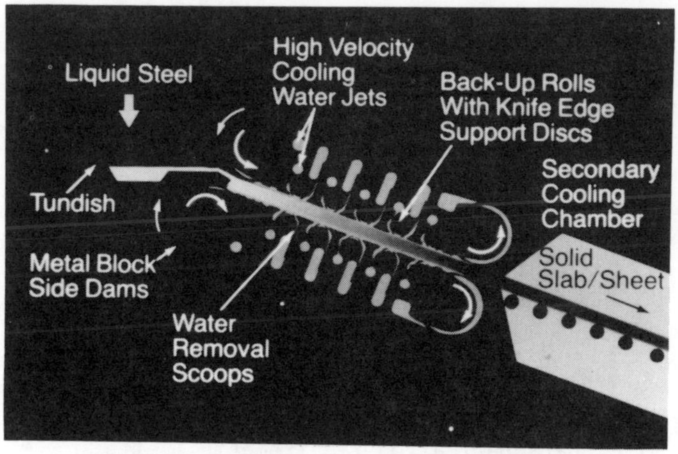

Fig. 4 – Schematic of moving-belt casting machine

The next step in our future steelmaking process is to homogenize the temperature of the cast thin section and direct roll it in a hot-strip finishing mill with controlled cooling of the strip to the proper coiling temperature.

From this stage on, our forming, heat treating, and surface treatment processes are like those now in use but they operate at higher speeds and under much tighter control. Here you see, Figure 5, a schematic of a hot finishing mill, pickling line, and cold finishing mill, all in tandem, to optimize transit time and eliminate materials handling. As you know, operations can only be placed in tandem effectively if there is a high degree of reliability to each operation between scheduled maintenance. So we must have reliability. Those of you unfamiliar with the operating speeds of such equipment may be surprised to hear that strip speeds can be typically 3500 to 5000 fpm in cold finishing operations.

Fig. 5 – Illustration of hot-finishing mill, pickling line and cold-finishing mill

Finally, we proceed to a continuous annealing line, temper mill, and surface treatment line, all operating in line and at high speed. If we are producing automotive sheet, we will probably have even one more operation -- an electrogalvanizing line. With this line we will apply a rather heavy zinc coating to one side of the sheet for corrosion protection and a lighter zinc-alloy coating to the other side to serve as a good base for painting, yet still provide good rust protection at scratched or flaked areas of the painted surface.

PROCESS CONTROL NEEDS

Now, to get back to the theme of this talk, process control needs and nondestructive characterization requirements, let's establish a format wherein we list the inputs to and outputs from a process, the control needs for the process and the measurements or characterizations that permit the inputs, outputs, and process parameters to be controlled.

The inputs to the 2-stage, high-pressure shaft-furnace, melter-gasifier process are iron ore (lumps or pellets), coal, oxygen, flux (lime and sand) and cooling water. The outputs of the process are hot metal, slag, hot fuel gas, dust, scrubber water, and desulfurizer sludge. The process control needs are input feed rates, combustion control, coolant flow and gas temperature and distribution. Thus the measurement needs are, particle size distributions for coal, ore and flux; raw material compositions for coal and ore; flow rates of the granular raw-material inputs, oxygen, sponge iron, hot metal, reducing gas, off gas, and coolant; pressures in shaft furnace and melter gasifier; pressure drop across fluidized bed; level of fluidized bed; bed temperature, shaft temperatures, and chamber wall temperatures; reducing gas composition and off-gas composition; dust loadings and hot metal and slag compositions. Because reactions are rapid at these temperatures and pressures, we would like to have continuous measurement of these variables.

For the desiliconizing, desulfurizing, and dephosphorizing stage the inputs are hot metal, mill scale, flux, and nitrogen; the outputs are slag, hot metal, fume, scrubber water, and sludge; and the process control needs are argon and oxygen flow rates, flux addition and slag removal rates, and duration of blows. The measurement needs then are hot metal composition, slag composition, gas flow, temperature, flux addition rate, and fume and dust loading. Again, these measurements, especially chemical compositions, should be continuous.

For the decarburizing vessel, the inputs are hot metal, oxygen, argon, inert gas shielding or vacuum; the outputs are liquid steel, slag and fume; the process control needs are argon and oxygen flow rates and blow durations. Measurement needs are bath temperature, continuous determination of C in liquid steel, gas flow rates, and degree of vacuum or protective atmosphere.

I won't belabor you with this detail for each processing step, but instead will refer you to the appendix. There we have listed in tabular format for each processing step, the inputs, the outputs, the process control needs, and the measurements needed to assure proper output from the process. We have further designated measurements as intermittent or continuous. Those continuous measurements are the big challenge. Without continuous measurement of liquid iron and steel composition, or at least the important elements, carbon, sulfur, silicon, phosphorus, hydrogen, oxygen, and nitrogen, we will probably not achieve truly continuous processes for iron making and steelmaking. You are setting the pace for us and we must count on your help.

Probably the most formidable measurement tasks are those for the final processing steps where the product is moving at speeds of 18 to 25 m/s (3500 to 5000 fpm or close to 60 mph). We would like to have structural information such as hardness, ductility, strength, grain size, preferred orientation, residual stress, composition gradients, and even magnetic properties. We would also like to have surface information such as a cleanliness, surface coating thickness and composition, reflectivity, surface roughness, and characterization of surface defects such as scratches, streaks from subsurface inclusions, or chatter marks.

Obtaining this information must not add significantly to cost. Because of the high surface-area rate of production, real-time scanning would require system time constants of the order of microseconds and would tax the information-handling capability of even large computers. Fortunately, going back to our quality management philosophy, real time scanning of 100 percent of the final product is probably not necessary. If we have controlled our process correctly in all previous steps, and if we properly control the final process, we must be producing material that meets the aim specifications. Statistical sampling of the final product, as a means of monitoring our input and process controls, should then be adequate.

MEASUREMENTS AND SENSORS

In general, process control and measurement needs for future steelmaking will be much more specialized than is the case now, because the entire process will be characterized by more continuous and higher speed operation, higher purity and more consistent quality requirements, and versatile and prompt response to demand with little in-process inventory. These trends will require, indeed may even be paced by, the development of sensors and instrumentation that will rapidly sense, digest, and act on large amounts of information in real time that would far surpass human capability to digest.

The development of these measurement and control systems will require more than isolated efforts. Rather, integrated and collaborative efforts will be needed that involve many disciplines across the fields of steel processing, process control, and sensor technology. I am encouraged along these lines by the process begun about two years ago, wherein steel company representatives working through the American Iron & Steel Institute Research Committee, with representatives of the Department of Energy, the National Bureau of Standards, the Defense Advanced Research Projects Agency and many instrument manufacturers, addressed the non-destructive characterization needs. This group, known as the Sensor Development Committee, after appreciable deliberation, established four development efforts with a task unit in charge of each. The four development efforts are: 1) Automatic Detection of Primary Pipe and Gross Porosity in Hot Slabs, Blooms and Billets; (2) On-Line Inspection for Surface Defects in Hot and Cold Strip; (3) Rapid In-Process Analysis of Molten Metal; and (4) The Rapid Measurement of Temperature Distribution Within a Solid or Solidifying Body of Hot Steel. You will hear more about these programs in the talk by Dr. Horvath.

Progress is being made and I am much encouraged, but I regret to say that each is still very much a challenge. If you review the list of measurement needs for each processing step in the appendix, you will find many other challenges. If you become bolder than I have been and assume that we have a continuous stream, chemical engineering type of steelmaking process in the future, you will have even more challenges. We will then have a set of cocurrent and counter current flows of liquid iron (hot metal), steel, and slags at high temperatures whose compositions and flow rates must be continuously determined.

Among the biggest challenges are those associated with the iron and steelmaking process where the temperatures are high, the environments are hostile, and the reaction rates are so rapid that quick-response sensors and fast-acting controls are prerequisites. There will also be many challenges in the thin-section casting step. Here we must deal with precise dimension, pressure, fluid flow, and temperature measurements, for automatic control systems that may need the sophistication of magneto-fluidynamic shaping and stirring and the sensitivity to respond to small changes in surface tension that affect the shape of the steel stream as it impinges on the chilled belt; that is, we must control the meniscus of the liquid steel.

COMPUTER CONTROL SYSTEM

So far, I have concentrated more on measurements as process control needs and indicated the measurement needs for each processing step. But there are other process control needs. There will be much more input for process control in the future, and the input data must be processed much more rapidly and more intelligently by the computerized control system than is current practice. The computer control system must function as an expert system with a complete model of the process in its program and a full set of responses to any of the "what-if" questions that the input data might dictate. And the expert system should learn from the responses to its control commands. It should embody advanced control algorithms with adaptation and optimization capabilities that continually update the control function.

The process control systems of the future, then, must embody new sophisticated, continuous measurement devices, new adaptive algorithms based on the development of appropriate control theory, and efficient engineering of both software and hardware to operate the process control system in real time.

I don't want to leave you with the feeling that by helping us solve our technical problems you will assure the revitalization of the steel industry. The development and implementation of technology is important and without modern technology we can't be competitive. But we must also have assurance that we can recover the cost of installing this technology that makes us competitive, and for that we need a fair market. Maintaining a fair market is a real challenge, but it is not your challenge.

Your challenge is to develop the means for measuring, on a continuous basis, the various inputs, outputs, and control functions listed in the appendix to this talk, and to develop the expert and adaptive systems that will continuously update the process models needed for process control. Those are formidable challenges. But I am sure this talented and knowledgeable audience will meet the challenge.

REFERENCE

1) Pabst, G., Steel Times 211 (7) 377-78 (1983)

APPENDIX

INPUTS, OUTPUTS, PROCESS CONTROL AND MEASUREMENTS FOR STEEL PROCESSING STEPS

I. Coal-Based Ironmaking

Inputs: ore, flux, coal, oxygen, water for cooling and gas cleaning.

Outputs: hot metal, slag, reducing gas, hot fuel gas, dust, cooling water, scrubber water, gas desulfurizer sludge.

Measurements (C = Continuous; I = Intermittent)

1. Coal to melter-gasifier unit - carbon (I), sulfur (I), volatile matter (I), and H_2O content (C), size consist (I), flow rate (C).
2. Ore to shaft reducer - chemical composition (I), size consist (I) and flow rate (C).
3. Lime to melter-gasifier unit - chemical composition (I), LOI (I), flow rate (C).
4. Oxygen to melter-gasifier unit - oxygen and nitrogen content (C), flow rate (C), pressure (C).
5. Reduced iron material from shaft reducer to melter-gasifier - chemical composition (C) and flow rate (C).
6. Feed gas to shaft reducer - gas composition (C) and temperature (C).
7. Off-gas from shaft reducer - temperature (C), pressure (C) and composition (C).
8. Off-gas from melter-gasifier unit - composition (C), amount of contained solids recycled to fluid bed (C).
9. Export off-gas - temperature (C), composition (C), flow rate (C).
10. Product pig iron (hot metal) - temperature (I), composition (I) and weight (I).
11. Product slag - weight (I), temperature (I) and composition (basicity, FeO and sulfur) (I).
12. Gas scrubber sludge - flow rate (C) and composition (I).
13. Heat losses from shaft reducer and melter-gasifier unit (C).
14. Pressure drop in fluid bed unit (C).

Process Control Criteria:

1. Temperature and composition of pig iron (hot metal).
2. Amount and composition of slag from melter-gasifier unit.
3. Flow rates of coal, ore, lime and oxygen.
4. Temperature and volume of feed gas to shaft reducer.
5. Gas pressure in shaft reducer and melter-gasifier unit.
6. Bed level in melter-gasifier unit.
7. Flow rate of reduced iron from shaft reducer to melter-gasifier unit.
8. Gas distribution in shaft reducer.
9. Cooling water flow rates.

II. Hot Metal Desiliconizing, Desulfurizing, and Dephosphorizing Station

Inputs: untreated hot metal, mill scale, flux reagent, nitrogen.

Outputs: slag, treated hot metal, fume.

Measurements (C = Continuous; I = Intermittent)

1. Untreated hot metal - weight (I), temperature (I) and composition (I).
2. Mill scale or ore concentrate - composition (C), flow rate (C), flow time (C), and particle size (I).
3. Flux reagent - composition (C), flow rate (C), flow time (C), and particle size (I).
4. Reagent carrier gas - flow rate (C), composition (C), pressure (C), and flow time (C).
5. Treated hot metal - composition (I), temperature (I) and weight (I).
6. Fume - composition (I) and weight (I).
7. Vessel refractories - thickness (I).

Process Control Criteria:

1. Reagent - flow rate and total amount.
2. Reagent carrier gas - flow rate.
3. Slag removal - before and after hot metal treatment.
4. Treated hot metal - chemical composition.

III. Decarburizing Station

Inputs: Desulfurized and dephosphorized hot metal, flux reagents, oxygen, argon/natural gas/nitrogen, cooling water for gas collection, scrap.

Outputs: Liquid "semi-steel," slag, waste gas, fume.

Measurements:

1. Desulfurized and dephosphorized hot metal - weight (I), composition (I), temperature (I).
2. Fluxes - composition (I), weight (C), flow rate (C).
3. Oxygen - flow rate (C), duration of flow (C).
4. Argon/natural gas/nitrogen (if used) - pressure (C), composition (CI), flow rate (C), duration of flow (C).
5. "Semi-steel" - weight (C), chemical composition (C), temperature (C).
6. Slag - weight (C), composition (C).
7. Waste gas - composition (C), temperature (C), flow rate (C).
8. Fume - weight (C), composition (C).
9. Refractory - condition (thickness) over time (I).
10. Bath depth in vessel (C).
11. Cooling water - temperature (C), pressure (C), flow rate (C).

Process Control Criteria:

1. Flux reagent - weight and flow rate.
2. Oxygen - flow rate and duration of flow.
3. Argon/natural gas (nitrogen) - pressure, flow rate and duration of flow.
4. Semi-steel - weight, composition and temperature.
5. Slag - weight and composition.
6. Waste gas - composition, temperature and flow rate.
7. Fume - weight.
8. Bath depth.
9. Cooling water - temperature, pressure and flow rate.

IV. <u>Grading Vessel - Composition/Temperature Adjustment</u>

Inputs: Liquid "semi-steel," alloying elements, electrical energy for heating, inert gas (argon) for shrouding, energy (steam) for vacuum generation.

Outputs: Liquid steel, fume.

Measurement Needs:

1. Liquid "semi-steel" - weight (I), composition (C), temperature (C).
2. Alloying elements - weight (C), composition (I), rate of addition (C).
3. Electrical energy - rate (C) and duration of input (C).
4. Inert gas (argon) - flow rate (C), pressure (C), duration of flow (C).
5. Composition of atmosphere (C).
6. Vacuum - energy for generation (C), (steam flow, etc.), chamber pressure (C), treatment time (C), number of steel cycles (C).
7. Liquid steel - weight (C), chemical composition (C), temperature (C).
8. Fume - weight (I), composition (I).

Process Control Criteria:

1. Alloying elements - weight and rate of addition.
2. Electrical energy - rate and duration of input.
3. Inert gas (argon) - flow rate, pressure and duration of flow.
4. Composition of atmosphere.
5. Vacuum chamber - pressure, treatment time, and number of steel cycles.
6. Liquid steel - weight, temperature and chemical composition.

V. <u>Thin Section Casting</u>

Inputs: Liquid steel, cooling water, casting fluxes, inert gas for shrouding (argon), mold lubricants.

Outputs: Coiled steel (1 inch thick) in coil box, cooling water.

Measurement Needs:

1. Liquid steel - weight (I), composition (I), temperature (C), flow rate (C), liquid level (C).
2. Casting fluxes - weight and level (C), composition (I), flow rate (C).
3. Mold lubricants - composition (I), usage rate (C), application pressure (C).
4. Inert gas (argon) - flow rate (C), pressure (C).
5. Cooling water (both in and out) - pressure (C), temperature (C), flow rate (C), make-up water additions required (volume) (C).
6. Caster - mold forces (C), roll forces (C), speed (C).
7. Coiled steel - weight (I), temperature (C), dimensions (C), cross section shape (C), surface quality (C), grain size, hot workability.

Process Control Criteria:

1. Liquid steel - weight, temperature, flow rate, liquid level.
2. Casting fluxes - level, flow rate.
3. Lubricants - usage rate, application pressure.
4. Inert gas (argon) - flow rate, pressure.
5. Cooling water - pressure, temperature, flow rate, make-up water requiements.
6. Caster - mold force, roll forces, speed.
7. Coiled steel - weight, temperature, dimensions, cross section shape, surface quality.

VI. Hot Rolling

Inputs: rolling mill finishing stand, 1-inch section steel coil, descaling and cooling water, lubricant, electrical energy, rolling schedules.

Outputs: hot band, scale, lubricant, cooling water.

Measurements:

1. Rolling stands - speed (C), interstand tension (C), roll force (C), roll gap (C), roll shape (C), temperature (C).
2. Descaling and cooling water - pressure (C), flow rate (C).
3. Lubricant - pressure (C), flow rate (C).
4. Rolls - profile (I), mechanical properties (I), wear rate (I).
5. Bearings - temperature (C), wear rate (C).
6. Steel strip - speed (C), temperature profile (C), contour (C), width (C), strip tension (C), microstructure (C), surface quality (C), coiling temperature (C), coiling tension (C).
7. Electrical motors - speed (rpm) (C), power requirements (C).

Process Control Criteria:

1. Mill stand speed.
2. Strip speed.
3. Mill stand temperature.
4. Roll temperature.
5. Roll gap.
6. Roll shape.
7. Roll cooling.
8. Roll lubrication.
9. Strip/roll friction.
10. Strip contour.
11. Strip temperature.
12. Strip width.
13. Reduction schedule.
14. Strip runout temperature and rate of cooling.
15. Table roll speed.
16. Coiling temperature.
17. Coiling tension.

VII. Pickle Line

Inputs: Steel hot band, acid, rinse water, heat.

Outputs: Scale-free steel, side trim scrap, contaminated water, spent acid.

Measurements:

1. Acid - temperature (C), concentration (C), bath level (C), flow rate (C).
2. Rinse water - temperature (C), acidity (C), bath level (C), flow rate (C).
3. Weld integrity - alignment (I), amperes (I), clamp pressure (I), weld soundness (I).
4. Trim knives - width setting (C), comparator (I).
5. Strip - speed (C), linear footage (C), width (C), surface cleanliness (C), presence of scale (C).

Process Control Criteria:

1. Weld cleanliness.
2. Weld alignment.
3. Weld amperes.
4. Weld clamp pressure.
5. Weld upset pressure.
6. Weld trim.
7. Acid temperature.
8. Acid concentration.
9. Rinse water acidity.
10. Rinse water temperature.
11. Weld tracking.
12. Edge trim integrity.
13. Strip width.
14. Strip surface cleanliness.
15. Strip speed.
16. Strip linear footage.

VIII. Accumulator

Inputs: oxide and scale-free steel strip

Outputs: oxide and scale-free steel strip

Measurements:

1. Strip speed (C).
2. Strip lineal footage (C).

Process Control Criteria;

1. Strip speed.
2. Strip lineal footage.
3. Weld tracking.

IX. Cold Reduction

Inputs: oxide and scale-free steel strip, lubricant, roll cooling medium.

Outputs: cold reduced steel strip with uniform gage, flatness, and stress level, lubricant, water.

Measurements:

1. Roll speed (C).
2. Strip speed (C).
3. Interstand tension (C).
4. Roll temperaure (C).
5. Interstand gage measurements (C).
6. Roll force (C).
7. Strip thickness (C).
8. Strip shape (C).
9. Lubricant - forward slip indicator (C).
10. Stand vibration sensor (C).
11. Roll surface texture (I).
12. Strip surface roughness (C).

Process Control Criteria:

1. Stand roll speed.
2. Interstand tension.
3. Strip speed.
4. Roll cooling.
5. Interstand strip gage.
6. Roll gap.
7. Roll crown adjustment.
8. Cooling water.
9. Lubrication.

10. Roll roughness.
11. Strip profile and surface texture.

X. Continuous Anneal

Inputs: cold reduced steel strip, surface cleaning reagent, water, heat, inert atmosphere.

Outputs: annealed steel strip, spent cleaning reagent, water.

Measurements:

1. Line speed (C).
2. Strip speed (C).
3. Strip tracking (C).
4. Zone temperatue (C).
5. Strip temperature (C).
6. Atmosphere dewpoint (C).
7. Atmosphere chemical composition (C).
8. Atmosphere pressure (C).
9. Atmosphere flow rate (C).
10. Cleaning reagent chemical composition (C).
11. Cleaning reagent temperature (C).
12. Strip shape (C).
13. Strip gage (C).
14. Strip surface roughness (C).
15. Strip mechanical properties (I).

Process Control Criteria:

1. Line speed.
2. Strip speed.
3. Strip temperature.
4. Furnace atmosphere.
5. Furnace temperature.
6. Cleaner concentration.
7. Cleaner temperature.
8. Product characteristics.

INTEGRATED MANUFACTURING TECHNOLOGY: AN ALUMINUM INDUSTRY OVERVIEW

Peter R. Bridenbaugh
Aluminum Company of America
Alcoa Center, PA

THE NATURE OF MATERIALS ENGINEERING is changing and with that change has come a demand for new materials and for traditional materials with properties clearly superior to those deemed acceptable only a few years ago.

Today, product designers are able to choose from a much broader range of materials than in the past. They desire, and properly so, to specify significantly superior levels of performance. The role of technology in meeting these demands is significant.

Today's challenge for such added-value, consistently-performing, and generally superior products requires that the scatterband of physical and mechanical properties, of dimensions, and of surface quality must be substantially reduced. To meet this challenge, the aluminum industry must have tightly controlled and properly integrated processing techniques characterized by on-line, real-time control of product characteristics and performance.

One product -- rigid container sheet -- has been chosen to illustrate the generic needs of the aluminum industry for advanced manufacturing technology. It should be pointed out, however, that the manufacture of all aluminum products follows a tortuous path, starting with mining, continuing through the production of semi-fabricated mill products and ending with the manufacture of a multitude of products from electrical wire to space shuttles. Each step in this sequence is comprised of a series of processes that must be both automated and integrated. The output of each step has a major impact on the next, and the characteristics of the final product are dependent upon the entire sequence.

The aluminum can is today the predominant container for the beer and soft drink industry. It is a rather common container. But it is also a sophisticated pressure vessel with extremely demanding material and processing requirements.

Tables I and II contain the manufacturing steps and the major property requirements for this product.

TABLE I - PROCESS STEPS FOR THE MANUFACTURE OF ALUMINUM RIGID CONTAINER SHEET

1. MELTING	6. HOT ROLLING
2. METAL TREATMENT	7. ANNEALING
3. CASTING	8. COLD ROLLING
4. SCALPING	9. FINISHING
5. PREHEATING	

TABLE II - PRODUCT REQUIREMENTS RIGID CONTAINER SHEET

GAUGE TOLERANCE

	SPECIFIED THICKNESS	TOLERANCE
END STOCK - 5182 SHEET H19 TEMPER	.01-.0136"	+.0005" -.0004"
BODY STOCK- 3004 SHEET H19 TEMPER	.011-.017"	±.0003"

GUARANTEED MINIMUM MECHANICAL PROPERTIES

	TENSILE (KSI)	YIELD (KSI)	% ELONGATION
*END STOCK - 5182 SHEET H19 TEMPER	56 MAX 51 MIN	47.5 MAX 42.5 MIN	6% IN 2"
BODY STOCK - 3004 SHEET H19 TEMPER	40	37	1

ANISOTROPY (EARING)

	MINIMUM PERCENT EARING
BODY STOCK - 3004 SHEET	6

\- - - - - - -
* AFTER HEATED FOR 20 MINUTES AT 400°F.

There are three elements of advanced manufacturing technology that are critical in the production of container sheet and are also generic to all of the steps in the manufacturing sequence of aluminum: 1) developing superior control systems for individual production centers, such as ingot casters, furnaces, rolling mills and finishing lines, 2) utilizing statistical procedures to achieve and maintain statistical control and to improve the operation of the individual production centers, and 3) knitting the total plant together to create a computer-integrated manufacturing complex.

Control system development follows an orderly sequence. It is built on process understanding. Initially, a qualitative understanding is employed to identify process variables. Quantitative understanding of these process variables is the next step. It involves two separate but related technologies; process modeling and process-metal structure relationships.

Process modeling is based on the conservation laws of continuum mechanics and results in mathematical descriptions of the processes. These models can predict various parameters -- stresses, strains and temperatures -- of both the product and the equipment. When these models are computationally correct, they can be utilized to develop real-time, computer-control algorithims of specific process and product parameters because they relate the effect of process variables to the response of both product and equipment.

Process-structure relationships utilize metallurgical understanding to control the microstructure, and therefore, the final properties of the product. Conceptually, this is the metallurgical equivalent of process modeling, but until recently, understanding has been essentially empirical. This is due to the multitude and heterogeneity of the microstructures produced. The relevant metallurgical processes, such as nucleation, diffusion, recovery and recrystallization, are sensitive to these local heterogeneities. Nevertheless, state-of-the-art technology is demonstrating that these metallurgical processes can be modeled successfully and compared against quantitative microstructural analysis. The major challenge in this area is to develop metallurgical process models that will express the kinetics of the most important microstructural features and their response to arbitrary process modifications and disturbances.

Quantitative understanding and resultant process models furnish the foundation for the construction of computer programs to simulate various process steps. These programs, in turn, make possible further testing and refinements of the model itself. Out of these further refinements emerges a control system design which can then be optimized through continued testing.

This control system development scheme permits many different variations to be tested, debugged and fine-tuned off-line, thus greatly expediting subsequent on-line implementation of the systems themselves.

Benefits achieved from technologically advanced control systems are quite substantial. In rolling mills, for example, they include more stable mill operation, higher speeds, fewer cobbles, much better gauge control, increased recovery and improved productivity. It's not surprising, that the cost/benefit ratio for such technology is quite favorable.

Today's improved mill control systems, good as they are, control only mill stability and dimensional characteristics. They do not control bulk metal properties such as strength, formability, workability and earing, nor do they control surface characteristics such as cleanliness, brightness and coating adhesion. And this brings us to our second element -- on-line data acquisition and analysis. The reason why some characteristics are controlled and others are not is largely a matter of available sensors. Present sensors are adequate to assure stable mill operation and good control of gauge and flatness but not for product and surface characteristics.

The superior method for overcoming existing sensor deficiencies lies in the combination of new sensor development with the correlation of related parameters to provide an inferential measurement of the desired product or property.

Programs are underway to develop needed sensors, such as non-contacting temperature and speed sensors and high-speed surface inspection devices. As the quantification of material-related constitutive relationships develops through progress in the process-metal structure models described earlier, the ability to parametrically represent properties such as strength, formability and earing, will make the model itself into an inferential sensor for these properties.

A third approach for process control is to always run the process exactly the same way, producing the same product all the time. But how do you know the process is consistent? The answer, lies in the application of statistical procedures to monitor both the process and product.

These procedures will not only lead to statistical control of the process, but equally important, they will also lead to increased process understanding and process improvement.

Some of the processes involved in the manufacture of container sheet are extremely complex with a multitude of variables. To effectively apply statistical analysis and control to such processes requires that both the data gathering and much of the data analysis be computer-driven. In addition, it requires critical analytical thought by personnel possessing deep process and metallurgical understanding. We believe that here is where powerful diagnostic tools will play a major role. Knowledge-based/expert systems which today embody artificial intelligence technologies will help identify and characterize process anomalies and will eventually combine quantitative process-metal structure understanding with experiential/heuristic process knowledge. Bringing the manufacture of container sheet into control is a difficult but essential endeavor and will combine statistical techniques, advanced sensor and control technologies and artificial intelligence.

The ultimate goal of process automation is computer-intergrated manufacturing wherein process control, product control and flexible manufacturing technologies are all merged into one plant-wide system which will improve productivity and product consistency while substantially reducing costs.

In the manufacture of container sheet, process control is defined as the parametric control of mill operation, statistical monitoring and automatic process diagnostics. Product control includes reduced tolerances for product characteristics such as dimensions, metallurgical properties and surface characteristics. Flexible manufacturing entails instant product optimization capabilities and a system permiting coil-to-coil tailoring. There appears to be no fundamental reason preventing the achievement of computer-integrated manufacturing. There are, of course, significant technical hurdles to be overcome.

1. We must continue to push forward our quantitative understanding of the effect of our processes on the product and hence on those properties and characteristics of the product deemed important by the users. In metals, for example, we must develop quantitative relationships between processing and metallurgical structure and between metallurgical structure and final properties and performance.

2. Today the control technology we employ in rolling is the most advanced in the world. But it must be applied to other processes and enhanced to include the new control theory concepts emerging from academia.

3. We must develop non-destructive sensors that allow us to evaluate the product we are producing and control its properties. These sensors can either measure the properties directly or provide related information that will allow us to infer the property based on our quantitative understanding. Some much needed sensors are those to provide: molten metal composition and inclusion content on the fly, metal and other fluid flow rates; non-contacting temperature measurement; and, as always, metallurgical properties in real-time and non-destructively.

4. We must apply statistical analysis techniques to our processes and then strive to understand and eliminate all sources of inconsistency. This requires the application of known technology as well as the development of new approaches.

5. The advent of high-speed, low-cost large data base number crunchers has allowed integrated manufacturing to become a gleam in our eyes. But a continued decrease in the cost of large-scale computing is required. This will undoubtedly require both advances in parallel processing as well as chip design.

6. We will never in my lifetime have precise algorithms to describe these extremely complex processes. So we will have to move to heuristic systems to interpret the data, to control the processes, and to provide the learning to advance both the degree of control and process understanding.

7. Our ability to model processes, while limited, is much more advanced than our ability to model, integrate and optimize entire systems of processes. Much must be done to advance this technology as well as to produce graduates who understand it, who can develop it and who can manage it.

The task of achieving superior products consistently is a formidable one. It will only be achieved through the cooperative efforts of scientists and engineers from a broad spectrum of disciplines. No single industry can be an island and hope to achieve integrated manufacturing. We are going to have to enter into the stream of creativity and intellect with universities, other industries and perhaps other nations to achieve our goal.

John Donne said three centuries ago, that "no man is an island, entire of itself: every man is a piece of the continent, a part of the main."

I can only speak for Alcoa ... but ... we are ready to join into a constructive dialog with anyone capable of assisting us in achieving integrated manufacturing.

APPENDIX I

Measurement and Sensor Needs in the Manufacture of Aluminum Rigid Container Sheet

Process Steps	Measurement Needs	Direct and Inferential Sensor Requirements
Melting, Metal Treatment Casting	Metal Temperature, Composition, Cleanliness Flow Rates Heat Transfer	On-Line Quantometric Analysis On-Line Sodium-Lithium Analysis Molten Metal Velocimeter Cooling Water Flow Rates
Thermal Treatment (Preheating, Annealing)	Metal Temperature Multi-Zone Temperatures	Cold-Spot Thermocouples Mathematical Models of Ingot Heat-up Characteristics
Hot Rolling Cold Rolling	Metal Dimensions • Gauge • Flatness • Profile Bulk Metal Properties • Strength • Anisotropy (earing) • Microstructure Rolling Mill Parameters • Roll Force • Roll Gap • Roll Temperature • Roll Speed Surface Characteristics • Cleanliness • Brightness • Roll Marks • Slivers	Mathematical Models of Metallurgical Processes • Nucleation • Diffusion • Recovery • Recrystallization Mathematical Models of Rolling Equipment and Product • Stress • Strain • Temperature Non-Contacting Sensors • Dimensions • Temperature • Surface
Finishing	Metal Dimensions • Gauge • Flatness • Profile Surface Characteristics • Cleanliness • Brightness	Non-Contacting Sensors • Dimensions • Surface On-Line Sensors • Oil Film Thickness • Coating Uniformity

THE IMPACT OF NEW TECHNOLOGY ON METALS PROCESSING A SYSTEMS PRODUCER'S PERSPECTIVE

Peter Cannon
Rockwell International
Science Center
1049 Camino dos Rios
1000 Oaks, CA 91360

As a major producer of aircraft, spacecraft, automotive products and electronics, Rockwell International uses a variety of processes which involve secondary processing and fabrication of metallic materials and alloy systems. These processes range from conventional ones such as machining, sheet metal forming, forging, casting, welding, and assembly, to new and emerging processes such as superplastic forming, diffusion bonding, powder metallurgy and fabrication of composite materials. Maintaining control of both the fabrication process and inspection of the parts produced is critical to integrity of the products and to success of the overall system.

The presentation will begin with a discussion of new directions in process control and advanced sensors for process control and will be followed by a survey of some areas of recent innovations in metals processing, including machining, casting, electroplating, laser welding and proessing, forming, superplastic forming, and diffusion bonding. An advanced system for inspecting diffusion bonded parts and the importance of processing and materials control from primary suppliers will conclude the talk.

Basic objectives of process control included increasing product and microstructure uniformity; automation process control to increase process uniformity and reliability, thereby eliminating human error and compensating for variabilities in incoming material. Reducing scrap loss and achieving near net shape processing are also important objectives of processing science and technology. While the processes are varied and the control systems are numerous, an appreciation of process control is gained through an examination of the individual processes.

Metal working processes can be classified broadly as those required to effect shape change and those used to improve properties. In many cases the two go hand-in-hand. Basically, shape change is produced through deformation, or melting and solidification, and through electrochemical plating. Thermal processing is used primarily for improved properties, such as strength or ductility, or in combination with shaping processes for synergism. At Rockwell, deformation and thermal processing covers approximately 70% of the materials usage, with solidification processing being utilized for about 25%. Deformation processes include machining, forging, sheet metal forming, extrusion, and rolling; while solidification processes include casting, welding, surface alloying, and crystal growth from liquid phase. Thermal processing treatments for annealing and phase transformations may be carried out using a variety of heat sources ranging from open flame to lasers and electron beams.

In the deformation processing area, the interaction between the tool and the workpiece -- the metal undergoing a shape change -- is controlled through friction and lubrication conditions of the interaction layer and has a profound influence on the uniformity of deformation in the zone of material undergoing deformation. For many years friction control has dominated metal working studies which involve machining, rolling, forging and sheet metal forming. The most important factor in deformation shape change is the geometry of the deformation zone itself, which determines the microstructure properties and the final part geometry. Since there are numerous possibilities for obtaining a final part shape from an initial geometry, the most desirable one maintains the material well below its processing limit so as to avoid internal microcracks and voids, and simultaneously minimizes scrap or trim loss. Computer analysis of metal deformation processes is being used in such optimizations and in the developments of die design and process flow. In forming process environments where in-process measurement and control are extremely difficult, computer-aided process simulations play a pivotal role in process control.

Metal response to the processing environment and finished component properties are influenced by chemical composition, microstructure, crystallographic texture, inclusions, and segregations. Control of these variables is essential at the primary metal producers end, and the effect on the success of processing and subsequent properties is important to the fabricator. In a solidification process, fluidity and porosity are controlled through melt superheat, agitation, cooling rate -- use of directional chills — and heat or applied pressure. Significant effort has been directed toward heat flow modeling and turbulence during solidification; however, the need for on-line control and sensor development in this area is much like that being felt in the deformation processing area.

No one will deny that process control is an important factor in achieving a profitable manufacturing operation, so I am not going to quote a lot of sources to prove that. However, the industry is at the threshold of a revolution in secondary metals processing; some of the aspects of that revolution can be seen clearly already, and the impact of that revolution over the next 20 years will be greater than it has been in the last 20 years. Before turning to the technology and application areas that the Science Center is involved in, those aspects of process control that are the most important at this time will be discussed.

The first side of the process coin involves all the factors that should be considered during the process itself, or the "real-time process control loop." Controlling the process precisely and in a timely fashion requires direct sensing of process parameters and affects quality directly. The sensor signals in turn require proper conditioning and processing to provide accurate measurements. The main difficulty connected with implementing real-time adaptive controls based on sensors, as found in the Science Center's Robotics Lab, is the integration of the sensors into the system. Since systems are built with several processors to avoid burdening any single processor, communication between the various processors is not standardized and represents a very difficult portion of the use of sensors in process control. Obtaining a high speed of information handling is much more important than is usually realized, because it affects the total throughput, and it limits the degree of sophistication in the feedback loop. Intelligent use of the process measurements requires the development of an accurate model so that the driving signals to the process actuators -- motors, pistons, and valves -- will be optimized. For example, in the current thrust toward automation of difficult welds on the Space Shuttle Main Engine, sensors under consideration for determining either position or weld process information include structured light, infrared detectors, acoustic emission and laser-induced time-of-flight signal analysis. However, even when the integration of sensors into the system is achieved for variations in weld seam position, the real-time modification of weld process parameters is very difficult since no agreed-upon model for the weld process exists. Finally, there must be a high degree of reliability, or fault tolerance, because breakdowns here affect the entire downstream portion of the operation. Systems such as those being developed for Autonetics Strategic Systems Division's "Factory of the Future" require extraordinary care for redundancy, supporting concepts such as those implemented by Tandem Computers.

The second and equally important side of the coin involves all the factors that should be considered to insure optimal planning of the process, to avoid bottlenecks in the critical path of production. This side becomes especially important when a significant fraction of the average batch count is made up of several different part numbers in the production line at the same time. One small-batch situation involves Space Shuttle main engine weld automation where 50% of current welds are manual; in the non-automated welds, "each inch of weld seam is different from every other inch;" thus, weld seams vary in production and during weld-processing. Another small-batch example occurs during circuit board inspection for "Peacekeeper," which involves a small number of boards (the total near 1000 for 100 systems). The concern is not so much with variation during production as with overall control for describing each board currently in production. Because this situation is common for the kind of systems Rockwell produces, it is being addressed at its most fundamental levels.

At this point, I would like to dig a little deeper and discuss some of the technological opportunities that we are excited about.

Few technologies are being discussed as widely these days as robotics and artificial intelligence. Advertised variously as the answer to the productivity problem, the knowledge explosion problem, and the potential drivers for a new level of leisure, these technologies are also among the most difficult to harness in a meaningful fashion. Some of you already may have discovered that the hard way, and may have a new meaning for the phrase "state-of-the-art." It means the machine will be down half the time; and it will be full of "bugs" when it happens to be up!

Fortunately, behind the hype associated with these topics, there are some nuggets of truth. Figure 1 shows a simplified diagram of the key technologies required for constructing intelligent machines. Current projects implementing use of Intelligent Machine development are cure-monitoring for composite materials, and weld-process control. Available support systems include general computer systems such as FORTRAN, whose development language is cumbersome for situations such as list-processing; or specially built systems, often known as LISP-MACHINES, which support programming and the use of large and varying knowledge bases.

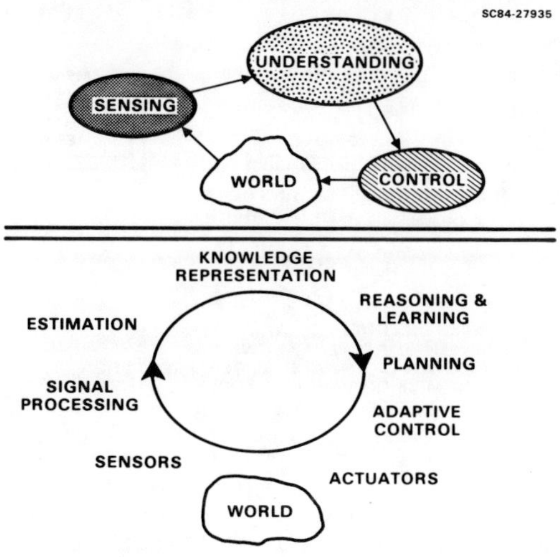

Fig. 1 - The technology of intelligent machines.

Figure 2 is a block diagram of the kind of adaptive processing system that can be implemented today, and shows the interrelationship of some of these elements. This represents a substantial advance over systems that could be built for manufacturing 10 years ago, and is economically realistic in many situations. Continuing progress is being made in learning how to implement and integrate these technologies so that they are useful to the real world. As a result, increasingly more successful demonstrations of the power of these disciplines, correctly balanced to an application, are happening.

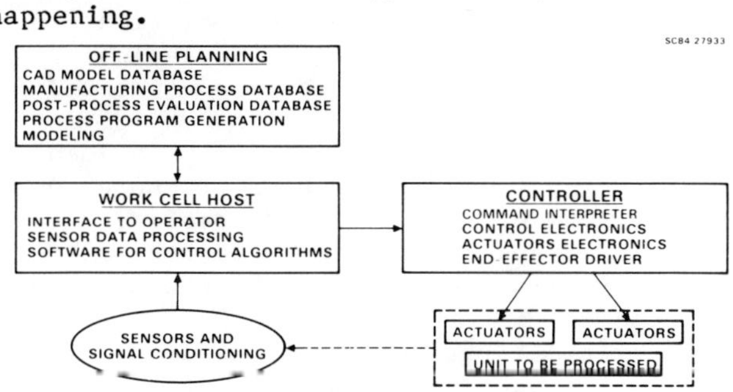

Fig. 2 - An architecture for adaptive processing.

For a large systems builder like Rockwell, the first of continuing challenges is that, unlike the popular uses of robotics such as materials handling, spot welding, and spray painting, each application requires a completely new analysis in order to make the correct tradeoffs. This yields a high degree of system uniqueness or "customization", which is made more difficult because very often there is a lack of detailed knowledge -- a model -- of the process. Therefore, it is important that integrating the technologies required for particular process control applications be done in such a way that the resulting mechanical system is still flexible. Right now, flexibility in the manufacturing process is maintained by close coordination among teams of skilled human beings who can react quickly to changes in part geometry or process specification. Any mechanical system designed to perform similar activities must be equally flexible.

For example, it is comparatively easy to build a riveting workcell that can handle one part geometry efficiently; it is an order of magnitude more difficult to build one that can handle 25 different part geometries efficiently, and still be able to easily accommodate "little things" like engineering change orders. Such a complex situation demands development of a hierarchical system, with specific tasks being accomplished by specific processors, and high speed communication existing between processors. A system of this type exists in the Science Center's Robotics Lab, where a VAX 11/780 acts as an overall host, or off-line programming station for a vision system based on multiple 68000 processors; an acoustic sensor based on an HP 9816, a single board computer for controlling real-time communications to an 11/23 based PUMA robot. All of these systems are dedicated to a particular task; it would be appropriate for a LISP-machine to serve as a decision-maker for prioritizing or ordering some of the sensor data, although that machine would not be appropriate to act as overall controller, due to its lack of real-time processing speed.

The second challenge at this point is that the more advanced tools in our computer science arsenal such as parallel processing and artificial intelligence must be incorporated into process control. Let's take a closer look at this last area.

Figure 3 is just one possible definition of artifical intelligence. Some of these techniques have been in use for some time; others are just being brought into the process control arena; and the implementation of others is much further out in time. Advanced techniques in pattern recognition and use of large knowledge bases are usually not thought of as artificial intelligence, but they are essential to successful machine intelligence. The excitement so apparent today is due mainly to the fact that perhaps for the first time in history, a machine can be programmed BY ITSELF to perform a desirable action. It can be given a set of rules, probability factors, and goals, and can actually create the detailed sequence of actions, or "program", for itself.

Figure 4 shows that not only the mechanization of mathematics must be considered, but also the mechanization of reasoning. The steps involved in building reasoning "systems" for manufacturing are just now becoming more clear to the research and development community. It is already possible to construct systems that can preserve and use the knowledge of skilled workers that have a lifetime of training in particular manufacturing processes. The knowledge of how to build such systems is spreading quickly, and has in fact become a branch of computer engineering. Therefore, we can safely

THE IMPLEMENTATION OF COMPUTATIONAL TECHNIQUES THAT HAVE THE POTENTIAL FOR EXHIBITING INTELLIGENT BEHAVIOR.

EXAMPLES OF INTELLIGENT BEHAVIOR:
- PATTERN RECOGNITION
- USE OF LARGE KNOWLEDGE BASES
- DEDUCTION AND REASONING
- USE OF "NATURAL" LANGUAGE
- LEARNING

Fig. 3 - What is "artificial intelligence?"

anticipate that, where it is economically sound, building expert systems to support manufacturing will be a standard activity of the information processing departments in our factories in the near future.

It is not so easy to predict when such reasoning capabilities will be integrated with real-time control so that a system can respond to unexpected events in a more intelligent way than just stopping the process and waiting for human intervention. A key problem is data processing speed: the number of data processing operations required to interpret an array of sensors and control a machining process, for example, is large enough that even the fastest LISP machines of today would produce unacceptable delays for satisfactory control. Other problems involve lack of I/O suitable for real-time control, and a general incompatibility with existing languages and computers. These latter problems will be solved in time and don't present the substantial barrier to implementting reasoning systems that can manufacture that basic data processing speed does.

Perhaps the most substantial barrier to use of automation and artificial intelligence techniques is that of understanding: in most cases, we lack good solid models of the process itself. The result is that we are forced to rely on phenomenology and heuristics, and these are extremely time-consuming ways to procure an advancement in technology.

Of course, some "travelers' advisories" are in order with respect to the entire area of robotics and artificial intelligence. In many cases the promise of individual technologies has not been realized in practice. The causes of disappointment are related mainly to the lack of reliability of these sophisticated machines, and the fact that methods that currently lie in the realm of "art" must be brought into the realm of engineering. To surmount the problems mentioned, those of us responsible for increasing productivity will have to exert a certain amount of courage. But despite the problems and apparent barriers, the promise of these concepts can only be termed extraordinary, and the rewards of our efforts will be much greater than the costs.

A recognized need for insuring the reproducibility and reliability of sophisticated microstructures is the development of "intel-

SC84-27934

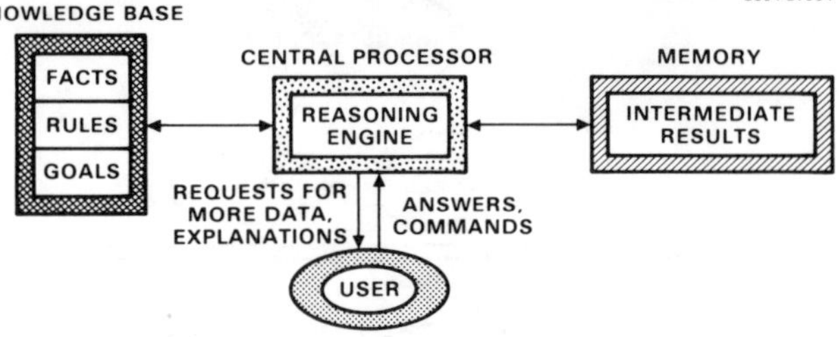

KEY ISSUES INCLUDE:
- EXISTENCE OF EXPERT KNOWLEDGE
- EFFICIENCY OF KNOWLEDGE REPRESENTATION
- ECONOMICS OF AVAILABLE TOOLS ($, SPEED)

Fig. 4 - Expert systems methods example.

ligent" computer-aided process control systems. Basic components of an "intelligent" materials process system include predictive process models relating process variables to microstructure/properties; sensors to characterize microstructures/physical properties during processing; intelligent feed-back control systems based on artificial intelligence (AI)/expert systems; and control theory.

Perhaps the weakest link in the "intelligent" processing of materials is the sensor technology. NDE techniques currently exist that in ideal conditions are capable of characterizing microstructures and process variables, such as temperature distribution. Some of this information has been processed successfully by computers to provide 3-dimensional displays. Although both the experimental and analytical tools still need considerable progress to provide quantitative information in real time, the difficult challenges lie in the adaptation of sensors to the complex geometries encountered in real parts, and the hostile environment of in-situ sensing of material consolidation. Such factors as high temperature, high pressure, corrosive environment, mechanical vibration, electrical interference, and non-stationary samples can turn a viable sensor approach into a meaningless one.

The critical role that sensors play in the "intelligent" control of materials processes is indicated schematically in Fig. 5 for the heating of a slab of alloy in a natural gas reheat furnace.[1] Here sensors must measure the temperature distribution during the heat treatment, the flow of gas, the furnace temperature, and oxygen/methane ratios. Irregularities in temperature profiles must be reported to a process supervisor; that is, an expert system which deduces the probable cause of the deviation from an "ideal" profile and issues commands to controllers for the necessary adjustments, for example the oxygen/methane ratios.

As shown in Fig. 6, sensor technology may be divided into two thrusts — the development of direct and indirect sensors. Direct sensors are those developed specifically to sense the process variables in real time and in-situ. These sensors are likely to be simple, rugged

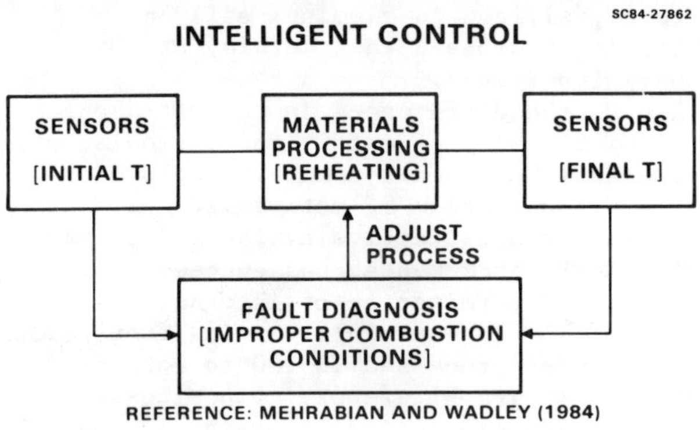

Fig. 5 - Intelligent control schematic.

TWO TYPES OF SENSORS

DIRECT

SENSORS WHICH CAN BE USED TO CONTROL PROCESSING DIRECTLY

INDIRECT

SENSORS WHICH CANNOT BE USED DIRECTLY FOR MATERIALS PROCESSING, BUT SERVE TO 1) CALIBRATE DIRECT SENSORS AND 2) AID DEVELOPMENT OF PREDICTIVE MODELS (e.g. SANS).

REFERENCE: MEHRABIAN AND WADLEY (1984)

Fig. 6 - Two types of sensors.

and miniaturized components capable of sending limited information to the process supervisor -- the expert system -- for inversion and interpretation. The development of direct sensors must go hand-in-hand with the development of indirect sensors, which would not be used directly for the process monitoring but for the calibration of the direct sensors and the development of predictive models.

One example chosen by Rockwell International Science Center for a more detailed treatment of the development of direct and indirect sensors is the problem of determining the porosity during powder consolidation. Applications to consolidation of ceramics and powder metal alloys are two cases of significant current interest. In collaboration with the National Bureau of Standards, for this application ultrasonic NDE techniques are being used for direct sensors and Small Angle Neutron Scattering (SANS) for the indirect sensor.

Machining, the most utilized manufacturing process for making finished parts, causes gradual wear, chipping, or catastrophic failure of the cutting tools. Generally this leads to surface finish deterioration and dimensional tolerance loss which may result in rework or scrapping and a consequent loss in productivity. Tool failure can also damage the machine tool itself. The value of efficient machining operations to Rockwell International can be illustrated by the B-1B program where machining constitutes about 25% of the manufacturing effort. Ongoing machining research at the Science Center is aimed at increasing the metal removal rate through improved tool materials and better utilization of the machine tools being made possible by on-line monitoring and control of the process.

Knowing when to change tools is particularly critical in unattended machining operations. Presently, this knowledge is based on empirical tool life data which, due to its scatter, is generally used conservatively to avoid premature tool failure. Thus, tools are not fully utilized. Inconsistent properties of the work and tool materials or other random events such as chatter can lead to unexpected catastrophic tool failure. Therefore, the major missing link needed to achieve fully automated machining is sensor systems capable of on-line detection of incipient tool failure and gradual wear.

Historically many direct and indirect sensor systems have been tried. Table I shows these and additional methods, some of which have worked successfully in the laboratory. So far, however, none works reliably in the shop floor machining environment.

Table I - On-Line Tool Wear Monitoring Techniques in Machining

DIRECT METHODS

- OPTICAL
  - REFLECTED LIGHT
  - FIBER OPTICS
  - LASER INTERFEROMETRY
  - TV CAMERA
- ELECTRICAL RESISTANCE
  - PRINTED
  - METAL FILM
- RADIOACTIVE
  - VOLUMETRIC
  - MICRO-ISOTOPE
- PNEUMATIC
  - AIRGAUGE

INDIRECT METHODS

- CUTTING FORCES AND TORQUE
  - STRAIN GAUGE
  - PIEZOELECTRIC
  - MAGNETIC
- VIBRATIONS
  - ACCELEROMETER
- ACOUSTIC EMISSION
  - PIEZOELECTRIC
  - ELECTROMAGNETIC/ACOUSTIC (EMAT)
  - CAPASITIVE
  - LASER

INDIRECT METHODS (CONT'D)

- SURFACE FINISH
  - LASER
  - FIBER OPTICS
  - HOLOGRAPHY
  - PHOTOCELL
  - PHOTO DIODE ARRAY
  - CONTACT GAUGES
- WORKPIECE DIMENSIONS
  - CONTACT
  - NON-CONTACT
    - PNEUMATIC
    - ELECTRICAL
    - LASER
- TOOL POST-WORKPIECE DISTANCE
  - ELECTRIC FEELER MICROMETER
  - ULTRASONICS
  - PNEUMATIC
  - LASER
- TEMPERATURE AND THERMOELECTRIC
  - TOOL-WORKPIECE THERMOCOUPLE
  - IMBEDDED THERMOCOUPLE
  - INFRARED
- ENERGY INPUT
  - ELECTRICAL POWER
  - POWER SPECTRUM
  - CURRENT

The high sensitivity and fast response time of acoustic emission appears to be the most promising. Figure 7 shows how easily tool chipping and fracturing can be distinguished from gradual wear by acoustic emission. Such systems are now being developed at Rockwell International Science Center and elsewhere.

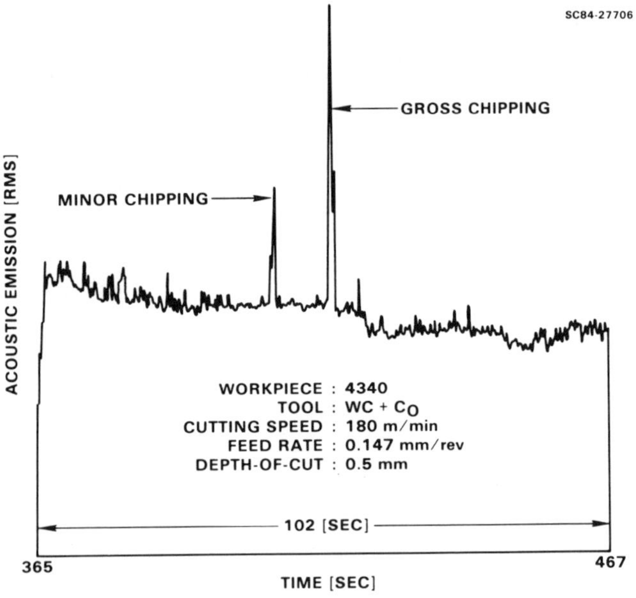

Fig. 7 - Tool chipping and cracking as distinguished from acoustic emission gradual wear.

Another area of renewed effort to achieve increased productivity is adaptive control of machine tools. Adaptive control for optimization and adaptive control with constraints are the two most common approaches to online adjustment of the machining parameters to obtain optimum and/or maximum metal removal rates without breaking the tools or overloading the machine tool. Once again, although systems have been demonstrated in the laboratory, the main impediments to industry acceptance and use are reliable, accurate online measurements which cover a wide range of machining parameters and inadequate control algorithms for the extremely complex cutting process.

To a large extent, no doubt machining of the future will be unmanned and will involve several levels of computer control, including positioning and velocity; adaptive control with constraints of cutting forces and torque; tool wear sensing; dimensional corrections; chatter control; automatic tool changing; and parts handling and inspection.

Cast materials often fail in service applications as a result of internal microdiscontinuities that are alien to the parent matrix. Such microdiscontinuities may be viewed as the weakest link in an otherwise homogeneous matrix and they may assume several forms. Microshrinkage cavities and/or gas porosities are typical initiators of failure of prime importance while morphological variation in second phase particles (e.g., silicon in aluminum-silicon or graphite in iron-based cast metals) are of secondary importance.

Due to the differences in the ultrasonic scattering response of matrix and microdiscontinuity, elastic waves are commonly employed in post facto examination of materials. With its expertise in nondestructive evaluation of materials the Science Center has succeeded in quantitative determination of internal microdiscontinuities in materials ranging from porosities in powder processed IN 100 to entrapped air bubbles in graphite/epoxy composites. Quantitative determination of grain size as well as carburized case thickness in steel has also been demonstrated.

Generally, microdiscontinuities in materials occur during early stages of manufacturing. Since Rockwell International is a manufacturing and process oriented corporation, it is to our advantage to be capable of monitoring the in situ development of porosity and alien particles and to control the microstructure of materials through use of elastic waves. Preliminary studies of the in situ use of high frequency, low intensity elastic waves for the early stages of process monitoring, and low frequency high intensity elastic waves to modify and control microstructures of cast aluminum-silicon alloys have been conducted. Some of these results follow.

The mechanical strength of aluminum-silicon castings, as well as machinability are enhanced when internal void segregation effects, grain coarsening, and dendrite arm spacings (DAS) are decreased. Voids formed by the exsolution and evaporation of hydrogen gas can usually be eliminated by chlorination of the melt. Grain refinement can be achieved using inoculants such as sodium or strontium. Rapid cooling using mold chills is also effective in producing fine-grained structures. However, there are still some internal porosities such as gas bubbles and microshrinkage cavities. There is also evidence for additional enhancement, both in mechanical properties and in fidelity of castings form, through the utilization of sound waves to modify microstructures of the silicon eutectic and primary aluminum morphology. While observations on the effect of vibrations or solidified textures date back to the 1950's,[2-8] very little research has been accomplished since that time and the mechanisms of grain refinement by sound waves are still not understood completely. Mechanisms involving early stages of nucleation and others advocating mechanical breaking of dendrites have been proposed. Preliminary research and experiments conducted at the Science Center have demonstrated that the application of elastic waves during the solidification of A 356 cast aluminum results in decreased grain size and dendrite arm spacing as well as a significant modification of the eutectic microstructure under certain conditions.

A series of experiments was conducted to investigate the influence of high intensity elastic waves on the solidification microstructure of cast aluminum-silicon alloys. Two mold configurations were employed; the first was a simplistic graphite-walled cylindrical mold while the second was a silica investment casting mold exemplar of production quality. In both series of experiments, high intensity elastic waves were generated using a commercially available 20 kHz system.

Following solidification of the aluminum-silicon alloy at a rate of approximately 10°C/min, representative samples were sectioned, polished and etched. In general, there were significant changes in microstructures in the samples insonified with high intensity compressional waves compared with the uninsonified samples.

Figures 8a and 8b are photomicrographs of the laboratory melts. The insonofied sample (8b) shows significantly finer DAS and a significant decrease in the mean free eutectic path compared with the uninsonified sample (8a). Figures 8c and 8e are photomicrographs of production cast, uninsonified samples while Figs. 8d and 8f show the corresponding microstructures of uninsonified samples. There is an obvious decrease in DAS, a significant reduction in dendritic texture, and a marked morphological change in the aluminum-silicon eutectic in the uninsonified samples. It should also be noted that Figs. 8c and 8d are micrographs of samples removed near the ultrasonic source while Figs. 8e and 8f are from samples approximately 8 in. (200 mm) from the source. Comparison of the uninsonified samples, near and far from the source, illustrates a potential problem with the use of a single elastic wave generator for microstructural modification of complex, production type castings. Far from the source, significant attenuation occurs causing a mixture of a refined and unrefined microstructure (Fig. 8f).

Subsequent experiments have reconfirmed our initial findings. We have also determined that continuous ultrasonic vibration is not required while the aluminum-silicon casting cools through the entire liquid plus solid regime of the phase diagram. Initial results suggest that elastic wave interaction more affects dendrite break-up than earlier stages of nucleation. At present, we have an on-going cooperative university-industry program to analytically model how elastic wave interaction affects the solidification process in cast metals. The performance of cast parts and uniformity of material will be enhanced by in situ monitoring and control of the microstructure, which should permit usage of cast components in primary aircraft structures that are presently forged or heavily machined. Depending on the present cost of forged or machined parts, replacements by casting could save hundreds to thousands of dollars. Since at the present time Rockwell International has no captive foundaries, current production volume is zero. However, if part production volume usage

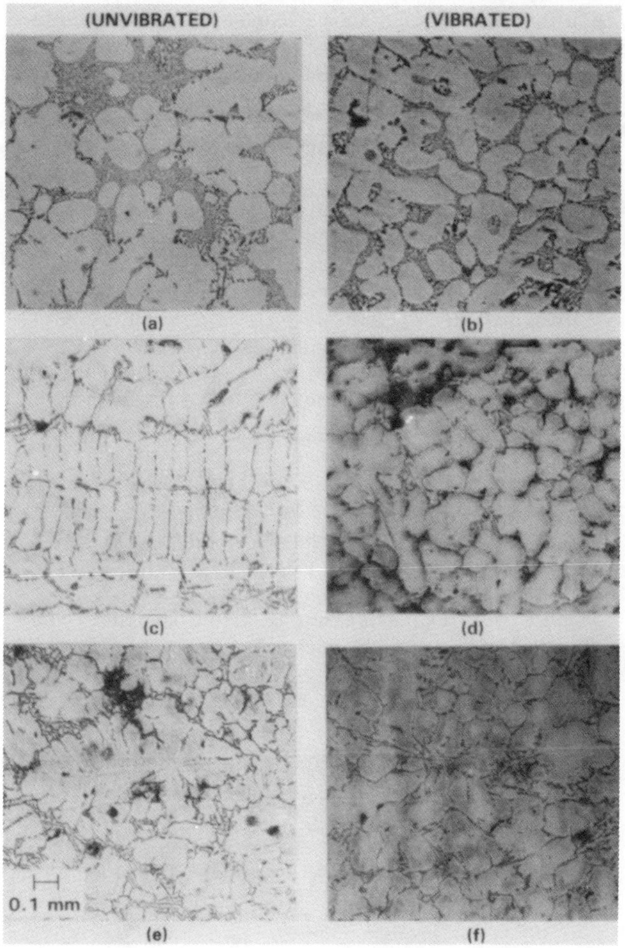

Fig. 8 - Photomicrographs of laboratory melts/production cast.

is considered, the volume of casting that can replace machined or forged parts is unlimited. Hypothetically, a "cast" frame with a Gr/Ep skin could be built! Realistically, approximately one-tenth of aircraft structure could be cast without safety being sacrificed. The payoff to the corporation would be a function of present costs mentioned above. Broad implications can be drawn for present work in process monitoring and control. Generically, at the Science Center we are attempting to exploit the fluid nature of a variety of materials, including colloid ceramics, polymers, and powder metals. Thus, the implication for research in general is certainly probable.

Although electroplating has been used in the fabrication of a wide variety of high technology products, from printed circuit boards and electronic components to rocket engines, until recently it has been considered a "black art" - largely uncontrolled and uncontrollable. One major difficulty has been lack of control of critical organic additives, which typically are present in the low ppm concentration range. Methods have also been needed to measure tensile strength and ductility accurately, and to determine the rate of electrolyte mass transport -- agitation -- in order to define optimum process conditions. The situation is further complicated by the build-up in production baths of contaminants which adversely affect deposit properties and interfere with bath analyses.

Within the last decade, however, electroplating process control methods have been developed, refined and implemented with dramatic results. A key milestone was the introduction of the cyclic voltammetric stripping (CVS) method[9] which measures the concentrations of bath additives, contaminants, and some minor bath constituents. This development was followed immediately by a rotating cylinder method[10] which precisely measures the tensile properties of electrodeposits and a method which evaluates bath agitation.[11] Concerted implementation of this technology for control of the process for plating high-reliability printed circuit boards[12] resulted in consistent process yields approaching 100%, whereas previously they had been below 70% at times.

Figure 9 shows instrumentation for performing CVS analyses, which recently has been introduced commercially.

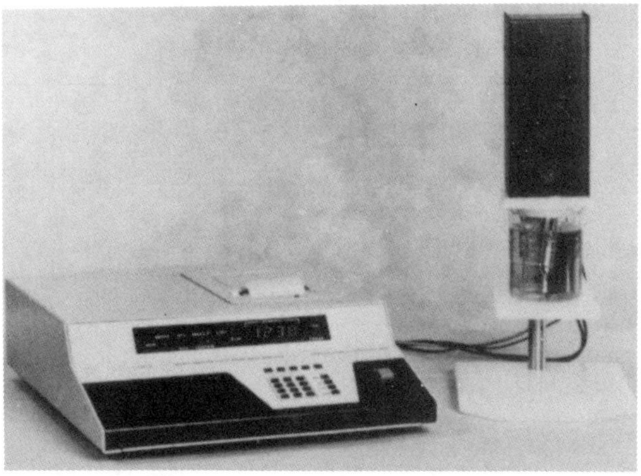

Fig. 9 Photo of CVS analyses instrumentation.

Currently there is considerable research activity directed toward controlling electroplating processes. Some of the more popular analytical methods being evaluated include polarography, high performance liquid chromatography (HPLC), x-ray fluorescence, and modifications of CVS. As might be expected since electroplating is an electrochemical process, the greatest successes in production have been with electrochemical techniques, CVS and polarography.

Most laser beam processing methods rely on laser radiation as a localized source of heat. By properly controlling the amount of heat input, processing methods such as surface heat treatment, welding, machining, and surface melting/rapid solidification, have been developed. Potential advantages of laser processing include automation and on-line process control; the ability to process a wide variety of alloys; fewer manufacturing steps; improved low-cost base material usage; low thermal distortion; processing under normal atmospheric conditions; and flexibility in the stand-off distance between the tooling and work surfaces.

Potential disadvantages of the method include high equipment and operating costs and the lack of scientific knowledge in areas such as energy transfer process during laser beam/material interaction, microstructure/property relationships, and process modeling involving heat and fluid flow. In spite of the high capital costs, laser beam processing is considered cost-effective in several applications. This presentation discusses application on heat treatment and welding.

Localized surface hardening through martensitic transformation in iron base alloys with laser beams provides increased resistance to wear and fatigue.[13] The majority of laser beam heat treating methods currently utilize a continuous wave CO_2 laser beam of 10.6 μ wavelength and beam power density of less than 10^4 W/cm^2. Normally a CO_2 laser beam with power density less than 10^4 W/cm^2 is not fully absorbed by machined iron base alloy surfaces. However, when iron base alloy surfaces are coated with infrared energy absorbing paints, these surfaces become highly absorbing for CO_2 laser beams.

Heat treating is accomplished by sweeping a paint coated workpiece at controlled speeds under a laser beam, as shown in Fig. 10. Upon absorbing the beam, the painted surface is heated to elevated temperatures localized only to those exposed surface areas. Therefore, steep temperature gradients extend from the surface into the substrate. Upon heat conduction from the work surface into the substrate, a finite volume of substrate is heated above the austenitizing temperature range and subsequently is cooled at rates rapid enough to yield martensite or bainite. When the workpiece is swept under the beam, a self-quenched heat-treated stripe is obtained along the surface. Dimensions of the heat treated stripe; that is, case width and depth, are controlled by absorbed laser beam power density, beam size, and travel speed. These three basic heat treating parameters control heating and cooling rates of work surfaces. Case hardness is influenced by composition of carbon and alloying elements and their distribution in the microstructure prior to laser heat treatment, and by the cooling rates along the austenitized zones.

In a project sponsored by the U.S. Army Tank-Automotive Command,[14] Rockwell International Science Center has developed the criterion for laser heat treatment. As shown in Fig. 11, by selectively modifying the laser beam's minor axis without altering its major axis, the profiles of the surface hardened patterns were changed. When the ratio of major to minor axis was high, the extent of heat flow was

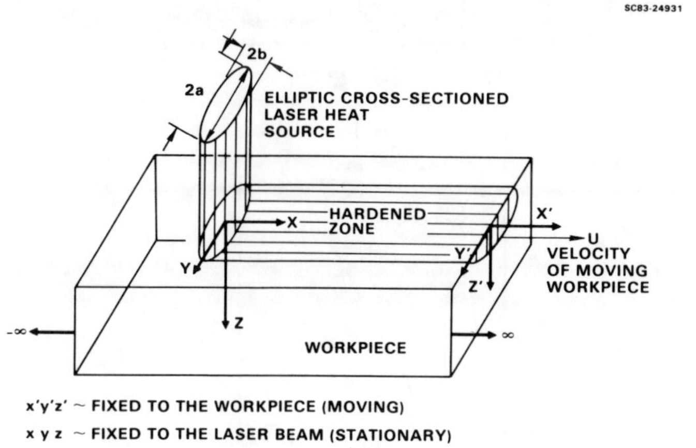

Fig. 10 - Model for laser beam heat treatment.

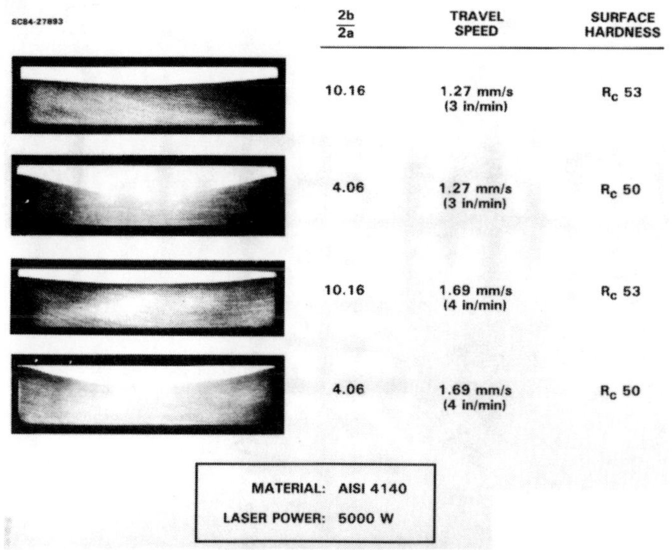

Fig. 11 - Cross-sectional views of laser heat-treated steel plates showing dependence of surface hardened patterns on laser beam's major (2b) and minor (2a) axes for identical laser power and travel speed.

confined to smaller depths below the surface; and therefore the hardened profile was relatively shallow. When the ratio of major to minor axis was low, heat flow extended to substantially deep regions below the surface, and the hardened profile was relatively thick. This potential to generate surface hardened patterns to required depths has broadened the use of laser heat treatment.

It seems appropriate to mention that through computational heat flow simulations, optimum selection of laser beam shape and size for improved process efficiency can be achieved.

Laser beam welding is done with a Nd:YAG (neodymium doped yttrium aluminum garnet) laser operating at a wavelength of 1,060 nm and CO_2 laser operating at a wavelength of 10,600 nm, each laser operating in the continuous wave or pulsed mode. When the laser beam power density is higher than 10^6 W/cm^2, the surfaces of alloys can be brought to the boiling point. The metal vapor interacts locally with the intense laser radiation and causes formation of a plasma. The beam penetrates the substrate so that laser energy is absorbed inside a plasma-filled pocket or keyhole and is localized only at the point of laser interaction.

Laser welding model in Fig. 12 shows that by moving the workpiece under the beam, the keyhole is translated and causes a slot along the weld path.[15] Molten material moves from the front to the back and around the sides of the keyhole, solidifying behind the laser beam, thus joining the two pieces together. Laser beam welding in general does not require filler material.

Several qualitative advantages over conventional welding processes are offered with laser welding. First, the welds are usually performed at higher speeds, therefore exhibiting finer solidification microstructures. Second, laser weldments have a higher depth-to-width ratio and narrower heat affected zones. Finally, part distortion is reduced due to the reduced total heat input of a laser weld compared to that of an arc weld.

A potential disadvantage of laser welding relates to its rapid solidification properties. Since the residual stress levels in the weld zone are normally higher in laser welded material, the process is somewhat more crack-sensitive. Laser beam parameters such as the power density and travel speed can overcome this disadvantage through control of cooling rates along the molten metal. As an example, Inconel 718 alloy plates in the mill-annealed condition were butt-welded using 5000 W of beam power and 19.6 mm/s travel speed.

Figure 13 shows cross-sectional and longitudinal views of the butt weld. The longitudinal view represents the plane along the cross-section's center line. The microstructures are free of cracking, and the weld-substrate interface does not show a heat-affected zone.

Due to the advantages of laser beam processing mentioned earlier, laser welding currently is being evaluated at Rockwell International for the manufacture of aerospace components. Laser welding process has the potential to cut manufacturing costs by 25 - 50% when compared to gas tungsten arc and electron beam welding processes. We anticipate that a laser welding application in Rockwell's business has potential for a 1 - 2 year payoff on the original investment in equipment and the necessary research and development.

Only two applications for laser beam processing, vis a vis surface hardening and welding, have been indicated. Besides these, there are a number of interesting and exciting avenues for research with laser processing, including innovative optoelectronic film processing; coatings for thermal protection; and fabrication of novel metallurgical microstructures through formation of gradients in microstructure for critical fatigue/fracture applications.

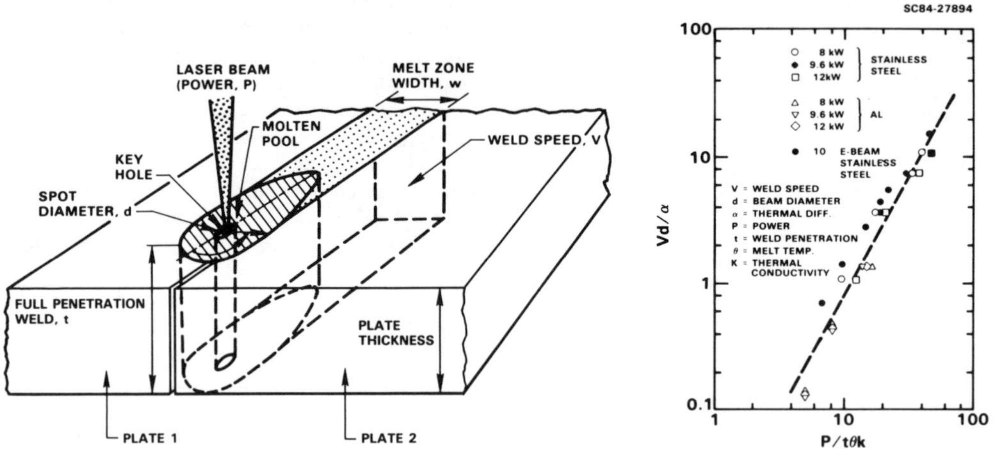

Fig. 12 - Model for laser beam welding and relationship between laser parameters, thermal properties and weld penetration.

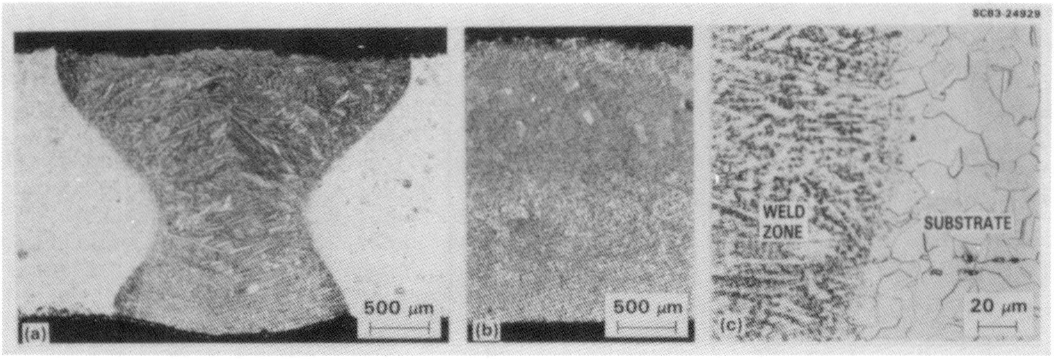

Fig. 13 - (a) Cross-sectional view, (b) longitudinal view, and (c) microstructure of laser weld zone/substrate interface in butt-welded Inconcel 718 alloy plates.

Conventional metal forming processes include bulk forming processes such as forging and extrusion and sheet metal forming processes such as drop hammer forming, press brake forming, and rubber pad forming. Basically, forming process control is important in the three different areas of quality control of the incoming stock; process control in the die tryout stage; and lubrication and tooling maintenance and control.

During cold forming of metals, the important material variables are strain hardening, strain rate sensitivity and fracture strain limit.[16] The latter is related to inclusion content and type, as well as to plasticity parameters. Strain concentration within the deformation zone and its proximity to the failure limit are critical considerations in selecting a material for a certain part. Strain hardening and strain rate sensitivity influence the plastic instability process which leads to the localization of flow in either macroscopic or microscopic manner. Quality control of raw materials is achieved by conducting tensile or compression tests to assess the material parameters of interest as well as simulative tests such as those which determine cup height[17] and forming limit diagrams.[18] Control of sheet metal drawability is assessed by conducting tests to determine plastic anisotropy parameter (ratio of the width-to-thickness strain in a tensile test).[19] The raw material supplier can control the above properties of the incoming material by suitably controlling his primary processing parameters. This subject will be discussed later.

Once the incoming stock meets the appropriate quality control tests, process control is achieved by varying such factors as the overall die and tool geometry and lubrication conditions for forming temperature, to produce a given component. Computer modeling of metal flow based on deformation and friction characteristics and strain limits are carried out to develop die design. This CAD process saves millions of dollars which would have been spent in die tryouts to determine the correct die geometry and the process sequence in multistage forming.

Both in the die tryout and production stages, production control is implemented by controlling press speed and lubricant clearance, as well as by maintaining tool and die geometry. Analyses of metal deformation is carried out using a grid circle analyzer capable of measuring the deformation of a preprinted grid on the specimen surface. Figure 14 shows a grid circle analyzer for sheet metal formability, with computer capability for determining major and minor strain levels in a sheet metal sample. Strain distribution measurements from critical regions of a part are analyzed and compared with formability limit diagrams. Alterations in lubrication or die geometry are made to bring a potentially failure-critical condition well under the formability limit diagram.

Superplastic forming (SPF) has emerged from a laboratory curiosity to an efficient, versatile manufacturing process which offers both design flexibility and reduced costs for fabrication of complex aerospace components. The application of the SPF process to production fabrication is growing as the number and variety of materials exhibiting superplasticity increases. An example of a superplastically formed/diffusion bonded titanium aluminide component is shown in Fig. 15. In addition to superplastic Ti alloys, wide varieties of alloys, including the high strength 7000 series Al alloys,[20] Al-Li alloys,[21] metal matrix components,[22] and high temperature alloys such as Inconel 718,[23] IN-100[24] and Titanium aluminides[25] have now been shown to be superplastic. Also, superplasticity has been demonstrated in the newly emerging low density AlLi alloys.[26] Clearly, SPF has an immediate future as an advanced manufacturing process for complex geometries made of metallic sheet as well as for material difficult to fabricate by more conventional techniques.

However, before this technology can reach full potential, advanced process control will be required to increase manufacturing efficiency and to simplify fabrication procedures. Superplasticity is a strain rate dependent behavior and the control of strain rate during superplastic forming is critically important to the success of this process. Since tensile elongation is maximized at a strain rate intermediate between dislocation and diffusional creep processes, it is necessary to maintain a constant strain rate throughout the forming cycle in order to achieve maximum forming while maintaining uniform sheet thinning. To achieve this, complex pressure/time (p/t) profiles are analytically determined based on experimentally measured materials characteristics and mathematical analysis of the part geometry. Initial parameters, however, change during forming. The sheet geometry changes, as well as a number of metallurgical characteristics of the SPF material such as grain size, flow stress, strain rate sensitivity, and sheet thickness. Due to

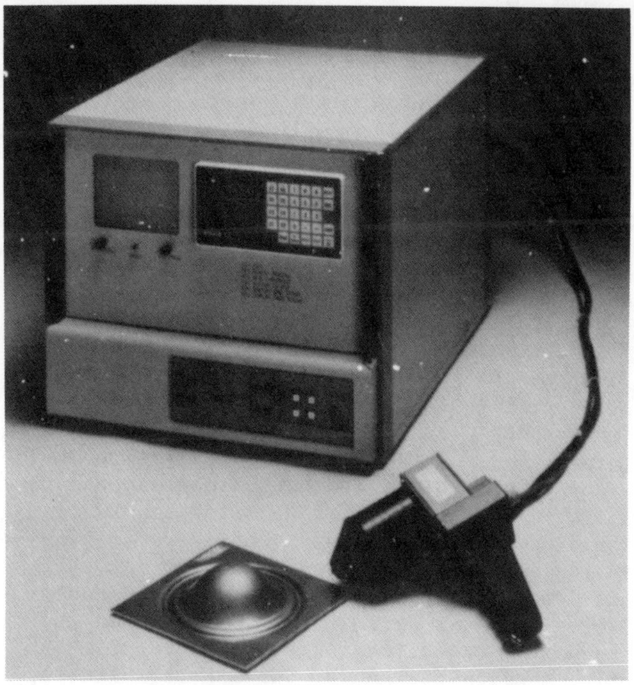

Fig. 14 - Grid circle analyzer.

Fig. 15 - Superplastically formed/diffusion-bonded Ti_3Al component.

the necessary extent of materials characterization, these material and geometry changes act synergistically to make the analytical determination of each p/t profile extremely complex and expensive.

Accordingly, this analytical procedure needs to be replaced with an "in process" monitoring system to determine in real time the extent of forming. The heart of this system would be either a proximity sensor (not in contact with the part); optical system or indirect and remote sensor (such as acoustic emission[27]) capable of operating in a severe environment.

Signals from the sensor would be processed to determine the extent of forming and thus the strain rate. In this manner, a closed loop system could be achieved wherein gas pressures would be adjusted continuously to maintain a constant strain rate throughout the forming cycle. This sensor development is considered crucial to continued implementation of SPF into aerospace manufacturing operations.

Diffusion bonding, a solid-state joining process in which a number of process and material variables are controlled, results in diffusion of elements from two either similar or dissimilar metals across an interface so that a metallurgically sound joint is created. As with the SPF process, analytical techniques are used to determine a relationship between time, pressure and temperature such that parameters can be established whereby complete bonding can be achieved. Complete bonding is often defined as achieving an interface strength equivalent to parent metal strength. In addition to process variables there are also material variables that change during the diffusion bonding cycle. For example, surface roughness, surface cleanliness, grain size and flow stress all influence the integrity of the final bond. Clearly a large number of variables must interact properly to create a high integrity diffusion bond. While process models for surface creep and prediction of diffusion bonding pressure and time are utilized, it would be both cost effective and more efficient to monitor the bond interface during diffusion bonding rather than to rely on indirect materials data that cannot allow for random differences between materials or process parameters.

During the bonding operation process monitoring of diffusion bonding requires NDE indentification of a signature that can be used to evaluate the integrity of the bond during real time. This signature could be different for different materials. For example, in some bonding operations it is required that significant metal deformation occur at the interface. At the bonding temperature this plastically deformed structure will recrystallize, so that often grain size at the interface is larger than in the matrix. This large grain structure could be used to evaluate the bond quality. Also, in almost all solid state bonding operations microvoid coalescence and diffusion occur until the bond line is no longer distinguishable from the matrix. An example of microvoids in the bond line of a titanium aluminide alloy is shown in Fig. 16. As shown, the voids are few, <1 μm in size, and most likely have little or no influence on bond line strength when they reach dimensions of this size. The absence of measurable sized voids could also identify the completion of a good bond cycle.

In addition to monitoring for metallurgical changes, acoustic emission (AE) techniques have been used to evaluate the quality of a diffusion bond during the bonding cycle.[27] To have a good bond, at the onset of the bonding cycle the

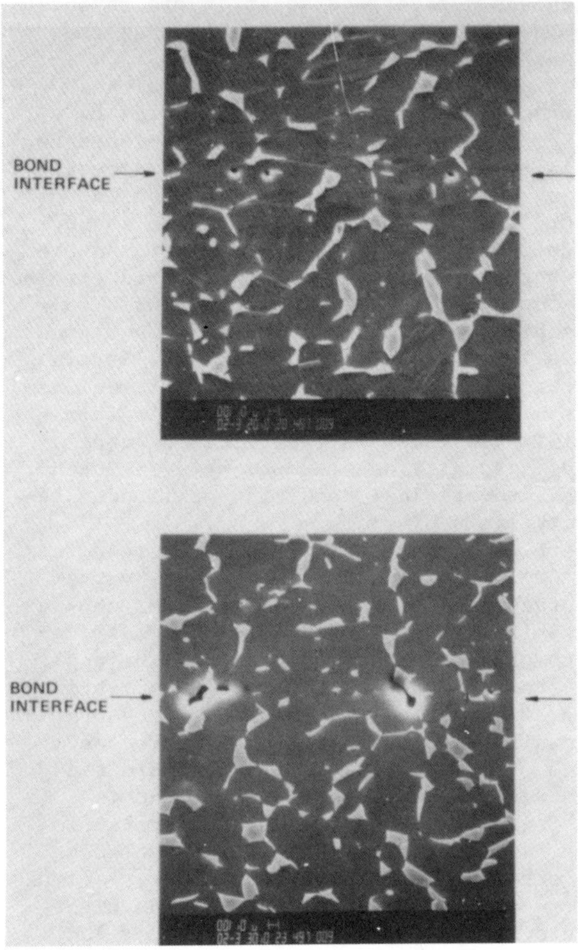

Fig. 16 - Microvoids in bond line of Ti_3Al alloy.

parts must be relieved from residual stresses, and then the materials must be deformed plastically.[28] Good bonds release AE on the onset of the bonding cycle. In contrast, unbonded materials behave as a spring, thus releasing AE. Weak bonds release a large amount of AE energy at the termination of the bonding cycle.

Clearly there are many possible routes for in situ monitoring of bond quality. Although the need is evident, advanced "in process" techniques have not yet been developed to the degree necessary for use in a manufacturing operation.

A number of advanced aircraft and spacecraft contain titanium structural parts made by the diffusion bonding (DB) process just described. As an example of this, Rockwell's B-1B bomber has 10 large diffusion bonded parts associated with the wing carry through structure -- the main structural element to which the wings of the aircraft attach and which provides the mechanism for the wings to pivot. The total titanium required to be purchased to support single B-1B aircraft is approximately 200,000 lbs, of which, after processing, 30,000 lbs goes into flight parts. Of the 30,000-lb flight weight, roughly 4,000 lbs is in diffusion bonded titanium parts. One of the principal advantages of the DB process for fabricating titanium parts is that it has a relatively low buy-to-fly weight ratio of about 7 versus a ratio of approximately 30 for machining an equivalent part out of titanium block. Although the DB process offers advantages for the fabrication of titanium structure, because of the critical role these parts play in aircraft structural integrity, it is necessary to provide 100% inspection of the diffusion bonds on these parts. The method of choice for inspecting titanium diffusion bonds nondestructively is pulse-echo ultrasonic inspection in a water tank in a frequency region of about 10 MHz. While it would be desirable to have in-process monitoring during the bonding cycle, presently there is no established on-line means to monitor the quality of the diffusion bond during DB processing. From the point of view of the aircraft assembler, ultrasonic inspection is an approach to in-process monitoring of the wing carry-through structure assembly. Each of the DB parts represents a very expensive piece of aircraft hardware (in the range of $100K to $200K). As aircraft structure assembly proceeds, the DB parts either are welded to each other or are bolted together. Thus, inspecting the integrity of each of these parts prior to assembly is an important step in in-process inspection. In fact, one of the DB parts of the B-1B, because of its complex geometry, goes through the diffusion bonding process twice, and it is ultrasonically inspected after each bonding step.

An enormous amount of effort is needed to inspect a typical DB part (Fig. 17), which generally contains more than half a dozen bond regions or entry surfaces requiring ultrasonic inspection. Because present commercially available ultrasonic inspection systems scan only along axes parallel to the rectangular dimensions of the water tank, it is necessary to adjust the orientation of the part in the tank so that the entry surface being scanned is parallel to a scanning plane. Sometimes this means realigning a part under water several times, which is very time-consuming. Other aspects of the inspection process are also very labor-intensive, and the net effect represents a significant cost element — something on the order of 2 - 3% of the cost of the part. There are corresponding manpower and inspection hardware requirements. For example, on the B-1B program, when production rate is reached, an average of 48 B-1B part inspections will have to be made each month.

Fig. 17 - Photo of B-1B DB part.

At Rockwell International Science Center an advanced Computer Aided NonDestructive Inspection System (CANDIS) for ultrasonically inspecting the diffusion bonded titanium parts of the B-1B aircraft is being developed (Fig. 18). The R&D for this project is being sponsored by the Air Force as a B-1B Tech Mod (Technology Modernization) Project. Rockwell's North American Aircraft Operations is providing the capital equipment funding for the project. CANDIS incorporates some of the latest advances in ultrasonic inspection techniques and flaw characterization methods as well as advanced approaches for user friendly man-machine interfaces.

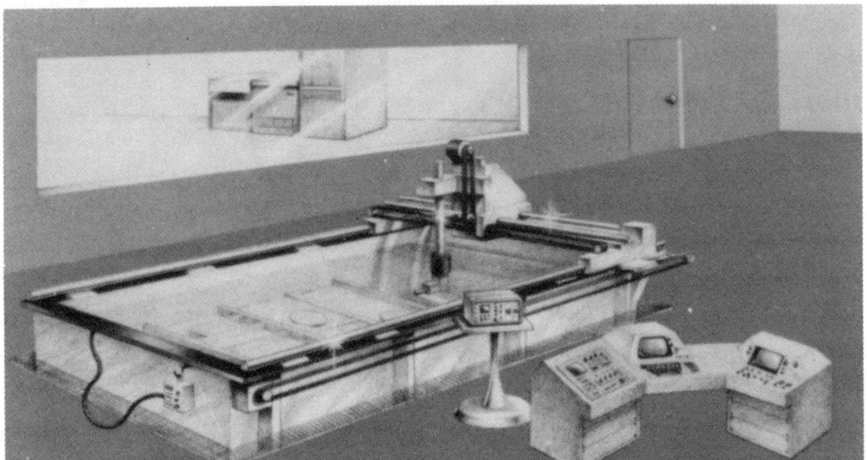

Fig. 18 - Photo of artist's sketch of CANDIS.

Moreover CANDIS will be fast -- much faster than conventional systems. Because it can scan in a plane of arbitrary orientation, the parts no longer have to be levelled and re-levelled -- a big time saver. It also scans rapidly, 20"/s, and accelerates rapidly to this speed at a rate of 40"/s/s. Other computer automated features result in significant time savings as well.

In CANDIS each ultrasonic pulse is analyzed and its peak amplitude is stored digitally, and eventually it is recorded permanently on magnetic tape. A color-enhanced image of an ultrasonic scan can be produced on a color monitor so that the operator can easily distinguish defective regions in the part (Fig. 19). The scan data can be recalled subsequently, long after the aircraft is flying, and reexamined under different conditions of background noise suppression and sensitivity adjustment, as well as under different color representations.

CANDIS is a large system -- a very large system. The tank, which weighs 21,000 lb empty, is 16 ft x 10 ft x 3-1/2 ft high and contains 3,600 gal of water. While the software and other components have been under development at the Science Center for nearly a year, the scanning tank and associated equipment were built by a vendor and recently were installed in our laboratory in Thousand Oaks, California (Fig. 20). Here the system will be completed for delivery to Rockwell's North American Aircraft Operations for use on the inspection floor in El Segundo, California.

This example presents an interesting illustration of how the producers of systems often have to become involved in a leadership role when the challenging problems of monitoring and controlling advanced materials processing result in new technologial advances being applied to a variety of disciplines. While it is very important to have in-process monitoring and inspection of parts after fabrication, because of the high value added in some of the advanced processing techniques discussed previously, it is essential that we have suitable starting materials. Post-processing inspection can prevent defective parts being delivered to the customer ... preprocessing inspection/control can have a major impact on the operation's bottom line.

Perhaps the most important factor in the success of secondary processing is the control of primary processing -- and systems producers may have only slight influence over this. Alloy chemistry and cleanliness are usually the most important variables at the primary metal processes end. Small variations in them can cause variations in material response to homogenization, rolling, and hot shortness as well as in all phases of secondary processing. Iron and silicon content in aluminum alloys, for example, must be controlled to extremely low values so that fracture toughness of these alloys is not adversely affected. Small inclusions become a serious problem in thin sheets such as in beverage can stocks.

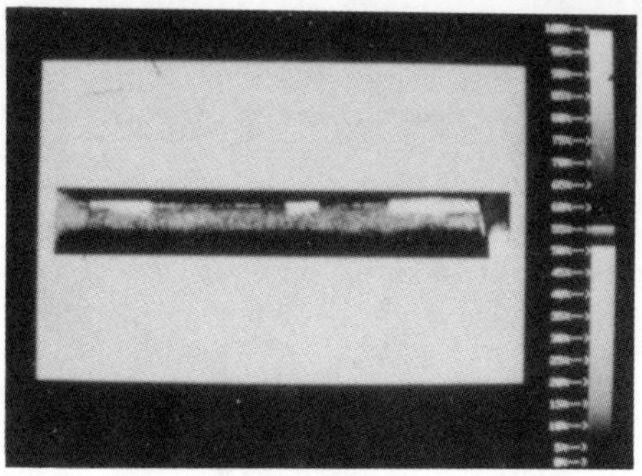

Fig. 19 - Photo of defective part scan.

Fig. 20 - Photo of CANDIS scanning system.

Segregation and piping are serious problems with castings which result from improper chemistry and poor control of cooling rate. Homogenization and proper breakdown of cost structures must be adequately conducted to provide an improved distribution of these constituents.

Wrought materials, which comprise nearly 90% of Rockwell's incoming material stock, are thermal and mechanically processed in a variety of steps. The breakdown of the cast structure and the extent of deformation introduced in the various stages are critical to the production of the appropriate microstructure and properties of the sheet or plate. Refinement of grain structure and production of a well recrystallized microstructure are usually the aim of primary material producers, since these enhance the subsequent strength, toughness, texture and forming characteristics of metals and alloys. Control of temperature and the amount of rolling reduction per pass have been found to be critically important to this end. For example, too high a beta processing temperature can cause such extensive grain growth in titanium alloys that no subsequent processing can correct this situation. Coarse grain size leads to high flow stresses at elevated temperature, thereby making superplastic forming or diffusion bonding practically impossible.

Microstructure control during primary metal processing is the key to grain refinement and superplasticity in structural alloys. In high strength aluminum alloys, for example, thermomechanical process to refine grain size generally consists of the following steps:[29]

1. Homogenization and solution treatment of the alloy;
2. Overaging to develop coarse precipitates;
3. Rolling to produce significant deformation around precipitates;
4. Recrystallization to develop fine grain size.

While the coarse precipitates (> 1 μ) concentrate deformation and aid in the nucleation of new grains, the very fine dispersoids existing in the alloy help retard grain boundary migration and achieve a fine grain size in the range 10-15 μ. A schematic illustration of this process appears in Fig. 21. The bimodal particle size distribution is a key to grain refinement in single phase alloys, and the control of the fine Cr, Co or Zr particles must be achieved from the casting stage.

In low density Al-Li alloys, which are emerging as strong candidates for future aerospace alloys, Zr is added as a grain refining agent, which must be dispersed uniformly by a rapid rate of heat extraction in direct chill cast molds. Control of cooling rate is also important in Ni-base superalloys used for turbine blades in various rocket engine and gas turbine engine applications. This enhances strength of blades with single crystal orientations, thus leading to desired properties.

Powder metallurgy of rapidly solidified alloys is becoming increasingly attractive for many aerospace structures applications. The control of powder particle size, atomizing medium, adsorbed gases, outgassing and working steps appears to strongly influence the product integrity and its mechanical properties.

Raw material control perhaps takes on the greatest challenge in epitaxial growth of various electronic film materials where Ga bearing solutions need to be free from trace elements to better than 10^{-12} parts in one. Control of materials processing for electronic applications and crystal growth has reached extreme sophistications and will continue to get even better.

Since its formation in the early 1960's, Rockwell International Science Center has been committed to maintaining its position as one of the leading industrial research laboratories in the country. Recent major technology thrusts in the materials synthesis and processing area have included superplastic metals and forming process development, near net shape technology, and electrochemical processing. In the automation sciences area, current emphasis at the Science Center involves developing the next generation of "intelligent" robots and machines. The availability of such technology can be expected to have an impact on the productivity of manufacturing operations by making possible applications in automation that are impossible with commercially available machines.

Continuing work on the development of superplastic forming and diffusion bonding offers a wide range of possible commercial applications.

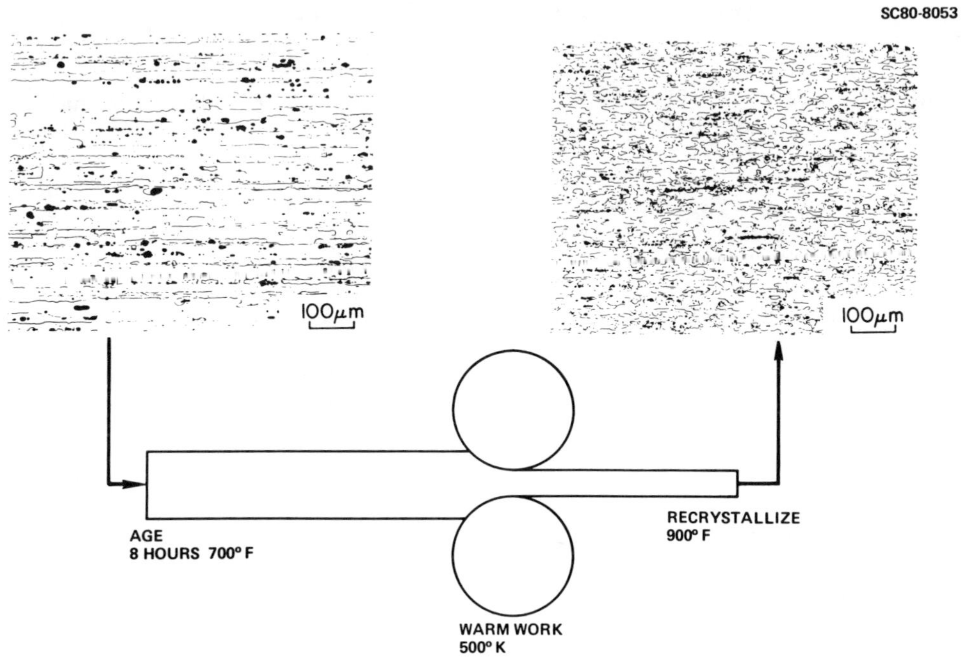

Fig. 21 - Schematic of coarse precipitates/grain boundary migration/fine grain size.

A new direction of effort focused on the synthesis of aluminum and titanium alloys for near net shape primary and secondary structural applications has an initial goal of enhancing mechanical ductility and ultimately increasing fatigue crack initiation and propagation resistance.

In view of the widespread use of electroplating processes in industry and the empirical way in which they have been developed, the opportunities for the application and extension of existing and future Science Center electroplating technology are tremendous.

In addition to the Computer-Aided Non-Destructive Inspection System described earlier, other potential automation programs presently in the planning and research stages will be integrated with state-of-the-art approaches in automated control, with the goal of yielding major improvements in the quality and consistency of materials. Thus, applications promise to be innovative and far-reaching.

Rockwell International Science Center is confident of the quality of its researchers and of its commitment to excellence. We will rely on the foundations of the past to welcome the challenges of the future. Thank you.

REFERENCES

1. R. Mehrabian and H.N.G. Wadley, "Needs for Process Control in Advanced Materials Processing," in Review of Progress in Quantitative NDE, ed. D.O. Thompson and D. Chimenti Proceedings of QNDE Conference, U.C. San Diego, July 1984, in press.
2. A.H. Freedman and J.F. Wallace, "The Influence of Vibration on Solidifying Metals," Trans. Am. Foundrymen's Society 65, 581-589 (1957).
3. E.A. Hiedemann, "Metallurgical Effects of Ultrasonic Waves," J. Acoust. Soc. of American 26, 5, 831-842 (1954).
4. D.W. Levinson, A.H. Murphy and W. Rostoker, "Influence of Vibration on Fluidity and Filling During Investment Casting of an Aluminum Alloy," Trans. Am. Foundrymen's Soc. 63, 683-686 (1955).
5. T.L. Mansfield, "Ultrasonic Technology for Measuring Molten Aluminum Quality," Materials Evaluation 41, 743-747 (1983).
6. D.P. Shukla, R. Agarwal, D.B. Gold and P.C. Pandey, "Effect of Vibration Imposed During Solidification of Castings," Indian Foundry Jour. 13-19, (1976).
7. N.F. Vil'danova, M.G. Gaydukov, M.I. Noskova and V.A. Pavlov, "Influence of Ultrasonic Treatment on the Microstructure and Creep Parameters of the Cast Alloy AL27," Inst. Metal Phys. Acad. Sci, USSR 39, 2, 367-374 (1975).
8. J. Seifert, F. Fischer, "Horizontal Continuous Casting of Metals with Mold Excited by Ultrasonic Waves," Ultrasonics, 154-158 (1977).
9. D. Tench and C. Ogden, J. Electrochem. Soc. 125, 194 (1978).
10. C. Ogden, D. Tench and J. White, Plating Surf. Fin. 67 (12), 50 (1980).
11. R. Haak, C. Ogden and D. Tench, J. Appl. Electrochem. 11, 771 (1981).
12. B. Lowry, C. Ogden, D. Tench and R. Young, Plating Surf. Fin. 70 (9), 70 (1983).
13. D.S. Gnanamuthu, C.B. Shaw, Jr., W.E. Lawrence, and M.R. Mitchell, "Laser Transformation Hardening," Laser-Solid Interaction and Laser Processing. Eds. S.D. Ferriss, H.J. Leamy, J.M. Poate, American Institute of Physics, AIP Conference Proceedings No. 50, 173 (1979).
14. D.S. Gnanamuthu, Laser Heat Treatment of Track Components in Combat Vehicles (Phase II-Task A), Contract No. DAAE 07-81-C-4069, U.S. Army Tank-Automotive Command, Final Report (1983).
15. D.T. Swift-Hook and A.E.F. Gick, "Penetration Welding with Lasers," Welding Research Supplement, 492-S (1973).
16. Z. Marciniak and K. Kuczynski, Int. J. Mech. Sci., 9, 609 (1967).
17. A.K. Ghosh, Metal Progress," 52 (May 1975).
18. S.P. Keeler, SAE Paper No. 680092, Detroit (Jan 1968).
19. G.M. Goodwin, SAE Paper No. 680093, Detroit (Jan 1968).
20. M.W. Mahoney and C.H. Hamilton, "Superplastic Aluminum Evaluation," AFWAL-TR-81-3051, June 1981.
21. J. Wadsworth, I.G. Palmer, D.D. Crooks and R.E. Lewis, Aluminum Lithium Alloys II, E.A. Starke and T.H. Sanders eds. TMS-AIME, 1983.
22. M.W. Mahoney and A.K. Ghosh, "Silicon Carbide (SiC) Reinforced Aluminum Structure," AFWAL-TR-82-3051, August 1982.
23. W.T. Chandler, A.K. Ghosh and M.W. Mahoney, "Superplastic Forming and Diffusion Bonding of Inconel 718," Journal of Spacecraft and Rockets, Vol. 21, No. 1, Jan. - Feb. 1984, p. 61.
24. R.G. Menzies, J.W. Edington and G.J. Davis, "Superplastic Behavior of Powder - Consolidated Nickel-Base Superalloy IN-100," Metal Science, May 1981, p210-216.
25. M.W. Mahoney and D.W. Schulz, "Production of Titanium Aluminide Products," Final Report, Contract No. F33615-78-C-5144, April 1981.
26. R. Grimes, "Aluminum Alloy Developments For Aerospace," Sheet Metal Industries, Dec. 1982, p. 885-888.
27. S.J. Vahaviolos, "Evaluating Superplastic Forming/Diffusion Bonding (SPF/DB) of Ti Strucutres with Acoustic Emission (AE)," Presented at 1983 ASMT Spring Conf., Orlando, FL.
28. O.L. Anderson, "The Role of Surface Shear Strains in the Adhesion of Metals," Wear, 3 (1960), pp. 253-273.
29. N.E. Paton and C.H. Hamilton, U.S. Patent 4,092,181. 1978.

NEW SENSOR DEVELOPMENT PROGRAMS IN THE STEEL INDUSTRY

Vincent V. Horvath
Bethlehem Steel Corporation
Research Department
Bethlehem, PA

ABSTRACT

To develop new sensors, American Iron and Steel Institute (AISI) task groups identified important new process control sensor needs. Cooperative R&D programs are underway by steel industry, equipment manufacturers, and government organizations to develop sensors and systems for: 1) measuring the temperature distribution inside hot metallic objects, 2) detecting internal defects in hot steel, 3) remotely analyzing liquid metals in situ, and 4) detecting and classifying surface imperfections in sheets moving at high speed. As a result of the positive response within the industry and participating organizations, other programs involving joint research and development are being organized.

INTRODUCTION

As one of its responsibilities, the American Iron and Steel Institute's (AISI) General Research Committee is charged with identifying and funding research and longer-range developments with a potentially favorable impact on steel production, processing, or products. In 1980, a study was carried out which identified "process control and sensor development" as a field with a large potential impact on steel quality and productivity. These needs were not being addressed by commercial suppliers because of the perceived limited size of the market, and the need for first-hand applications experience to develop and evaluate the equipment. In addition:

- The need was recognized for a focused effort and the avoidance of overlapping technical development programs.
- Many metals companies were under financial pressure and were forced to consider reducing longer-range research and development costs.
- The Federal government recognized that selected cooperative developments could help the U.S. steel industry in its efforts to compete internationally.

These factors led to the recognition that the development of sensors for process control was an appropriate candidate for cooperative efforts. A research task force was subsequently organized under AISI sponsorship to identify and prioritize specific sensor and process control needs. Thirteen steel companies contributed to the effort in which several hundred process control and sensor needs were defined, and 18 were identified as "most significant". Of these 18, four were selected as high priority on the basis of a large potential payback and a moderate to high risk. It was felt that low risk developments were not well suited to a cooperative development program, since individual companies would likely pursue those activities on their own. Furthermore, the nature of any cooperative development entails a certain amount of organizational overhead more appropriately absorbed by longer term developments.

Once the four high priority sensor needs were selected, individual working task groups, or "units" were organized by part time volunteers from the various steel companies. Their objectives were to establish task priorities, define the potential economic justification for the sensors, discuss technical approaches to solving the problems, and to organize the approaches for attacking the problems. A workshop was held in mid-1982 at the National Bureau of Standards (NBS) under sponsorship of NBS, Defense Advanced Research Projects Agency (DARPA), and AISI. The workshop brought together representatives from government, industry, and academia, and led to in-depth technical discussions of the selected sensors. As an outcome[1], the four task groups received numerous proposals for research work. The four "key" sensor topics included:

- Rapid measurement of temperature inside solid bodies of hot steel. (Unit 5-4)

- o Automatic detection of pipe and porosity in hot steel. (Unit 3-4)
- o Rapid in-process analysis of molten metal. (Unit 5-2)
- o On-line inspection for surface defects in hot and cold strip. (Unit 4-5)

From this beginning, a broad-based cooperative program evolved as shown in Figure 1. AISI remains as the focus of the effort, but significant technical work is underway at NBS, Department of Energy laboratories at Los Alamos and Battelle/PNL, and several private and academic organizations. Funds and researchers are being provided by the steel industry as well as by NBS and DoE. The list of sponsors has also expanded to include aluminum companies as well as steel companies. In the remainder of this report, we will discuss the objectives and status of the programs underway, and some thoughts on the process of cooperative sensor development.

RAPID MEASUREMENT OF TEMPERATURE DISTRIBUTION WITHIN A SOLID OR SOLIDIFYING BODY OF HOT STEEL

There are several potential applications in which the direct measurement of internal temperatures could lead to improved energy utilization, productivity, and quality. For example, to maximize productivity with the continuous casting process, it is desirable to cast steel as fast as possible, consistent with the need to maintain product quality. In doing this it is important to maintain surface and internal temperatures within certain limits to reduce chances of rupturing the cast shell or causing thermal cracks. Models can predict the temperatures based on casting speed, liquidus temperature, and other factors, but these factors change during the casting process. To achieve optimum performance, therefore, it is desirable to have temperature measurements of the surface and interior of the product as it is cast.

A second example of a process which could benefit from having internal temperature measurements is that of "hot-charging", where hot slabs are transported directly to a rolling mill for further processing. Many North American steel mills were designed when energy was cheap and when the process technology did not support the concept of hot charging. As energy prices rose, and continuous casters were added to existing steel plants, the concept of hot charging became more important. Because of plant logistics, however, the time needed to transport and roll the hot cast steel is not consistent. To optimize the processing and reheating of the "hot" steel, it is desirable to measure its interior temperature. There are several other potential applications for direct internal temperature measurements, such as controlled-temperature rolling, where the metallurgical properties of the finished product depend on the thermal state of the steel during processing.

To serve as a suitable tool for addressing these needs, a program was undertaken to:

Develop a system for the rapid direct measurement of temperatures and temperature distribution within a solid or solidifying body of hot steel.

An initial assessment was made of the specifications for an "ideal" sensor:

- o Measure internal temperature with an accuracy of ±10C.
- o Maintain accuracy at metal temperatures to 1400C.
- o Associate each temperature value to a known location within ±2mm inside the sample.
- o Complete a measurement within seconds.
- o Operate reliably in severe environments.

It was recognized that a sensor need not meet all these requirements for it to be useful for specific applications.

The task force established to initiate this sensor first reviewed possible technical approaches. They concluded that ultrasonic (high-frequency sound) methods had the potential for a non-invasive approach to internal temperature measurement, since the velocity of propagation of an ultrasonic wave in steel varies with its temperature. It was further suggested that ultrasonic velocity measurements might be processed mathematically to provide a complete temperature profile of the interior of the steel. This concept can be demonstrated by heating a steel cylinder to a uniform temperature and allowing it to cool. As it cools, a radially symmetric temperature distribution develops. If an ultrasonic wave is then introduced at one point on its surface, as shown in Figure 2, the sound velocity can be measured along paths one through four. If the continuous temperature distribution is approximated by concentric constant-temperature rings, then ray four passes only through one ring of temperature T_4. Ray three passes through rings of temperature T_4 and T_3, and so forth. Using computer methods, it is possible to calculate the temperature in each of the rings, if the ultrasonic velocity along each path is measured. Figure 2 also shows the calculated temperature distribution on the basis of this concept using the four rays. The stepwise temperature distribution can then be "smoothed" to create a more-realistic smoothly varying function.

Work is now underway at the National Bureau of Standards (NBS) to investigate these and other methods of internal temperature measurements. A major goal of this work is to develop "practical" ways to generate ultrasound without physically contacting the hot steel. Steel industry researchers are working with NBS scientists through special assignments, supported by AISI, in which they serve as Research Associates at NBS's Gaithersburg facilities.

Figure 3 shows how lasers can be used to generate the ultrasonic waves for the heated cylinder concept described above. A high-powered pulsed laser is used to generate a well-characterized ultrasonic wave at the point at which the laser strikes the surface. No physical contact with the material is required. In this case, the ultrasonic waves are detected using quartz rods in contact with the steel, and conventional piezo-electric detectors. For steel mill use, non-contact receivers are also desirable. Work is also underway to develop non-contact electro-magnetic-acoustic transducers (EMATS) suitable for use on hot steel. The work at NBS also includes another important aspect of the program: characterizing the ultrasonic properties of the various steel alloys as a function of temperature so that temperature can be calculated on the basis of ultrasonic velocity measurements.

A complementary effort is underway at Battelle's Pacific Northwest Laboratory (PNL) to develop a system for measuring the temperature inside hot steel. This Department of Energy (DoE) and AISI sponsored program also relies on the use of ultrasonic measurements as a means to indicate temperatures. The initial work at PNL is aimed at designing and constructing prototype hardware and software for demonstrating the technology. Laboratory tests under carefully controlled conditions indicate that ultrasonic velocity measurements have sufficient sensitivity for measuring internal temperatures to the ±10C accuracy needed. However the tests were performed at steel temperatures near ambient, so work is underway to extend the technology, and the tests, to the high temperature range of interest.

AUTOMATIC DETECTION OF PIPE AND POROSITY IN HOT STEEL

The objective of this sensor development activity is to:

> Develop instrumentation for the detection of primary pipe, gross porosity and fishtailing in hot primary rolled materials and to provide a map showing such anomalies during cropping.

By "pipe" we refer to the shrinkage cavity which is formed as liquid steel solidifies in a mold to form an ingot. The shrinkage cavity can be pictured as a conical open void extending inward from the end of a solidified ingot. During the rolling process, pipe is removed by shearing or "cropping" the material from the parent metal. It is necessary to completely remove the piped material with minimal removal of the base material. Visual inspection, the technique most commonly used, can lead to quality problems through cropping too little, or yield loss through the removal of excessive sound material. Automatic detection of pipe and subsequent control can therefore lead to improvements in both yield and quality. Continuously-cast steel also shrinks during solidification, but any voids are usually dispersed in the cross-section in the form of pores. The problem of pipe/porosity is usually not as serious in continuously-cast steels as in ingot-cast steels, but if there are problems in the casting process, excessive porosity can occur.

Considerable effort has been expended over the years to detect pipe in hot steel. Figure 4 shows some tests carried out at Bethlehem[2] several years ago aimed at detecting pipe in hot steel blooms. The technique used pressure contact to couple ultrasonic waves from conventional transducers into the hot steel. Although the system could successfully detect pipe in hot steel, it was judged that the mechanical equipment needed for pressure contact methods would have high maintenance costs and questionable reliability when in full production use. Others have carried out similar tests which demonstrated that ultrasonic methods can detect pipe and porosity. The main problem to overcome is similar to that in the measurement of temperature; mainly, the need to transmit ultrasonic waves into the hot steel and to receive them with high reliability and sensitivity. With regard to reliability, noncontact methods can have considerable advantages over the need to physically contact the hot steel.

The focus of the pipe/porosity detection program is twofold: (1) develop improved non-contact cost effective ultrasonic sensors suitable for inspecting hot steel, and (2) investigate means for improving the potential reliability of contact ultrasonic methods. Specifications for the ideal system include:

- Locate pipe to within 2.5 cm.
- Detect pores greater than one cm.
- Inspect slabs at least 30 cm thick at temperatures to 1400C.
- complete the inspection process in less than 30 seconds.
- Operate reliably in severe environments.

The initial focus of this program is similar to the internal temperature measurement work, since both are aimed at ultrasonic measurements in hot steel. Therefore it was judged useful to coordinate the development of the ultrasonic transducers at NBS. Research is underway by NBS staff and AISI-supported Research Associates to explore laser methods to introduce the sound into hot steel, and non-contact methods using electro-magnetic-acoustic transducers (EMATS) as receivers. The use of the EMATS on hot steel has been demonstrated by several researchers[3,4], but a truly cost-effective, low power, reliable system has not been available. Experimental high-temperature EMAT receiver transducers are being evaluated. The actual EMAT receiver coil is less than one centimeter in width, but a large magnet is required to generate a magnetic biasing field, and this increases the total size of the receiver assembly. Initial results with the EMAT receiver and laser excitation of the ultrasound appear promising.

In addition to the work being done at NBS, researchers at Magnasonics have been working to develop and test improved EMAT transducers for use on hot steel. In addition, Magnaflux Corporation under AISI support, has been experimenting with contact-wheel ultrasonic transducers to inspect hot steel samples. The sensitivity is high with the contact wheel transducers, but additional work is needed to improve signal levels at reduced contact pressures.

RAPID IN-PROCESS ANALYSIS OF MOLTEN METAL

In-process or "in-situ" analysis of molten metal is very important because of the potential it offers for improving productivity. Quality improvements and energy savings can also be considerable. The specific objective of this task force is to:

> Develop a system for the rapid in-process chemical analysis of molten iron and steel for compositional control during: refining in basic-oxygen vessels, melting in arc-furnaces, and alloy additions in the ladle.

An "ideal" system should be able to:

- o Analyze for C, Mn, P, S, Al and other elements.
- o Complete sampling and analysis in one minute or less.
- o Analyze metal at temperatures up to 3800F.
- o Be cost effective.
- o Operate in severe environments.

The concept of in-situ analysis of molten metal has been the subject of numerous publications and patents. However none has been completely successful in a production environment. Spectrometric methods are effective for analyzing solid samples, and most published techniques for in-situ analysis have been aimed at: (1) generating an electric spark plasma at the liquid sample, and analyzing the emitted light, or (2) rapidly acquiring a sample and transporting it to a spectrometer for analysis. Newer methods employ high-powered laser excitation to generate a plasma at the liquid metal surface and subsequent spectral analysis of the light emitted. Ozaki[5], et al, reported results obtained in using the laser-spectrometry concept. The principle was demonstrated successfully, but accuracy and precision were not sufficient due to the need to very accurately focus the laser beam on the liquid metal surface.

Another approach using a high-power pulsed laser is shown schematically in Figure 5. The purpose of the laser in this case is to explosively generate, or ablate small particles which can be transmitted to external equipment for rapid analysis. This technique is not as sensitive as the laser-induced plasma method to changes in the optics focus of the laser system. Questions must be resolved, however, regarding the size and shape of the ablated particles, potential chemical changes during transport, and the relationship between the chemistry of the particles and the bulk molten metal.

Work is underway to investigate laser analysis methods at Los Alamos under DoE and AISI sponsorship. The metals industry is actively contributing to the technical development and supports a Visiting Scientist at Los Alamos. In addition to laser methods, rapid sampling and analysis methods are being developed as potential near-term solutions.

ON LINE INSPECTION FOR SURFACE DEFECTS ON HOT AND COLD STRIP

The surface quality of steel sheet is very important to both steel producers and customers. An on-line assessment of surface quality, however, is extremely difficult because of the processing speeds up to several thousand linear feet per minute, and the wide size and variety of surface blemishes which may be encountered. In addition, minor surface imperfections may be perfectly acceptable for some end uses, so the problem is not only one of detecting surface imperfections, but classifying them as well. A task group was recently initiated to organize a program with the objective to:

> Develop instrumentation to detect, characterize and classify surface imperfections on hot or cold strip during processing.

Specifications for the "ideal" instrument include the need to:

- o Detect imperfections as small as one mm.
- o Classify imperfections
- o Process information in real-time, consistent with inspection at surface speeds up to 30 meters per second
- o Operate in severe environmental conditions
- o Be economically justifiable for multiple installations

In addition, it is important to detect the more significant defects early in the process to prevent the production of large quantities of flawed material. The initial work to organize this program was carried out by steel industry representatives, but the problem is also important to aluminum producers, and they have joined in the present effort.

Some variations of optical imaging or optical scanning appear to be the only approaches with the potential to inspect sheet metals at high speeds. Laser scanners are available and have been used successfully in continuous processing operations such as paper mills. They have also been successfully applied to a few metals processing applications such as coating lines. For many high speed lines, however, they have not been completely successful because of the subtle nature of many imperfections, and the wide range of imperfection types encountered. The need to detect and classify the imperfections at production speeds involves the need for very high speed data acquisition and pattern recognition systems. For example, to completely inspect a two meter wide strip moving at 30 meters per second with a scanned laser spot of one mm can require a signal bandwidth of 100 MHz. Processing large volumes of data at this rate is impractical even with today's state-of-the art electronics, and it is necessary to pre-process the information to reduce the bandwidth. It seems reasonable to consider the pre-processing concept, since a large part of the sheet is usually defect-free. The pre-processor would, in effect, sort through the incoming high-speed data stream and reduce the information to only that representing the "imperfect" regions. Subsequent processing and pattern recognition would be used to further analyze the data at reduced data rates and classify the imperfection by type. A program is just getting underway in conjunction with Westinghouse aimed at developing instrumentation based on these concepts. A laser scanner will provide input to high speed pre-processor and pattern recognition hardware.

STATUS AND CONCLUSIONS

The process of initiating cooperative programs aimed at addressing metals industry sensor and process control needs has resulted in four programs now underway. Initial technical results appear promising in the work aimed at detecting pipe and measuring temperatures inside hot steel. Progress has also been made in the in-situ analysis of molten metal. In all three programs, additional work is needed to implement practical technology. The work to develop instrumentation to assess surface quality in metal sheet is still in the initial laboratory stages. In addition to the programs underway, several new cooperative programs are being considered:

- development of sensors to measure molten metal cleanliness
- development of instruments to measure wall thickness of seamless pipe
- work by DoE, industry, and other laboratories aimed at developing new steelmaking and "near net shape" production processes

Although the ultimate degree of technical success of these cooperative programs has yet to be determined, participants have reached some conclusions:

- Cooperative programs are a successful means to apply scarce technical resources.
- Concentrated, focused efforts are helpful in developing visibility and in attracting additional resources from government, industry, and universities.
- Scientists, engineers and managers involved have benefitted from new associations.
- It is necessary to understand and adapt to the differing goals, objectives, and driving forces of the organizations involved.
- Technology transfer and implementation are major hurdles yet to be overcome.

The last of these items will ultimately determine the success or failure of the cooperative approach to research. It is clear that the active involvement of metals industry personnel is essential in providing focus to ongoing research efforts. Industry personnel will also play key roles in putting into practice the technology developed at government and university laboratories.

REFERENCES

1. Mehrabian, R., R. L. Whiteley, E. Van Reuth and H. N. G. Wadley, "Process Control Sensors for the Steel Industry", Report of Workshop held July 27-28, 1984, NBSIR 82-2618, (1984).
2. Droney, B. E., and T. J. Pfeiffer, "Ultrasonic Inspection of Hot Steel Blooms to Detect Internal Pipe", Materials Evaluation, 31-36, June (1980).
3. Whittington, K. R., "Ultrasonic Inspection of Hot Steel", British Journal of NDT, 242-247, September (1978).
4. Kawashima, K., S. Murota, H. Soga, "On-Line Detection of Mechanical Pipe Patterns in Hot Steel Slabs", 99th ISIJ Meeting, Lecture No. S285, April (1980).
5. Ozaki, T., T. Takahashi, Y. Iwai, K. Gunji, and E. Sudo, "Giant Pulse Laser-Direct Spectrochemical Analysis of C, Si, and Mn in Liquid Iron", Trans. Iron and Steel Institute of Japan, July (1984).

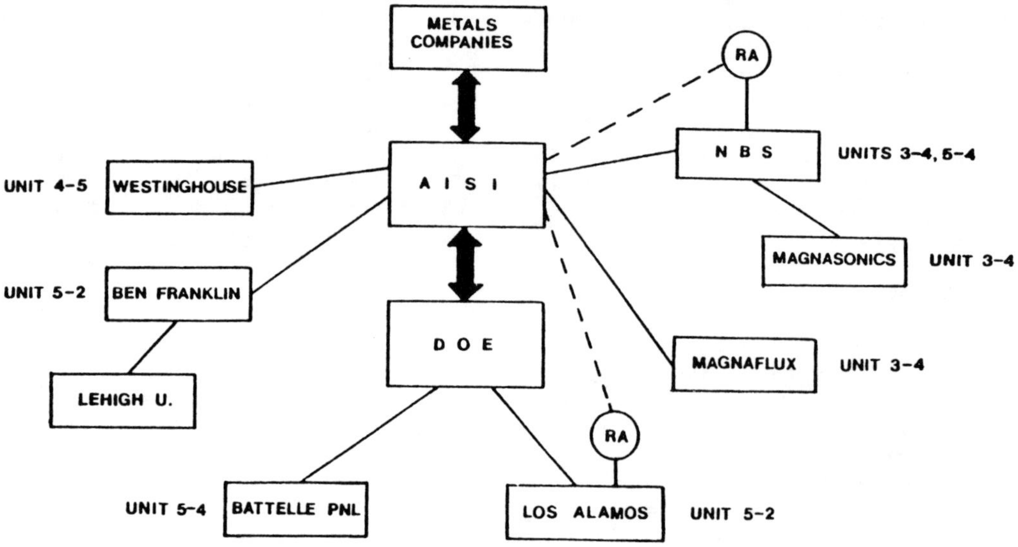

Figure 1. The AISI process control sensor development programs involve strong support from many diverse organizations.

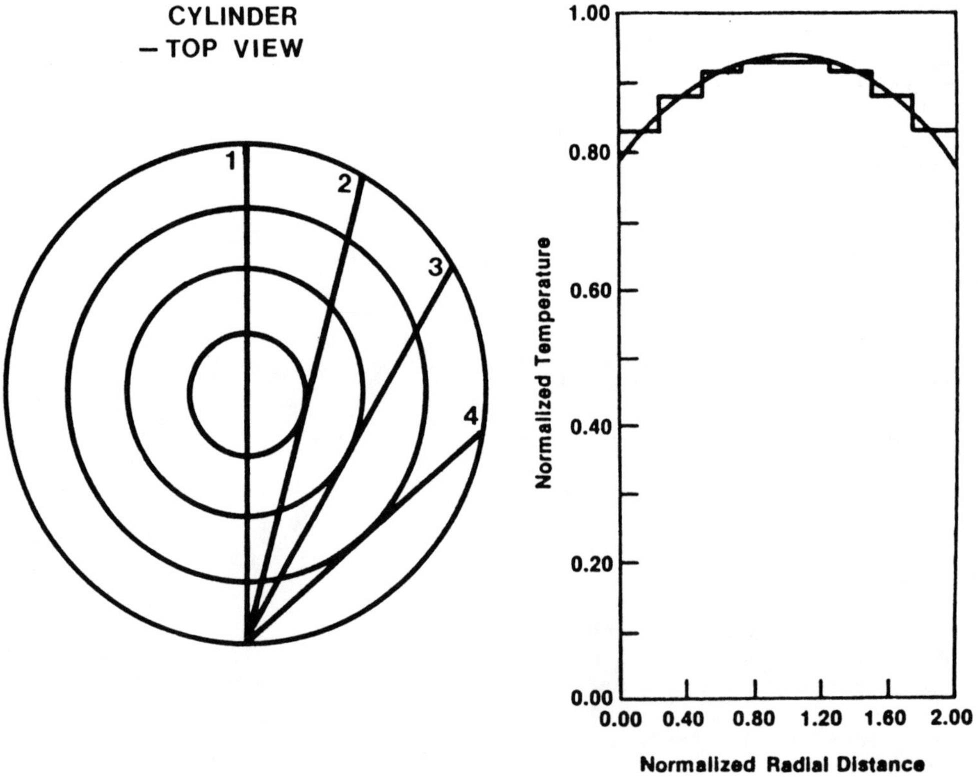

Figure 2. The velocity of ultrasonic waves propagating along various paths through a cooling steel cylinder can be used to calculate its internal temperature distribution. (Courtesy-NBS)

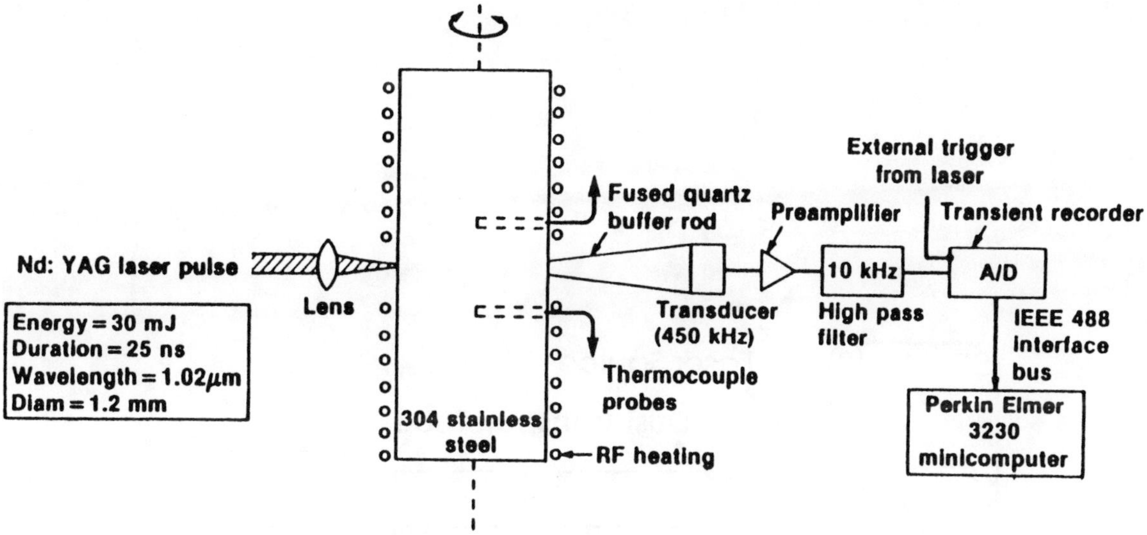

Figure 3. Laser generation of ultrasonic waves appears promising as a means to measure internal temperatures without physical contact. Contact transducers are shown here as receivers, but noncontact receivers are also feasible. (Courtesy-NBS)

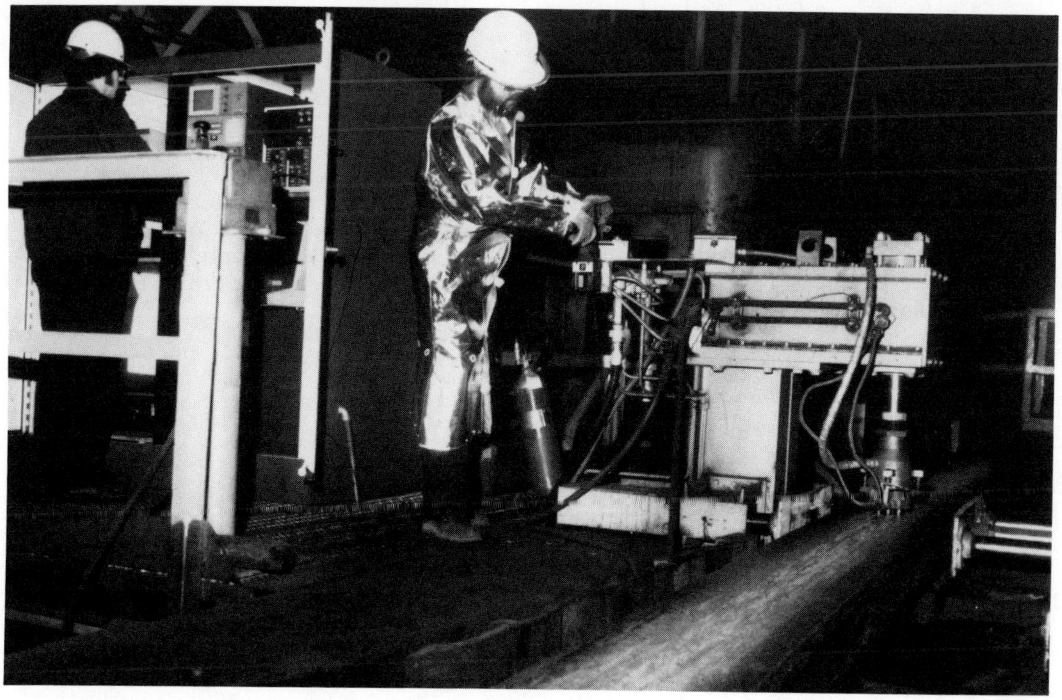

Figure 4. Contact ultrasonic inspection methods are feasible for detecting pipe in hot steel. This is a prototype unit developed and evaluated by Bethlehem Steel several years ago.

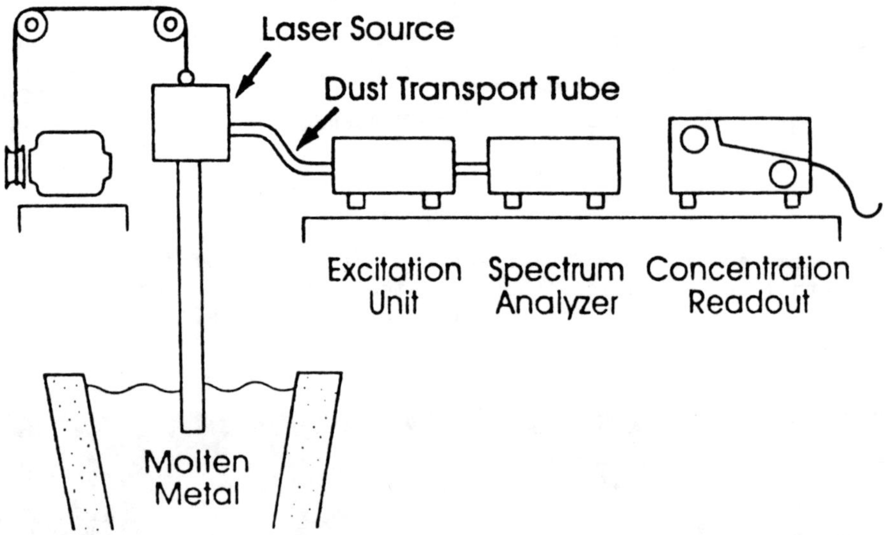

Figure 5. High-power pulsed lasers can cause particles to be expelled from liquid metal surfaces at the point of impact of the laser beam. The particles can then be transmitted to an instrument, such as the inductively coupled plasma spectrometer shown here, for analysis.

INVERSION OF ATTENUATION MEASUREMENTS IN TERMS OF A PARAMETERIZED AUTOCORRELATION FUNCTION

Fred E. Stanke
Schlumberger-Doll Research
Ridgefield, CT 06877-4108

ABSTRACT

Ultrasonic attenuation measurements in single-phase polycrystalline materials can be used to obtain information about the grain size and anisotropy factor of the grains. This abstraction of information requires an accurate, general theory for the attenuation of ultrasound in such samples. The unified theory of Stanke and Kino provides calculations which agree with attenuation and grain-size measurements in copper and 304 stainless steel, and should lead to inversions with at least ±20% accuracy.

INTRODUCTION

The unified theory of Stanke and Kino[1] accurately predicts the attenuation of longitudinal waves in single-phase, polycrystalline materials and can therefore be used as a means of inverting ultrasonic measurements to obtain information about the microstructure. The present paper compares measurements in samples which have known microstructures, to theoretical predictions calculated for a range of parameters in an attempt to ascertain the accuracies which might be obtained from such an inversion. For the cases examined, the grain size and anisotropy factor (of the grains) could be determined to an accuracy of 20% or better.

PREDICTIONS OF THE UNIFIED THEORY

The unified theory has four key attributes which make it ammenable for inversion of measured attenuation coefficients. First, it is valid for all ratios of grain size to ultrasonic wavelength, i.e. at all frequencies. This allows the ultrasonic frequency to be chosen for samples of given thicknesses and microstructures so that the attenuation coefficient can be accurately measured. Accurate measurements require that the actual attenuation through the sample be larger than reflection losses and instrument error, and small enough so that all of the ultrasonic signals of interest are larger than the measurement noise. Theories which are valid only when the grain size is much smaller or much larger than the wavelength (i.e., at high or low frequencies) do not allow measurements to be optimized with nearly as much flexibility. Second, the theory accounts for the elastic anisotropy of the individual grains, which is often the dominant cause of attenuation in single-phase metals at ultrasonic frequencies. Theories which attempt to simplify the problem by using an analogous, locally isotropic medium have errors which far exceed the differences between the unified theory and measurements. Third, the theory accounts for the immediate proximity of the grains to one another, and for their irregular (i.e. non-spherical) shape. This is related to the first attribute because spherical particles exhibit resonances when the grain size is comparable to the wavelength (i.e., in the intermediate frequency range) while real grains generally do not exhibit such resonances. Fourth, the theory characterizes the grain size in a way which corresponds in a simple way to measurements made along lines drawn on photomicrographs, so that the theory can be easily verified, and will hopefully provide information in a useable form.

Figure 1 shows the predictions of the unified theory for the attenuation of ultrasonic waves in aluminum, and compares them to the results of some of the other available theories. The vertical scale is the logarithm of the attenuation coefficient, α, normalized by the grain size, $\bar{d}$, and the horizontal scale is the logarithm of the ratio of the grain size to the wavelength, which ratio can be considered a normalized frequency. All of the theories assume that the bulk materials are macroscopically homogeneous and isotropic, i.e. that the statistics of the microstructural properties are independent of position and direction. In addition, the unified theory assumes that the lengths of cords between grain boundaries on a line through the medium have Poisson statistics, as described and justified below. At low frequencies the unified theory agrees with the equivalent Rayleigh theories of Bhatia and Moore[2] and Lifshits and Parkhomovski[3]. Of the various low-frequency theories which have been presented, these are the most accurate because they account for mode and the anisotropy of the grains. At

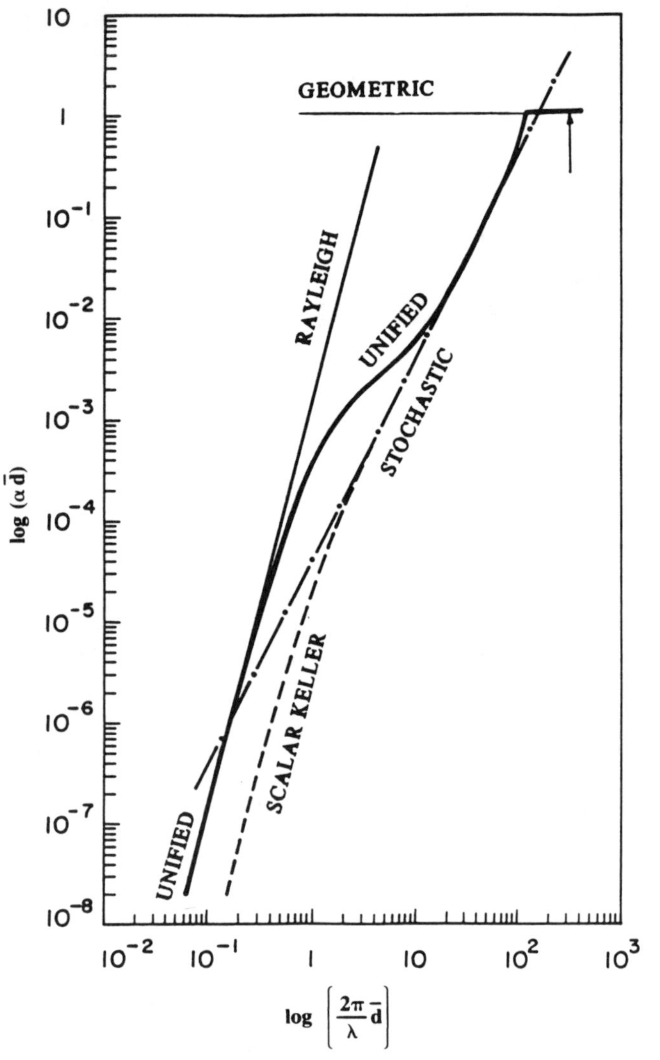

Figure 1 - Predictions of the attenuation coefficient α for the unified theory and simpler theories versus normalized frequency $2\pi\bar{d}/\lambda = 2\pi f\bar{d}/v$.

(moderately) high frequencies when the grain size is somewhat longer than the wavelength, the unified theory agrees with the stochastic theories of Huntington[4] and Lifshits and Parkhomovski[3]. In this frequency range the anisotropy of the grains can be modelled by isotropic grains with varying longitudinal wave velocities, as was done by Huntington. At very small ratios of grain size to wavelength the unified theory agrees with the predictions of geometric acoustics, and is independent of frequency and the elastic properties of the grains. The Scalar Keller[5] theory is closely related to the unified theory. Like the unified theory it is valid for all ratios of grain size to wavelength for the physical situation to which it applies, i.e. compressional waves in a fluid with varying compressibility. At low frequencies the elastic and scalar cases are clearly not analogous, because most of the loss in the elastic case is caused by mode conversion.

CHARACTERIZATION OF GRAIN GEOMETRY

Calculations based on the unified theory require knowledge of the probability that two points separated by any given distance are in the same grain. We will denote this probability as W(r), where r is the distance separating the two points. The use of this particular function to characterize grain geometry is a direct consequence of the second-order perturbation analysis used in the unified theory to describe an inhomogeneous medium. W(r) can be determined for real samples from measurements of the lengths of the cords (i.e. segments) which lie between grain boundaries on a line drawn across a photomicrogragh. The formula[6] for such a determination is:

$$W(r) = \sum_{\substack{n=1 \\ c_n > r}}^{\infty} \frac{c_n - r}{\bar{c}} \qquad (1)$$

where c_n is the length of the nth cord, $\bar{c}$ is the average cord length, and only the positive terms of the sum are included in its calculation. This equation assumes only that the statistics of the grain sizes are homogeneous and isotropic, and that the grains are convex. If it is known or assumed that the cord lengths have Poisson statistics, it can be shown that the result is $W(r) = \exp(-r/\bar{c})$, which is the assumption used in calculations based on the unified theory. In general we will specify grain size as the mean free path, denoted $\bar{d}$, which is the average length of cords which pass through randomly chosen points in the medium. In general, $\bar{d}$ is larger than $\bar{c}$ because a random point is more likely to fall on a long cord than on a short one, whereas a line through the medium discriminates less against short cords. For Poisson statistics, $\bar{d} = 2\,\bar{c}$.

COMPARISONS OF MEASUREMENTS WITH THE UNIFIED THEORY

DESCRIPTION OF SAMPLES - The two samples for the present examination are composed of copper and 304 stainless steel, respectively. The copper sample is very pure, and was made by deforming a cylindrical bar of copper along its axis in compression. The aim of this preparation was to reduce the effects of the sample anisotropy which was introduced when it was initially formed as a bar. The 304 stainless steel[7] is also taken from a round bar, but has been measured in its as-rolled condition, which has a 75mm diameter. Grain size measurements were performed from micrographs of surfaces perpendicular to the bars' axes, and acoustic attenuations were measured parallel to the bars' axes. In all of the cases discussed herein, the measured W(r) is very closely approximated by the inverse exponential form assumed by the unified theory. Also, in all cord-length measurements, twin planes have been regarded as grain boundaries, since the acoustic properties vary across them as for the case of real grain boundaries. Calculations based on the unified theory require knowledge of the elastic constants for the individual crystals. The values for the present calculations have been taken from the literature, and are shown in Table 1. There are numerous published sets of elastic constants for copper, and one[8] was chosen because it leads to upper and lower Hashin and Strikman[9] bounds for the low-frequency, longitudinal-wave velocity which are above and below the measured value for the sample. Only one set[10] of published, single-crystal constants for 304 stainless steel is known to the author, and it has been used.

Table 1. Elastic constants of copper and 304 stainless steel used in unified theory to calculate the attenuation coefficients.

	Elastic Constants (10^{10} N/m)			Density (kg/m^3)
	C_{11}	C_{12}	C_{44}	p
Cu	16.83	12.11	7.57	8.392
304SS	20.05	13.30	12.50	8.010

RESULTS FOR COPPER - The attenuation coefficients in the copper sample were measured from 7 to 80 MHz, and are shown in Figures 2 and 3. In both figures there are three theoretical curves. The central theoretical curve is based on the measured mean free path ($\bar{d}$=22.9 microns) and the published elastic constants. The agreement between this "best" theoretical predication with the measured values is very good, especially considering that the large majority of the data is at frequencies were both the high frequency and the low frequency theories are an order of magnitude in error. The frequency at which $2\pi\bar{d}/\lambda = 1$ is 33.7 MHz.

If the elastic properties of the grains are known, a single-frequency attenuation measurement is sufficient to determine the grain size. In Figure 2, the two outer theoretical curves are based on mean free paths which differ from the measured value by ±20%. Since all of the measured values fall within the two bounds of these calculations, any single measurement could be used, along with the unified theory and the single-crystal constants, to predict a grain size with less than 20% error. An inversion technique which employed all of the data points could obtain an even better estimate of the grain size. The relative accuracy of measurements is limited by the ratio of the attenuation to reflection losses and instrument inaccuracy at low attenuations, and by noise at high attenuations. An inversion algorithm could take these variations in accuracy into account by weighting the data more heavily when it is in a range with higher expected accuracy. Also, the sensitivity of the attenuation coefficient to variations in grain size is higher at the lower frequencies, as indicated by the spreading of the bounds. This sensitivity variation can also be used to weight averages of measurements taken over a range of frequencies.

If the elastic anisotropy of the single crystals is unknown,

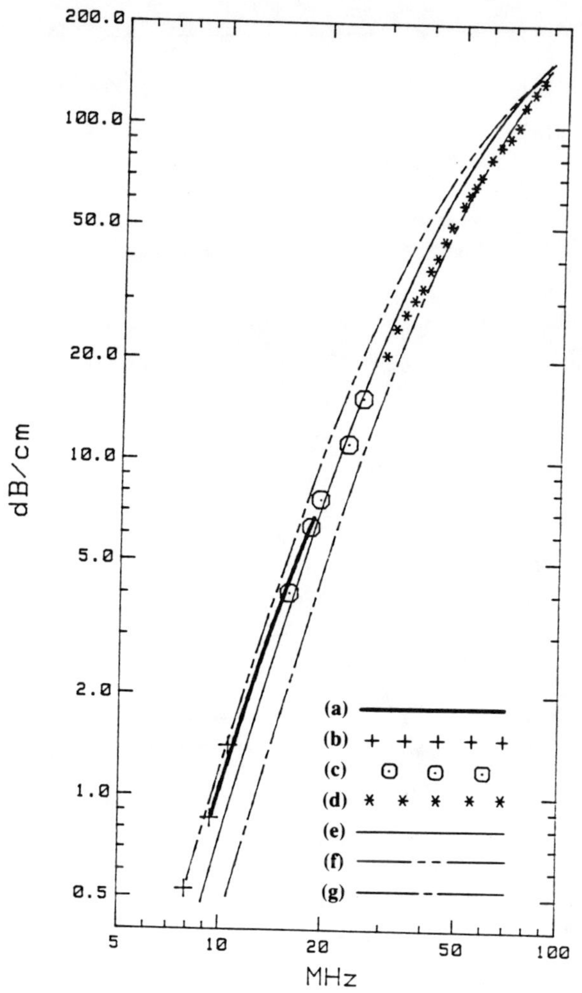

Figure 2 - Comparison of measured attenuation in copper with predictions based on different grain sizes. (a) Spectroscopically measured attenuation. (b), (c) and (d) Quasi-CW measurements. (e) Theory based on published elastic constants and independently measured grain size. (f) Grains 20% larger than measured. (g) Grains 20% smaller than measured.

but the shear and longitudinal velocities are known, then it may be possible to determine both the anisotropy and the grain size from broadband measurements of the attenuation coefficient. The anisotropy factor used here is defined as C_{11} - C_{12} - 2 C_{44}. In Figure 3 the central theoretical curve is based on the published anisotropy factor, and the other two are based on anisotropy factors which are different by ± 10%. The sensitivity of the attenuation coefficient to changes in the anisotropy factor is greatest at the higher frequencies, as indicated by the spreading of the curves. The region of highest sensitivity to changes in grain size and is therefore different from the region of highest sensitivity to changes in anisotropy factor. This distinction is the key to the possibility of obtaining both parameters from the attenuation coefficient. The inversion algorithm should heavily weight the data at high frequencies when estimating the anisotropy factor,

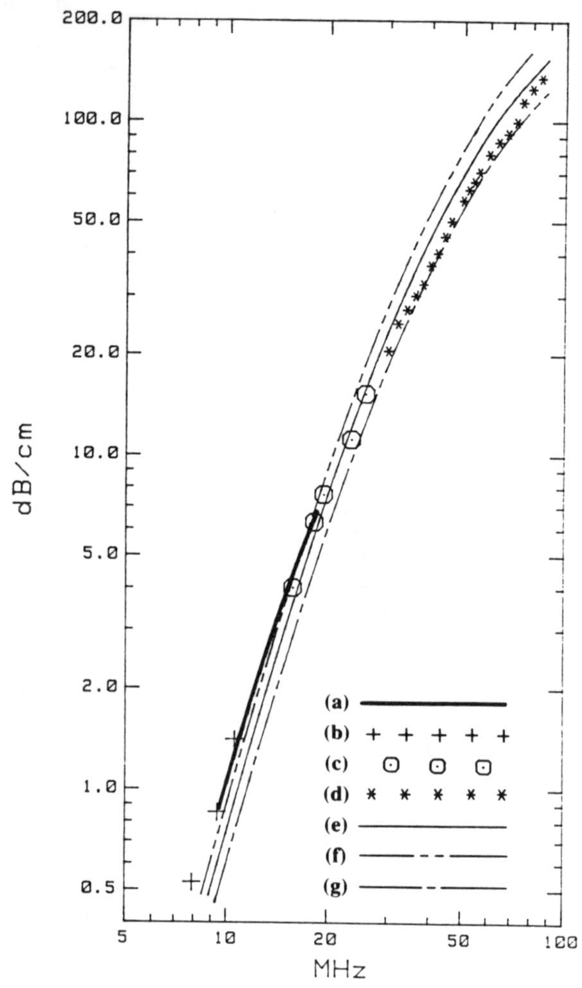

Figure 3 - Comparison of measured attenuation in copper with predictions based on different anisotropy factors. (a) Spectroscopically measured attenuation. (b), (c) and (d) Quasi-CW measurements. (e) Theory based on published elastic constants and independently measured grain size. (f) Anisotropy 10% higher than published. (g) Anisotropy 10% lower than published.

and heavily weight the data at low frequencies when estimating the grain size. The precision of the estimates will depend on the spectrum over which there are measured attenuation coefficients because there must be data both in the grain-size-sensitive region and the anisotropy-sensitive region. On the log-log plot in Figure 1, the Rayleigh approximation appears as a line with a slope of four, corresponding to a fourth-order frequency dependence. The unified theory is indistinguishable from the Rayleigh approximation at very low frequencies and initially diverges very gradually as the frequency increases. Attenuation coefficients which are in this region of nearly fourth-order frequency dependence have high sensitivities to changes in grain size. Then, in the transition region near $2\pi\bar{d}/\lambda = 1$, the unified theory begins diverging much more rapidly, and this is where the attenuation coefficient is least sensitive to grain size and most sensitive to the anisotropy factor. Inversion requires enough data to verify that some is in the nearly Rayleigh region and that some is in the transition region. This implies that the data must cover at least a decade. The spectrum of the measurements shown for the copper specimen is probably adequate.

RESULTS FOR STAINLESS STEEL - Figures 4 and 5 show results for 304 stainless steel that are similar to those for copper shown in Figures 2 and 3, respectively. However, there are two major difference between the results for stainless steel and those for copper. The statistics of the grains in the stainless steel sample are inhomogeneous, being larger in the center ($\bar{d}$=57.2 microns) than they are near the edge ($\bar{d}$=36.4 microns), as a result of the forming process. Consequently, the attenuation coefficients measured at the center and at the edge are shown separately and compared with the results of the unified theory. The second

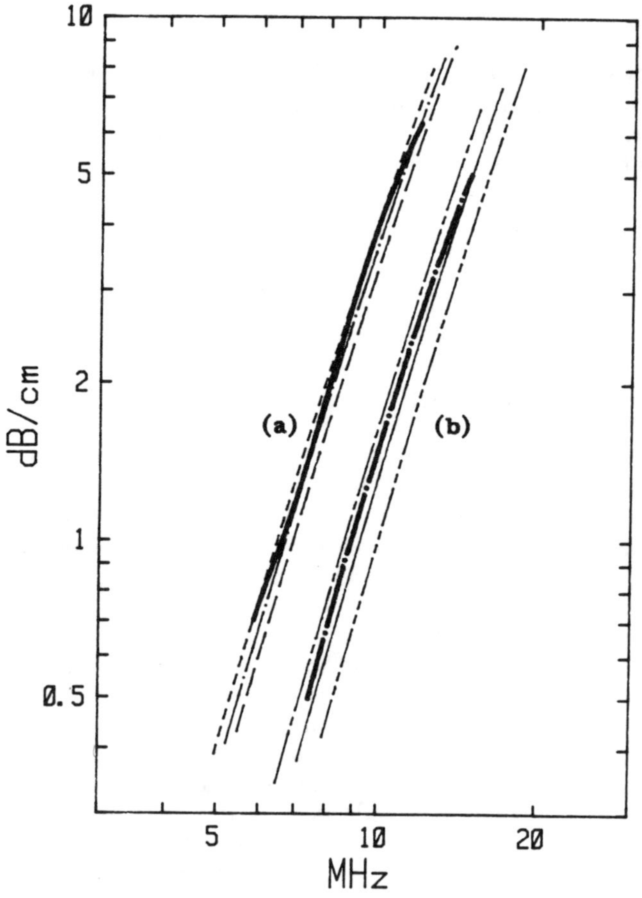

Figure 4 - Comparisons of spectroscopically measured attenuation in stainless steel with theoretical predictions based on different grain sizes. The heavy lines are measurements and the lighter lines are predictions. (a) At the center of the sample, the grain sizes have been varied by ±5%. (b) At the edge of the sample, the grain sizes have been varied by ±5%.

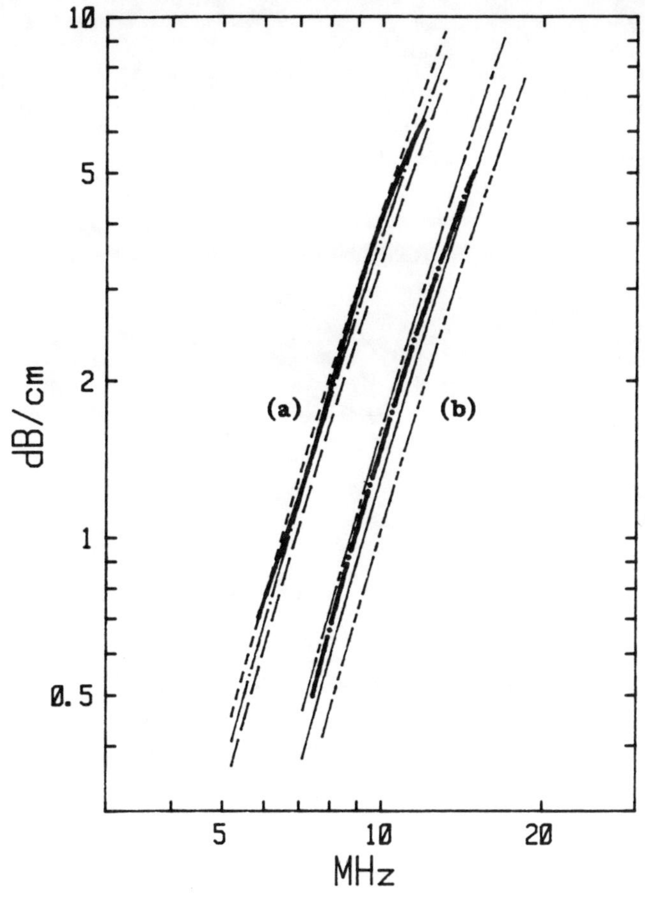

Figure 5 - Comparisons of spectroscopically measured attenuation in stainless steel with predictions based on different anisotropy factors. The heavy lines are measurements and the lighter lines are predictions. (a) At the center of the sample, the anisotropy factors have been varied by ±5%. (b) At the edge of the sample, the anisotropy factors have been varied by ±5%.

difference is that the data covers a much narrower frequency range, approximately one octave at both positions. In both figures, the two sets of measured values are accompanied by three sets of calculated values as before. The central set is based on the published elastic constants and the measured grain size. The attenuation coefficients measured for the stainless steel specimen fall within bounds which represent errors in the predicted grain size of ±5% at its center and ±10% at its edge, as shown in Figure 4. The measured attenuation coefficients in 304 stainless steel are bounded in Figure 5 by theoretical calculations which use anisotropy factors differing from the measured value by ± 5% in the center of the bar and by ± 10% at its edge. The spectra of the measurements in stainless steel are not adequate for inverting both the grain size and the anisotropy because the deviations in grain size and anisotropy factor in the two figures lead to virtually indistinguishable curves. For measurements having a spectrum of approximately an octave, only one of the two parameters can be inverted and the other must be known. Inversion of both parameters requires measurements over at least a decade in frequency.

SUMMARY AND CONCLUSIONS

Any inversions of the measured acoustic properties depend on knowledge that the material at least approximately fulfills the assumptions of the unified theory. Therefore, acoustic measurements cannot replace the various existing analytical techniques used for studying microstructure in materials science. However, they may serve as a useful tool for process control and quality control of manufactured materials which are already well characterized but have some unknown parameters. For example, both materials discussed, here, have primary phases with cubic symmetry grains, neither was strongly textured from severe cold working, and both have grains that are well described by Poisson statistics. In these cases there are only four parameters needed to fully describe any (location in a) sample: shear-wave and longitudinal wave velocities, the relative anisotropy and the mean free path, which are all accessible with acoustic measurements made along one direction of propagation. It seems reasonable to expect predictions of grain size based on attenuation coefficients measured at a single frequency and inversion with the unified theory to be accurate to at least 20% for such materials if the elastic constants are known. Measurements over a band should allow better accuracy to be obtained, and, if the band is very broad, should allow the anisotropy to be inverted as well. Shear-wave and longitudinal-wave velocities can be measured directly. Even when there is not a rigorous theory for the particular case of interest, or there are more parameters than can be conveniently measured, purely empirical methods may be useful.

ACKNOWLEDGEMENT

The author wishes to thank Dr. Hayden N.G. Wadley of the National Bureau of Standards for preparing and micrograghing the copper sample which was used for this work.

REFERENCES

1. F.E. Stanke and G.S. Kino, "A Unified Theory for Elastic Wave Propagation in Polycrystalline Materials", J. Acoustical Soc. Am. **75** 665-681 (1984).
2. A.B. Bhatia and R.A. Moore, "Scattering of High Frequency Sound Waves in Polycrystalline Materials. II," J. Acoustical Soc. Am. **31**, 1140-1141 (1959).
3. E.M. Lifshits and G.D. Parkhomovski, *Zh. Eksperim. i Theoret. Fiz.* **20**, 175-182 (1950).
4. H.B. Huntington, "On Ultrasonic Scattering by Polycrystals," J. Acoustical Soc. Am. **22**, 362-364 (1950).
5. F.C. Karal, Jr. and J. B. Keller, "Elastic, Electromagnetic, and Other Waves in a Random

Medium," Journal of Mathematical Physics **5**, 537-547 (1964).

6. F.E. Stanke, Unified Theory and Measurements of Elastic Waves in Polycrystalline Materials", Ph.D. Thesis, Stanford University, Department of Electrical Engineering (1983).

7. N. Grayeli, F. Stanke, G.S. Kino, and John Shyne, "Effect of Grain Size and Preferred Crystal Texture on Acoustic Properties of 304 Stainless Steel," Proceedings of the ARPA/AFML Review of Progress in Quantitative Nondestructive Evaluation, LaJolla, California, July (1980).

8. S.G. Epstein and O.N. Carlson, "The Elastic Constants of Nickel-Copper Alloy Single Crystals," *Acta Metallurgica* **13**, 487-491 (1965).

9. Gene Simons and Herbert Wang, "Single Crystal Constants and Calculated Aggregate Properties: A Handbook," p.22, The MIT Press, Cambridge, MASS.

10. David S. Kupperman and Karl J. Reimann, "Ultrasonic Wave Propagation and Anisotropy in Austenitic Stainless Steel Weld Material," IEEE Transactions on Sonics and Ultrasonics, **SU-27**, No. 1, January (1980).

MORPHOLOGICAL SEVERITY FACTOR FOR GRAPHITE SHAPE IN CAST IRON AND ITS RELATION TO ULTRASONIC VELOCITY AND TENSILE PROPERTIES

Emmanuel P. Papadakis
Ford Motor Co.
Redford, MI

Leslie Bartosiewicz
Ford Motor Co.
Dearborn, MI

John D. Altstetter (Retired)
Ford Motor Co.

Gilbert B. Chapman, II
Ford Motor Co.
Redford, MI

ABSTRACT

An expression has been derived to quantify the effects of shape and size of all graphite particles in a specimen of cast iron. The expression yields a number termed the Morphological Severity Factor (MSF) when quantitative metallographic measurements are used as data in the expression. The theory used in deriving the expression indicates that the MSF is related to the increase in elastic compliance (decrease in modulus) provided to the iron by the graphite particles. Experiments on metallography, ultrasonic velocity and tensile properties have been performed on a set of specimens covering the range nodular, CG, and gray iron to define the relationships between the MSF and the physical properties. The principal result is that ultrasonic velocity is linearly related to the MSF with a negative slope. Yield and tensile strengths are also monotonic decreasing in MSF as long as the pearlite/ferrite ratio is constant. Apparently both ultrasonic velocity and MSF can be used for assurance of cast iron strength as long as the matrix microstructure is known.

BACKGROUND

GRAPHITE MORPHOLOGY is the most influential of the various metallurgical factors determining the strength of cast iron. (The other principal factors are the pearlite-to-ferrite ratio and the presence or absence of carbides.) Graphite morphology determines the difference between nodular iron, compacted graphite (CG) iron, and gray iron. Therefore it is vital to be able to assure the morphology of the graphite in cast parts to be used for various purposes, particularly in parts to be placed under stress.

At present; there are two methods which are widely used to assure graphite morphology; visual (microscopic) and ultrasonic velocity. At their present state of development, both have limitations. The visual method is carried out by skilled operators and employs visual comparisons with standard micrographs, point-count methods, and quantitative microscopes. Specimen preparation and examination are time-consuming and judgmental in part. In addition, the visual method examines only a surface. Ultrasonic velocity examines a volume of material and can be measured automatically. Generally the material with the highest nodularity exhibits the highest ultrasonic velocity. At lower nodularity values (CG iron), the velocity shows considerable scatter when plotted versus nodularity. See Fig. 1. Gray iron has very

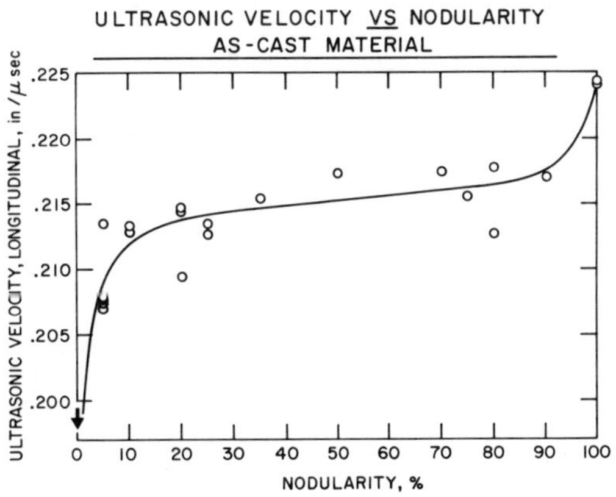

Fig. 1. Ultrasonic velocity plotted versus nodularity in as-cast nodular, CG, and gray iron. In the central region of the curve, there is a low slope and considerable scatter.

low ultrasonic velocity. For nodular iron of moderate to high nodularity, the ultrasonic velocity can be used as an approximate predictor of tensile and yield strength (other metallurgical factors being constant). See Fig. 2.

Ultrasonic velocity holds the potential on both theoretical and practical grounds for being developed into a predictor of graphite morphology in cast iron over the entire range of possible morphologies. Theory predicts that the ultrasonic velocity should decrease if the continuity of the iron is decreased by the graphite. The continuity of the iron is maximum in 100% nodular iron, intermediate in CG iron, and minimum in gray iron. The missing link is a quantitative measure of the shape of the graphite. The measure-of-shape would of necessity account for the possibility of several graphite types within one specimen.

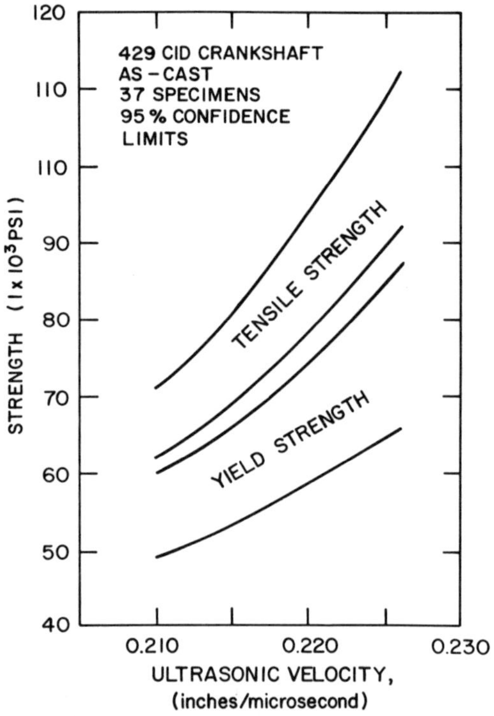

Fig. 2. Tensile and yield strength plotted versus ultrasonic velocity in as-cast modular iron made by the "sandwich" process with some fading to lower the velocity and nodularity.

In the present program, the theory was developed for a measure-of-shape for the graphite. This measure is termed the Morphological Severity Factor (MSF) in the present paper. A methodology was devised for gathering the requisite data by means of a quantitative television microscope (QTM), and a program was written to calculate the MSF from these data. Ultrasonic velocity measurements and QTM measurements were made on a group of cast iron specimens including tensile specimens to develop correlations among the ultrasonic velocity, the MSF, and the tensile data.

The present paper will give the theory of the MSF, the methodology for collecting QTM data, and the mathematics of calculating the MSF from it. The paper will also present the experimental work and the results of the correlations. In particular, the paper will show that the ultrasonic velocity is a linear decreasing function of the Morphological Severity Factor calculated for the graphite.

THEORY

An ultrasonic wave propagating through a solid is an elastic wave[1]. As such, it stresses the solid and produces strains in the elastic region. For the traveling waves used in ultrasonic testing and scientific measurements, the strains are smaller than 10^{-5} in magnitude[2]. Nevertheless, there is a gradient of stress and strain which accompanies the wave front as it passes through the solid.

The gradient of stress acts upon the graphite particles in cast iron to cause different degrees of strain at the forward and rear faces of the graphite (relative to the direction of propagation). This is because the modulus of the graphite is much lower (up to 20 times lower) than the modulus of the iron[3]. Visualize, as in Fig. 3, an extended graphite particle of idealized shape like a thick irregular disc viewed edgewise. The stress wave in

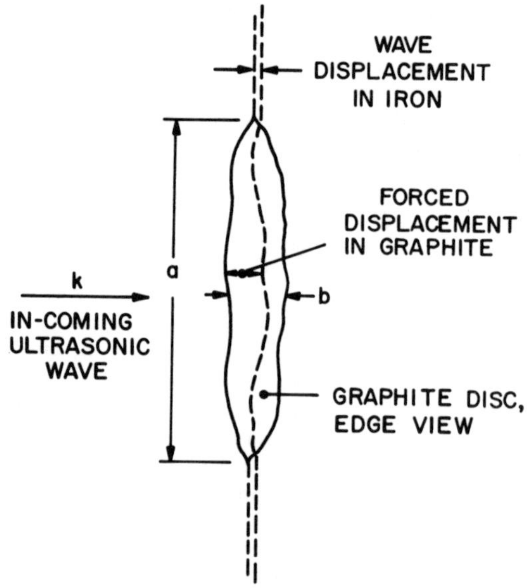

Fig. 3. An idealized disc-shaped graphite particle viewed in cross-section with an ultrasonic wave impinging from the left. The left face, being more compliant than the iron, has a larger displacement as shown by the dashed line. The surface moves analogous to a drum head or a membrane in a pressure wave in air.

the iron can move the front face of the disc more than the adjacent iron outside the perimeter of the disc because of the low graphite modulus. This process is analogous to a pressure wave moving a drum head or the membrane of a microphone. The motion of a drum head in acoustics has been worked out theoretically[4]. In that case, the displacement of the drum head is proportional to the applied pressure and to the area of the membrane. In terms of the analogous disc of diameter a in Fig. 3, the displacement is proportional to a^2. The cross-sectional area of the disc viewed edgewise, A, is proportional to ab where b is the thickness of the disc.

The displacement of the major face of the graphite disc, being larger than the displacement of the adjacent iron, adds to the compliance of the composite iron/graphite structure. Adding compliance implies lowering the modulus. Since the ultrasonic velocity is proportional to the square root of the modulus, the graphite lowers the velocity relative to pure iron. Discs of graphite will have an effect proportional to their area. Thick discs up to and including spheres will have an effect proportional to their projected area in the direction of propagation.

This analysis shows the need to define both the shape and size of each graphite particle in the cast iron. A measure of the shape and size

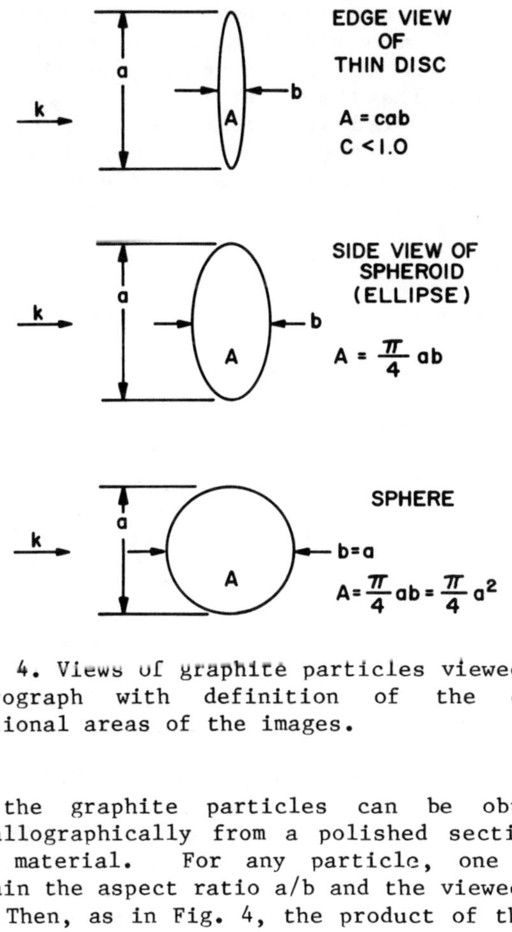

Fig. 4. Views of graphite particles viewed on a micrograph with definition of the cross-sectional areas of the images.

of the graphite particles can be obtained metallographically from a polished section of the material. For any particle, one could obtain the aspect ratio a/b and the viewed area A. Then, as in Fig. 4, the product of the two quantities is:

Thin Disc

$$(a/b)A = (a/b)cab = ca^2 \qquad (1)$$

where c is smaller than 1.0 if b is the maximum thickness.

Ellipse

$$(a/b)A = (a/b)\ (\pi/4)ab = (\pi/4)a^2 \qquad (2)$$

Sphere

$$(a/b)A = (a/b)\ (\pi/4)ab = (\pi/4)a^2 \qquad (3)$$

where a=b for the sphere, making a simplification. In each case, the result is proportional to the projected area of the object in the direction of propagation of the ultrasonic wave. Thus, a measure proportional to (a/b)A is required to characterize the particle. Particles may have any orientation in space, so an average or summation of

$$F_s = \frac{a}{b}\ A \qquad (4)$$

over a representative sample of them is desired. F_s is the definition of the shape factor for one particle; the Morphological Severity Factor (MSF) will be the sum of these for all the particles in a certain area of a micrograph.

From a QTM one can obtain data on areas A and perimeters P of images. The parameter $P^2/4\Pi A$ is dimensionless and is proportional to a/b as follows:

Thin Disc

$$\frac{P^2}{4\pi A} \sim \frac{(2a + 2b)^2}{4\pi ab} \sim \frac{4a^2}{4\pi ab} = \frac{a}{\pi b} \qquad (5)$$

Ellipse

Complicated total elliptic integral which is proportional to a/b. (6)

Sphere

$$\frac{P^2}{4\Pi A} = 1, \text{ the limit of this dimensionless parameter} \qquad (7)$$

Hence, $P^2/(4\Pi A)$ is the first factor in F_s, while A itself is the second factor.

To systematize the theory for data acquisition, a matrix was drawn up using ten rows and eleven columns. The rows represent the reciprocal aspect ratio b/a given by $4\Pi A/P^2$ in units of 0.1. In the first row long particles with ratios 0.0 to 0.1 appear; in the second row ratios 0.1 to 0.2 are recorded; and so on through 0.9 to 1.0. The columns represent the area class interval A in $(\text{microns})^2$, and run from 101-152 through 5838-8757. A blank data acquisition form appears in Fig. 5, and shows the magnitudes of the classes in each row and column. The table is later filled in with the count of particles conforming to the classes.

Since the theory requires the reciprocal $P^2/4\Pi A$ and the area in each class, the averages of these quantities for the 10 x 11 matrix are used in the calculation. The corresponding form appears in Fig. 6, and can be used for hand calculations. The form shows the area A_j as the label of column j, the aspect ratio C_i as the label of row i, and the number of objects B_{ij} filling any category.

QTM DATA ACQUISITION FORM FOR MSF

Sample: ______ Date: ______

Shape Factor $4\pi A / P^2$	Frequency of Occurence										
0.0 - 0.1											
0.1 - 0.2											
0.2 - 0.3											
0.3 - 0.4											
0.4 - 0.5											
0.5 - 0.6											
0.6 - 0.7											
0.7 - 0.8											
0.8 - 0.9											
0.9 - 1.0											
Area Class Interval, Microns ^{2}	101-152	152-228	228-342	342-512	512-769	769-1153	1153-1729	1729-2595	2595-3892	3892-5838	5838-8757

Total Feature Count: ______ Nodularity, % ______
Area Scanned: ____ mm x ____ mm Frames: ______

Fig. 5. Tabular form for collecting data with the QTM.

QTM DATA COMPUTATION FORM

Sample: ______ Date: ______

$C_i = P^2 / 4\pi A$	B_{ij} = Frequency of Occurence										
20											
6.67											
4.00											
2.86											
2.22											
1.82											
1.54											
1.33											
1.18											
1.05											
A_j = Area Microns ^{2}	126	190	285	427	640	961	1441	2162	3244	4865	7298

Fig. 6. Tabular form for data with the variables transformed for computation by Eq. (8).

The final form of the Morphological Severity Factor is

$$MSF = 10^{-7} \sum_{i=1}^{10} \sum_{j=1}^{11} C_i \, A_j \, B_{ij} \qquad (8)$$

One can see that this is the proper form since C_i is a/b, A_j is A, and B_{ij} is an integer from the frequency distribution. The factor 10^{-7} was introduced to scale the results on the basis of the size and number of fields used in the QTM as explained below under Experiments. In the present work, the factor 10^{-7} gave MSF in the range 0.1 to 1.5. A natural way to normalize the function would be to divide by the total area examined by the QTM. This was not done in the present paper.

EXPERIMENTS

A. SPECIMEN - Most specimens were in the as-cast condition. Four annealed specimens (ferritic) were included for comparison. Nodular, CG, and gray iron specimens were obtained. Most of the specimens were tensile bars, some 0.50 in diameter and some 0.40 in diameter. Of these, some were from production parts and some from arbitration bars, brick-shaped bars, and other shapes cast in our experimental foundry. A few specimens from arbitration bars were only 1.25 in long, and Oould not be used for tensile tests.

B. ULTRASONIC VELOCITY - Ultrasonic velocity was measured in three specimen configurations and with three test systems. For most of the measurements, a pulse-echo-overlap system[5], the Panametrics 5054*, was used in the through-transmission mode with NDT transducers to measure travel time. Micrometers were used for measuring length. To account for the time delay in the wearplates of the two transducers, the travel time in a steel step block was measured at several steps of thickness, and the wear plate time increment was computed[6]. Measurements in tensile specimens were made along the axis of the tensile bars where possible. Two cases required measurements through the diameter within the gage length: (1) gray iron with high attenuation**, and (2) previously pulled tensile bars, a few of which were still available. These were measured in undisturbed areas away from the necking region. Measurements in the 1.25 in sections of arbitration bars were performed with a Branson 1207D automatic velocity system***. One subset of specimens was measured with a velocity-calibrated thickness gage, the Panametrics 5221****. Figure 7 shows the arrangement of the transducers for the various measurements. None of the measurements were corrected for difraction (beam spreading)[7] because of the complicated geometries. It is estimated that the maximum errors in any measurement is 1/2%, and that the precision from part to part of identical geometry is better than 1/4%.

C. YIELD AND TENSILE STRENGTH - Yield and tensile strengths were measured on tensile machines calibrated according to ASTM-E4-72(78). While the iron exhibits a smooth curve of stress versus strain with no definite yield point, the 0.2% offset definition was used for yield.

---

* Panametrics, Inc., Waltham, MA

** Some gray specimens of rectangular cross-section were measured similarly.

*** Krautkramer-Branson, Stratford, Conn.

**** Panametrics, Inc. Waltham, MA

D. OPTICAL MEASUREMENTS: QTM - For the tensile bars, the optical specimens were cut from one end of the bar near the end of the threads nearest the gage section. The optical face was perpendicular to the bar axis. For the 1.25 in. long cylinders from the arbitration bars, one end (and sometimes both ends) was polished.

The QTM data were taken on 56 fields at 500X in a 7 x 8 rectangular array distributed over a 7mm x 7mm area of each iron specimen. (Some specimens were too small; smaller areas were used.) This area was slightly smaller than the cross-section of the ultrasonic beam. Where possible, the optical area was placed within the ultrasonic beam position (specifically for the arbitration bar pieces and for the tensile bars ultrasonically measured along their length.)

Typical results of the QTM measurements are shown in Tables 1, 2, and 3 for a nodular, a CG, and a gray iron specimen, respectively. One can see the progression from a preponderance of low aspect ratios (low C_i) in nodular iron to high preponderance of high aspect ratios in gray iron.

Complete data on the specimens are given in the Appednix.

E. CALCULATIONS - For each specimen, the QTM measurements were used as data B_{ij} to calculate the MSF by Eq.(8). Some specimens had two QTM areas (both ends, for instance) and/or two ultrasonic velocity measurements on different regions. When these data are plotted under Results, below, each point is treated as an independent datum.

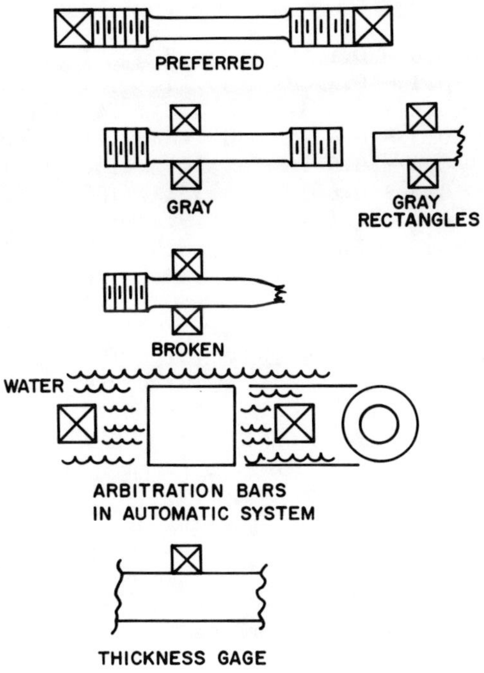

Fig. 7. Transducer locations for ultrasonic velocity measurements.

Table 1 - QTM Data for Nominally Nodular Specimen

Sample: C455A - Red 1 | Date: 5-5-82

Shape Factor $4\pi A/P^2$	Frequency of Occurrence										
0.0 - 0.1	0	0	0	0	0	0	1	0	0	0	0
0.1 - 0.2	0	0	0	0	0	1	1	0	0	0	0
0.2 - 0.3	0	1	1	0	3	2	0	2	0	0	0
0.3 - 0.4	5	3	3	0	3	8	5	2	0	0	0
0.4 - 0.5	7	10	0	10	4	9	5	2	1	0	0
0.5 - 0.6	15	16	18	15	14	14	12	4	1	0	0
0.6 - 0.7	10	17	21	22	32	21	20	6	1	0	0
0.7 - 0.8	15	20	21	39	52	33	21	5	2	0	0
0.8 - 0.9	26	17	31	34	63	81	60	27	2	0	0
0.9 - 1.0	15	21	31	47	111	131	146	72	5	0	0
Area Class Interval, microns^2	101 152	152 228	228 342	342 512	512 769	769 1153	1153 1729	1729 2595	2595 3892	3892 5838	5838 8757

Total Feature Count: 1478 | Nodularity, %: 62
Area Scanned: 7.0mm x 7.0mm | Frames: 56

Table 2 - QTM Data for Nomically CG Specimen

Sample: C454-2A - Green 1 Date: 5-5-82

Shape Factor $4\pi A/P^2$	Frequency of Occurrence										
0.0 - 0.1	3	1	1	3	6	11	5	1	1	0	0
0.1 - 0.2	1	4	23	48	58	31	17	7	2	1	1
0.2 - 0.3	20	46	107	134	82	40	14	1	1	0	0
0.3 - 0.4	48	123	169	106	48	20	2	1	1	0	0
0.4 - 0.5	109	166	121	45	26	18	4	0	0	0	0
0.5 - 0.6	131	106	38	21	13	8	3	3	0	0	0
0.6 - 0.7	91	58	21	16	6	6	3	3	1	0	0
0.7 - 0.8	63	40	21	11	11	8	4	2	1	0	0
0.8 - 0.9	39	26	14	12	13	13	14	3	2	0	0
0.9 - 1.0	50	22	13	15	23	17	11	10	1	0	0
Area Class Interval,	101	152	228	342	512	769	1153	1729	2595	3892	5838
microns^2	152	228	342	512	769	1153	1729	2595	3892	5838	8757

Total Feature Count: 2664 Nodularity, %: 11
Area Scanned: 7.0mm x 7.0mm Frames: 56

Table 3 - QTM Data for Nominally Gray Specimen

Sample: C463A - Orange 1 Date: 5-5-82

Shape Factor $4\pi A/P^2$	Frequency of Occurrence										
0.0 - 0.1	8	17	18	22	40	68	50	27	17	6	3
0.1 - 0.2	18	50	112	180	177	124	58	24	5	1	1
0.2 - 0.3	111	224	295	204	126	44	11	5	2	1	0
0.3 - 0.4	263	280	222	88	48	25	7	3	1	1	0
0.4 - 0.5	243	188	114	47	19	8	2	1	1	0	0
0.5 - 0.6	175	113	44	25	4	5	0	0	0	0	0
0.6 - 0.7	119	73	22	13	9	2	2	0	0	0	0
0.7 - 0.8	51	18	8	8	7	2	1	1	0	0	0
0.8 - 0.9	45	13	12	4	3	2	0	0	0	0	0
0.9 - 1.0	42	29	17	17	10	7	4	0	2	0	0
Area Class Interval,	101	152	228	342	512	769	1153	1729	2595	3892	5838
microns^2	152	228	342	512	769	1153	1729	2595	3892	5838	8757

Total Feature Count: 4519 Nodularity, %: 5
Area Scanned: 7.0mm x 7.0mm Frames: 56

RESULTS

Ultrasonic velocity is plotted against MSF in Fig. 8. The result is linear with negative slope from nodular iron through gray iron. The range of MSF is from 0.1 to 1.5 in the present formulation, while the range of velocity is from 0.229 to 0.155 in inches/microsecond (or 0.582 to 0.394 cm/microsecond). The data in Fig. 8 encompass the range from 85% pearlite to 15% pearlite with respect to pearlite-to-ferrite ratio. None of the specimens were carbidic. The best line through the points calculated by orthogonal regression[8] is

$$v = 0.22808 - 0.052520 \ \text{MSF} \tag{9}$$

For v as the running variable, the equation is

$$\text{MSF} = 4.3426 - 19.040 \ v \tag{10}$$

The correlation coefficient in both cases is -0.9035 or better. The negative sign indicates the slope of the lines in Eqs. (9) and (10). The magnitude above 0.9 indicates the good correlation achieved. 81% of the variability is explained by the correlation. The result in Fig. 8 can be summarized as follows: ultrasonic velocity and the MSF are linearly related with good correlation in as-cast cast iron independent of pearlite/ferrite ratio in structures which are not carbidic. There is at present no data on carbidic structures.

An interesting point to note is that plain carbon and low alloy steel with velocity around 0.232 inches/microsecond falls on the line if one assumes MSF = 0 (no graphite) for it. Also, the annealed specimens fall on the same line. The steel was not included in the calculation of the regression line.

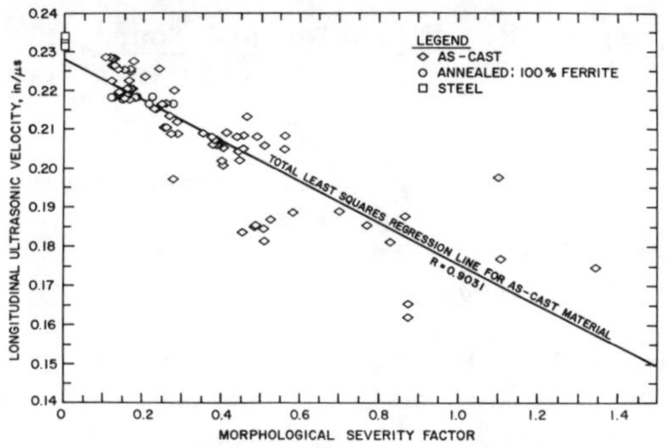

Fig. 8. Ultrasonic velocity plotted versus the Morphological Severity Factor (MSF) calculated from Eq. (8). A linear relationship obtains over all iron grades. MSF is in the vicinity of 0.15 for 100% nodularity and generally above 0.8 for gray iron.

Results of yield strength and tensile strength versus the MSF are plotted in Figs. 9 and 10, respectively. There is a general decrease in strength with increasing MSF. However, pearlite-to-ferrite ratio is very influential. At constant MSF, 40-85% pearlite specimens are stronger by a factor of as much as 1.5 than ferritic specimens. The lower strength and higher ductility of ferritic specimens is well known. There is a large spread of values among the dominantly pearlitic specimens, however.

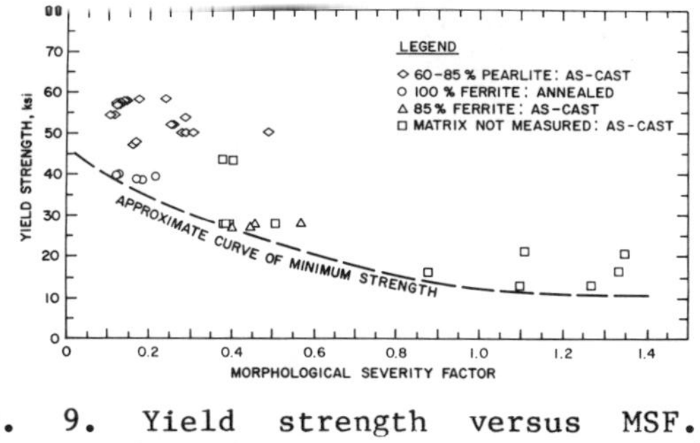

Fig. 9. Yield strength versus MSF. The strength depends on MSF and pearlite/ferrite ratio.

To complete the data presentation, yield and tensile strength are plotted against ultrasonic velocity in Figs. 11 and 12, respectively. There is a general increase in strength with increasing velocity as Fig. 8 would predict, but again the pearlite-to-ferrite ratio is a strong influence. As found in other studies, it seems that the minimum strength can be inferred from the ultrasonic velocity.

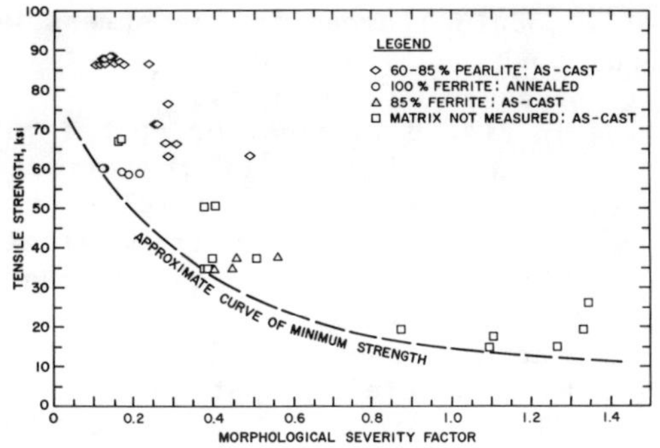

Fig. 10. Tensile strength versus MSF. The strength depends on MSF and pearlite/ferrite ratio.

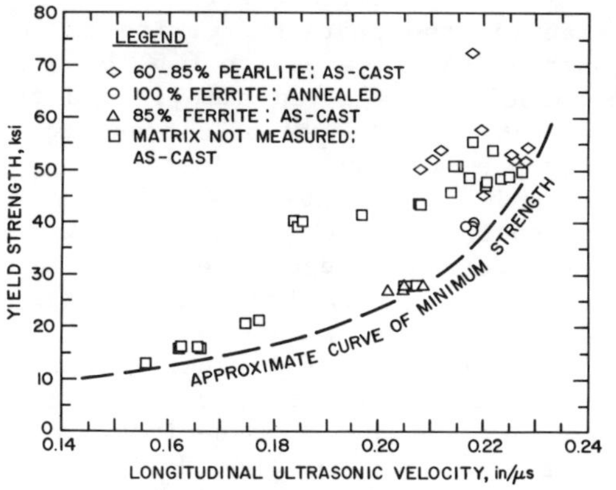

Fig. 11. Yield strength versus ultrasonic velocity. At constant pearlite/ferrite ratio, the velocity shows the strength.

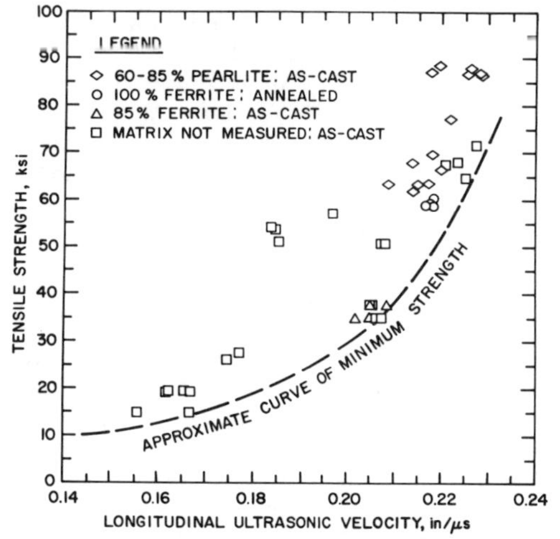

Fig. 12. Tensile strength versus ultrasonic velocity. At constant pearlite/ferrite ratio, the velocity shows the strength.

DISCUSSION

The strong correlation with negative slope between ultrasonic velocity and the Morphological Severity Factor provides evidence that the theory of added compliance caused by the graphite is correct. The surfaces of the graphite particles appear to act as high compliance (low modulus) areas in the stress wave (ultrasound) path analogous to membranes in an atmospheric pressure wave (sound).

The line in Fig. 8 is superior to the curve in Fig. 1 for the evaluation of the effect of graphite morphology on ultrasonic velocity and for the prediction of morphology from velocity. Nodularity (Fig. 1) is inadequate to characterize the ultrasonic velocity in CG iron. The effect of the actual shape and size of the non-nodular graphite particles must be accounted for. The MSF performs this function adequately.

Figures 9 through 12 indicate that the strength parameters should be predictable from combined knowledge of the MSF and the pearlite/ferrite ratio or equally from the ultrasonic velocity and the pearlite/ferrite ratio. In foundry practice the pearlite/ferrite ratio is fairly constant (e.g., a 0.4 to 0.6 for a nominal 0.5 value) for a given alloy, mold type, and cooling rate, so the use of ultrasonic velocity to assure strength is justified. The proper approach is to choose a velocity high enough to assure strength no lower than some value at a chosen confidence level. The 95% level is shown in Fig. 2. The accuracy will be determined empirically in each case.

There are two areas of further research on the MSF. One is to measure the pearlite/ferrite ratio in more of the specimens in the present set to improve the definition in Figs. 9 through 12. The second is to study the MSF and corresponding ultrasonic velocity in a broader range of annealed specimens made ferritic deliberately. It is expected, by analogy with steel[10], that the matrix will have little effect upon velocity, and that the major effect will come through the MSF. It is known[11] that annealing does have a considerable effect upon velocity, but the mechanism accounting for the change is not known at this time. The present modicum of data indicates that annealing results in points which fall on the v vs. MSF curve for as-cast material.

ACKNOWLEDGMENTS

The authors wish to acknowledge the careful and painstaking QTM work of S. LaRouche at the Ford Motor Company Central Services Laboratory. After the QTM methodology was devised and explained by one of the authors (L.B.), he performed many of the measurements. Advice and suggestions from B. V. Kovacs proved helpful, also.

REFERENCES

1. Mason, W. P., Physical Acoustics and the Properties of Solids, Van Nostrand, Princeton, N.J., 1958.

2. Lindsay, R. B., Mechanical Radiation, McGraw-Hill, N.Y., 1960, pp. 249-250.

3. Papadakis, E. P., and Bernstein, H., "Elastic Moduli of Pyrolytic Graphite," J. Acoust. Soc. Amer. 35, 521-524 (1963).

4. Morse, P. M., Vibration and Sound, McGraw-Hill, N.Y., 1948, pp. 183-208 (particularly 202).

5.Papadakis, E. P., "New Compact Instrument for Pulse-Echo-Overlap Measurements of Ultrasonic Wave Transit Times," Rev. Sci. Instrum. 47, 806-813 (1976).

6. Papadakis, E. P., and Petersen B. W., "Ultrasonic Velocity as a Predictor of Density in Sintered Powder Metal Parts," Mater. Eval. 37 (5), 76-80, (April 1979).

7. Papadakis, E. P., "Ultrasonic Diffraction Loss and Phase Change in Anisotropic Materials," J. Acoust. Soc. Amer. 40, 863-876 (1966).

8. Pearson, K., "On Lines and Planes of Closest Fit to Systems of Points in Space," Phil. Mag. 2, 559-572 (1901).

9. Papadakis, E. P. and Chapman, II, G. B., "Computer Method for Linear Regression Analysis Applicable to Situations With Errors in Both Variables," manuscript in preparation.

10. Papadakis, E. P., "Ultrasonic Attenuation and Velocity in Three Transformation Products in Steel," J. Appl. Phys. 35, 1474-1482 (1964).

11. Henderson, H. E. "The Effect of Heat Treatment on Ultrasonic Velocity of Ductile Iron Castings," The Iron Worker 40 (3), 23-25 (Summer 1976).

Appendix - Data on Specimens

Specimen	MSF	Ultrasonic Velocity in/μs	Strength, ksi Tensile	Strength, ksi Yield	% Ferrite	Annealed	Source	Velocity Test Set	Nom.
11G	0.1171 0.1609	0.2241 0.2241	NA	NA	45	No	Lab	1207D	Nod.
41J	0.2468 0.2550	0.2167 0.2167	NA	NA	30	No	Lab	1207D	Lo.Nod.
61L	0.2662 0.4637	0.2134 0.2134	NA	NA	18	No	Lab	1207D	CG
81J	0.5090 0.8303	0.1811 0.1811	NA	NA	0	No	Lab	1207D	Gray
21J	0.1513 0.1369	0.2243 0.2243	NA	NA	NA	No	Lab	1207D	Nod.
41N	0.2329 0.2317	0.2153 0.2153	NA	NA	23	No	Lab	1207D	Lo.Nod.
71B	0.2897 0.2727	0.2080 0.2080	NA	NA	16	No	Lab	1207D	CG
C455A Red-1	0.1692	0.2207	67.7	48.0	NA	No	Lab	5054	Nod.
C455B Yellow-1	0.1612	0.2207	67.3	47.2	NA	No	Lab	5054	Nod.
C454-1B Brown-1	0.4040	0.2081	50.7	43.4	NA	No	Lab	5054	CG
C454-2A Green-1	0.3790	0.2080	50.7	43.6	NA	No	Lab	5054	CG
C456A White-1	1.1092	0.1771	27.8	21.3	NA	No	Lab	5054	Gray
C456A Black-1	1.3462	0.1745	26.2	20.8	NA	No	Lab	5054	Gray
C463A Pink-1	0.8769 0.8769	0.1656 0.1624	19.5	16.4	NA	No	Lab	5054	Gray
C463A Orange-1	1.3337 1.3337	0.1669 0.1622	19.4	16.2	NA	No	Lab	5054	Gray
C430-2	0.1403 0.1450	0.2196 0.2196	88.5	57.9	40	No	Lab	5054	Nod.
C430-4	0.1665 0.1523	0.2179 0.2179	87.1	72.4	40	No	Lab	5054	Lo.Nod.
C430-6	0.2865	0.2120	76.9	53.9	40	No	Lab	5054	CG
C430-8	0.2559 0.2570	0.2103 0.2103	71.4	52.0	40	No	Lab	5054	CG

Specimen	MSF	Ultrasonic Velocity in/μs	Strength, ksi Tensile	Yield	% Ferrite	Annealed	Source	Velocity Test Set	Nom.
C430-10	0.2900	0.2081	63.4	50.3	40	No	Lab	5054	CG
	0.4910	0.2081							
C460-2	NA	0.2174	63.5	48.7	40	No	Lab	5054	CG
C460-4		0.2180	69.7	55.5	40	No	Lab	5054	CG
C469-6		0.2140	61.8	45.9	40	No	Lab	5054	CG
C460-8		0.2219	77.2	53.9	40	No	Lab	5054	Nod.
C460-10		0.2140	67.9	50.9	40	No	Lab	5054	CG
C460-12		0.2150	63.3	48.9	40	No	Lab	5054	CG
105494A									
"A	0.4877	0.1852	51.1	40.1	NA	No	Comm.	5054	CG
"B	0.4518	0.1836	54.1	40.3	NA	No	Comm.	5054	CG
"C	0.5060	0.1844	53.7	39.1	NA	No	Comm.	5054	CG
"D	0.2784	0.1971	57.1	41.4	NA	No	Comm.	5054	CG
C234-G0	1.0997	0.1976	NA	NA	NA	No	Lab	5054	Gray
C234-G1	0.7021	0.1890	NA	NA	NA	No	Lab	5054	Gray
C234-G2	0.5821	0.1887	NA	NA	NA	No	Lab	5054	Gray
C234-G3	0.8675	0.1875	NA	NA	NA	No	Lab	5054	Gray
C234-G4	0.5246	0.1869	NA	NA	NA	No	Lab	5054	Gray
CCP-A	0.1736	0.2275	71.7	49.8	NA	No	Comm.	5054	Nod.
CCP-B	0.1699	0.2251	64.7	48.9	NA	No	Comm.	5054	Nod.
CCP-C	0.2034	0.2235	68.3	48.6	NA	No	Comm.	5054	Nod.
1-26	0.4576	0.2051	37.6	28.0	85	No	Lab	5221	CG
	0.5619	0.2051							
	0.4576	0.2086							
	0.5619	0.2086							
2-10-1	0.4466	0.2020	34.9	27.0	85	No	Lab	5221	CG
	0.4020	0.2020							
	0.4466	0.2049							
	0.4020	0.2049							
2-10-3	1.2675	0.1557	15.0	13.0	85	No	Lab	5221	Gray
	1.0971	0.1557							
	1.2675	0.1668							
	1.0971	0.1668							
2-17-1	0.3874	0.2062	35.0	28.0	85	No	Lab	5221	CG
	0.3855	0.2062							
	0.3888	0.2062							
	0.3874	0.2074							
	0.3855	0.2074							
	0.3888	0.2074							

Specimen	MSF	Ultra-sonic Velocity in/μs	Strength, ksi Tensile	Yield	% Ferrite	An-nealed	Source	Velocity Test Set	Nom.
2-17-2	0.3976	0.2051	37.4	28.1	85	No	Lab	5221	CG
	0.5082	0.2051							
	0.3976	0.2059							
	0.5082	0.2059							
421-1	0.2802	0.2200	66.4	50.2	40	No	Lab	5221	Nod.
	0.3083	0.2200							
421-2	0.1277	0.2284	86.8	56.9	40	No	Lab	5221	Nod.
	0.1257	0.2284							
421-3	0.1164	0.2286	86.4	54.5	40	No	Lab	5221	Nod.
	0.1068	0.2286							
421-4	0.1265	0.2262	88.0	57.1	40	No	Lab	5221	Nod.
	0.1227	0.2262							
421-5	0.1750	0.2256	86.8	58.1	40	No	Lab	5221	Nod.
	0.2406	0.2256							
C122-7	0.2150	0.2167	59.0	39.5	100	Yes	Lab	5054	Nod.
	0.2784	0.2167							
C121-18	0.1202	0.2183	60.3	40.1	100	Yes	Lab	5054	Nod.
	0.1229	0.2183							
	0.1781	0.2183							
C127	0.1852	0.2184	58.8	38.6	100	Yes	Lab	5054	Nod.
	0.2216	0.2184							
GM	0.1695	0.2179	59.4	39.0	100	Yes	Comm.	5054	Nod.
	0.1495	0.2179							
Steel* SAE 4150	0	0.2340	-	-	Pearlite & Ferrite	No	Lab	5054 Prototype	-
	0	0.2329	-	-	Bainite	No	Lab	5054 Prototype	
	0	0.2315	-	-	Martensite	No	Lab	5054 Prototype	
	0	0.2325	-	-	Tempered Martensite	Yes	Lab	5054	

* E. P. Papadakis, "Ultrasonic Attenuation and Velocity in Three Transformation Products in Steel," J. Appl. Phys. 35, 1474-1482 (1964).

INFERENCE OF STRESS AND TEXTURE FROM THE ANGULAR DEPENDENCE OF ULTRASONIC PLATE MODE VELOCITIES

R.B. Thompson, J.F. Smith, S.S. Lee
Ames Laboratory
Iowa State University
Ames, Iowa

ABSTRACT

The theory for the angular dependence of the ultrasonic wave velocity in a symmetry plane of an orthorhombic, stressed material is presented. The two waves having polarizations in this plane are shown to have velocities which can be estimated from measurements of the SH_0 and S_0 guided modes of a thin plate: the relationship being exact for the SH_0 mode and requiring a 10% correction for the S_0 mode at long wavelength. It is then shown how stress and texture can be independently inferred from various features of the angular dependence of these two velocities. From the SH_0 data, the ability to determine the directions and differences in magnitudes of principal stresses is described and supported by experimental data on several materials. From a combination of the SH_0 and S_0 data, a procedure is proposed for determining the coefficients W_{400}, W_{420}, and W_{440} of an expansion of the crystallite orientation distribution function in terms of generalized Legendre functions. Possible applications in process control are indicated.

INTRODUCTION

Successful process control requires the nondestructive characterization of a number of structural properties. Included are hardness, grain size, ductility, strength, preferred orientation, and stress.[1] This paper discusses techniques whereby significant information related to the latter two quantities in rolled plate can be inferred from the angular dependence of the velocity of ultrasonic waves propagating in the plane of the plate.

THEORY OF THE ANGULAR DEPENDENCE OF ULTRASONIC VELOCITY

CONSIDER A ROLLED PLATE and a coordinate system in which the 1, 2, and 3 axes coincide, respectively, with the rolling direction, the transverse direction, and the thickness direction, as shown in Fig. 1. Assume that the material can be represented as an elastic continuum with orthorhombic symmetry, and imagine that a biaxial stress is present in the plane of the plate with principal values σ_a and σ_b and orientation angle Ω as shown in Fig. 2. The velocity of ultrasonic waves propagating in the plane of the plate, i.e. guided modes of the plate, must be obtained by solving the nonlinear, anisotropic equations of dynamic elasticity subject to the stress free boundary conditions on the two surfaces of the plate. The formidable job can be simplified by first neglecting the plate surfaces and treating plane wave propagation in an unbounded medium. Using an analysis to be reported elsewhere,[2] the solutions for the longitudinal and transverse modes, to second order in the elastic anisotropy, are found to be of the form

$$\begin{aligned}\rho V_L{}^2 &= \bar{C}_L + T \\ &+ \frac{\bar{\alpha}}{2}\bar{C}_L \cos 2\theta - \frac{\bar{\beta}}{2}\bar{C}_T (1-\cos 4\theta) \\ &- \frac{\bar{\beta}^2 \bar{C}_T{}^2}{4(\bar{C}_L-\bar{C}_T)}\left[(1-\cos 4\theta)^2 - 2(1-\cos 4\theta)\right] \\ &+ \frac{\bar{\alpha}^2}{32}\frac{\bar{C}_L{}^2}{(\bar{C}_L-\bar{C}_T)}\left[1-\cos 4\theta\right] \\ &+ \frac{\overline{\alpha\beta}}{4}\frac{\bar{C}_L\bar{C}_T}{(\bar{C}_L-\bar{C}_T)}\left[(1-\cos 4\theta)(\cos 2\theta)\right] \\ &+ (\bar{C}_{16}+\bar{C}_{26})\sin 2\theta + \frac{1}{2}(\bar{C}_{16}-\bar{C}_{26})\sin 4\theta\end{aligned} \qquad (1)$$

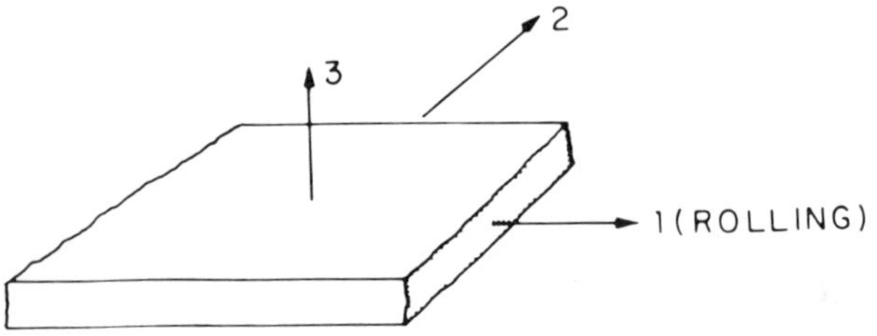

Fig. 1. Coordinate system.

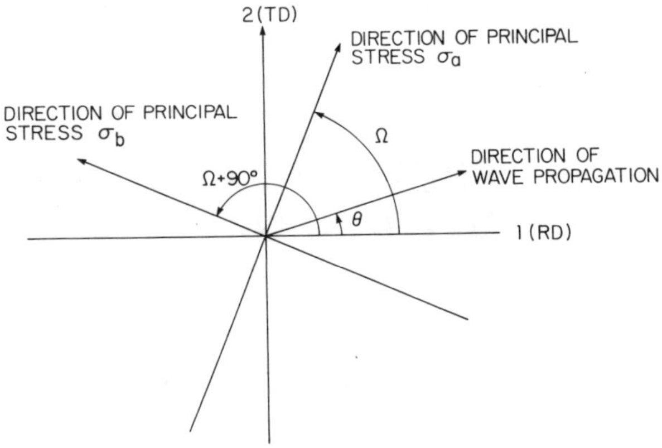

Fig. 2. Orientation of stress and wave propagation direction.

$$\rho V_T{}^2 = \overline{C}_T + T + \frac{\overline{\beta}}{2}\overline{C}_T\,(1-\cos 4\theta) + \frac{\overline{\beta}^2\,\overline{C}_T{}^2}{4\,(\overline{C}_L-\overline{C}_T)}\left[(1-\cos 4\theta)^2 - 2\,(1-\cos 4\theta)\right] - \frac{\overline{\alpha}^2}{32}\,\frac{\overline{C}_L{}^2}{(\overline{C}_L-\overline{C}_T)}\left[1 - \cos 4\theta\right] - \frac{\overline{\alpha\beta}}{4}\,\frac{\overline{C}_L\overline{C}_T}{(\overline{C}_L-\overline{C}_T)}\,(1-\cos 4\theta)\,(\cos 2\theta) - \frac{1}{2}\,(\overline{C}_{16} - \overline{C}_{26})\sin 4\theta \qquad (2)$$

Here ρ is density, $V_{L,T}$ is the wave velocity of longitudinal and transverse waves, θ is the angle of propagation with respect to the 1-direction, $\overline{C}_T = \overline{C}_{66}$, $\overline{C}_L = (\overline{C}_{11} + \overline{C}_{22})/2$, $\overline{\alpha} = (\overline{C}_{11} - \overline{C}_{22})/C_L$, $\overline{\beta} = \left[(C_L - \overline{C}_{12})/2 - \overline{C}_T\right]/\overline{C}_T$. The latter two parameters are measures of longitudinal and shear wave velocity anisotropies, respectively. In addition,

$$T = (\sigma_a + \sigma_b)/2 + \left[(\sigma_a - \sigma_b)/2\right]\cos 2\,(\Omega - \theta) \qquad (3)$$

where σ_a and σ_b are the principal stresses with the former acting along an axis inclined at an angle Ω with respect to the 1-direction. In the absence of stress, the $\overline{C}_{ij}$ are equal to the anisotropic elastic stiffnesses of the orthorhombic continuum, for which $\overline{C}_{16} = \overline{C}_{26} = 0$. Hence there are nine independent values.[3]

When stress is present, there will be small shifts in the values of these nine constants.[4] If $\Omega \neq 0$ the constant $\overline{C}_{16}$ and $\overline{C}_{26}$ will no longer vanish due to the lowering of the effective symmetry of the problem from orthorhombic to triclinic.

The solution V_L corresponds to a quasi-longitudinal wave which is polarized in the 1-2 plane, while the solution V_T corresponds to a quasi-transverse wave polarized in the same plane. A third solution, polarized along the 3-direction, also exists which will not be discussed here.

The relationship of these plane wave solutions for an unbounded medium will next be related to wave propagation in a plate. Consider first the shear wave solution given by Eq. (2). Examination of the full solution shows that all stress components vanish on planes Z = constant. Hence the plane wave results are also exact solutions in the plate geometry. Since they are uninfluenced by the position of the plate faces, they are non-dispersive. This solution is known as the SH_0 mode of the plate.[5]

The plane waves whose velocities are given by Eq. (1) are not exact solutions in the plate geometry since the dynamic stress component σ_{33} does not vanish for these solutions. However, a simple approximation exists when the frequency is sufficiently low that the wavelength is large with respect to the plate thickness. In isotropic, linear elasticity, the fundamental, symmetric Lamb mode[5], S_o, is known to propagate with a velocity of $V_{S_o} = V_{L_o}\left[1-\left(\frac{\nu}{1-\nu}\right)^2\right]^{1/2}$ in the low frequency limit. Here V_{L_o} is the plane wave longitudinal velocity and ν is Poisson's ratio. For a Poisson's ratio of 0.3, $V_{S_o}/V_{L_o} = 0.90$.

In an orthorhombic material, this relationship generalizes to

$$V_{S_o} = V_{L_o}(1-C_{i3}^2/C_{ii}C_{33})^{1/2} \qquad (4)$$

for propagation along the "i" symmetry axis. Thus the correction relating the plate to plane wave velocities will depend on direction. The correction is the order of 10% as noted above, and it depends on elastic constants which only vary by a few percent in modestly textured materials. Hence, in many cases, the angular dependence of the correction may be negligible and its isotropic value can be used. This point will be discussed further in the second section following.

As an experimental test of Eq. (2), consider the plot of $\Delta V/V$ versus for θ for 6061 Al in Fig. 3.[6] In the upper plot, $\sigma_a = \sigma_b = 0$ and the data are fitted well by the first and third term on the right hand side of Eq. (2). When $\sigma_a \neq 0$ and $\Omega = 0$, the results shown in the lower plot were obtained. Here the theory includes the first three terms. $\overline{C}_T$ was shifted slightly to fit the zero degree data, σ_a was determined from load cell readings, and $\overline{\beta}$ was held equal to the unstressed value. Although some minor differences between theory and experiment are observed, the agreement is believed to be quite good. This is consistent with the theoretical expectation that the remaining omitted terms vary as a higher order in the anisotropy.

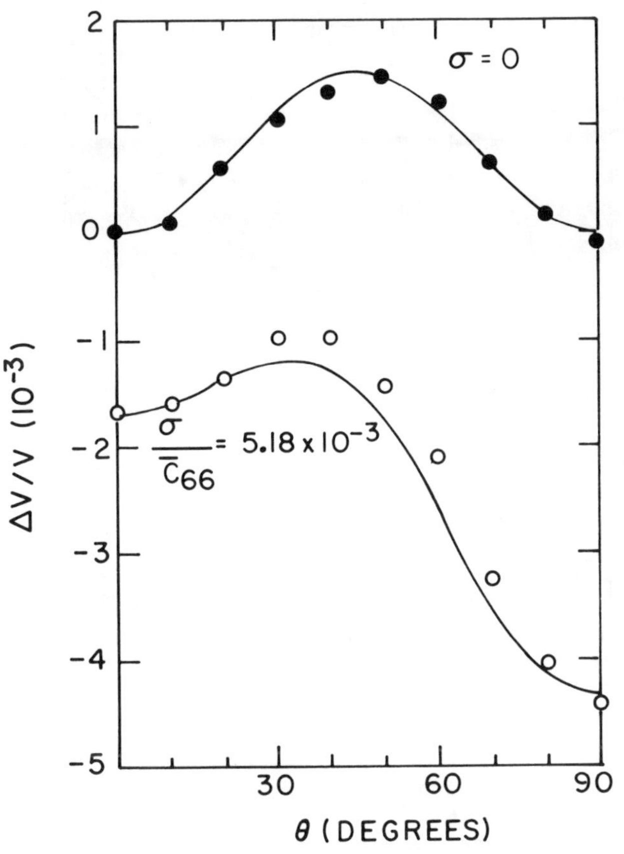

Fig. 3. Angular dependence of SH_o velocity in a rolled, 6061-T6 aluminum plate.

DETERMINATION OF STRESS

Stress may be determined from measurements of the shear wave velocity as predicted by Eq. (2). As the results are discussed in detail elsewhere,[6-10] only a brief summary will be given. Consider first the expression for T in Eq. (3). Note that no material constants influence the dependence of T on σ_a and σ_b. Thus, if the contributions of the terms following T in Eq. (2) could be suppressed, the difference in principle stresses, $\sigma_a-\sigma_b$, could be determined from measurements of the angular dependence of V_T. This can be accomplished by noting that all but one of those terms has four-fold symmetry. Thus, if one measures $V_T(\theta)$ and $V_T(\theta + 90°)$ and takes the difference in ρV^2, it follows that

$$[V_T(\theta) - V_T(\theta + 90°)]/\overline{V}_T = \left(\frac{1}{2\rho\overline{V}_T^2}\right)\Big[(\sigma_a-\sigma_b)\cos 2(\Omega-\theta) - \frac{\overline{\alpha}\,\overline{\beta}\,\overline{C}_L\,\overline{C}_T}{4(\overline{C}_L - \overline{C}_T)}(\cos 2\theta - \cos 6\theta)\Big]. \qquad (5)$$

where $\overline{V}_T \triangleq [V(\theta) + V_T(\theta + 90°)]/2$. For small anisotropies, one can approximate $\rho\overline{V}_T^2$ by $\overline{C}_T$ for further simplification.

The final term in Eq. (5) is second order in the elastic anisotropy. In cases in which it is negligible, the principal stress difference may be deduced by varying θ until the magnitude of the left hand side of Eq. (5) is maximized and setting that value equal to $(\sigma_a-\sigma_b)/2\,C_T$. As an example, Fig. 4 presents a comparison of measured and predicted stress in a second 6061-T6 aluminum plate having an unstressed SH_o wave velocity anisotropy of 0.5%, as distinct from the 0.15% anisotropy of the first sample shown in Fig. 3. The stress was applied at 45° with respect to the rolling direction, $\Omega = 45°$. The maximum value of velocity shift between the θ and θ+90° directions was observed to occur when $\theta = 45°$, and a plot of predicted stress versus applied stress (not shown) closely followed the ideal,

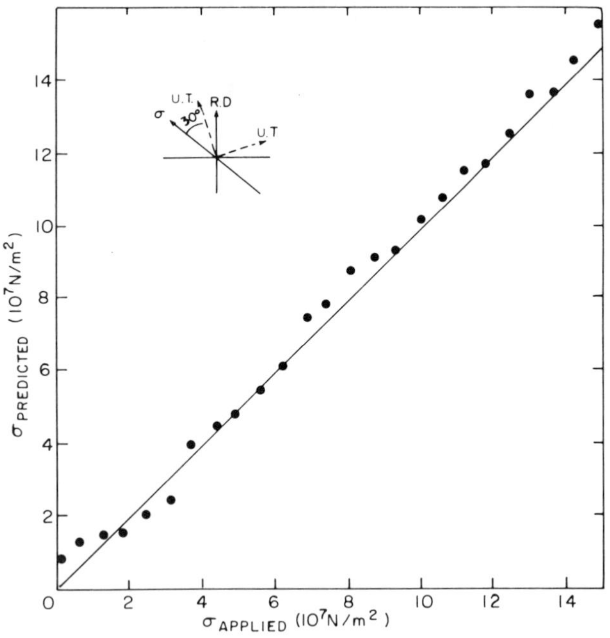

Fig. 4. Comparison of predicted and applied stress in 6061-T6 aluminum when stress is not parallel to material symmetry axes. Experimental configuration is shown in the inset and the method of comparison is described in the text.

unity slope line.[10] However, that was not a critical test of the smallness of the final term in Eq. (5) since cos 2θ = cos 6θ = 0 for θ = 45°. Hence the measurement direction was then changed to θ = 15° so that the final term would no longer vanish identically. In the plot shown in Fig. 4, the predicted stress was deduced on the basis of inserting the known values of Ω = 45°, θ = 15° into Eq. (5) and neglecting the final term. The excellent agreement is consistent with the postulated small value of this second order anisotropy term, at least in the particular aluminum sample studied.

If the anisotropy term is not small, a procedure is still available for removing its influence on the data. Note that Eq. (5) contains only one term with a 6θ angular variation, which has as its coefficient the second order material anisotropy parameters. Determination of that coefficient from the value of the 6θ term in a Fourier series representation of $[V_T(\theta) - V_T(\theta + 90°)]$ should allow the entire anisotropy term to be subtracted from the data. The previously described procedure could then be applied to the corrected data.

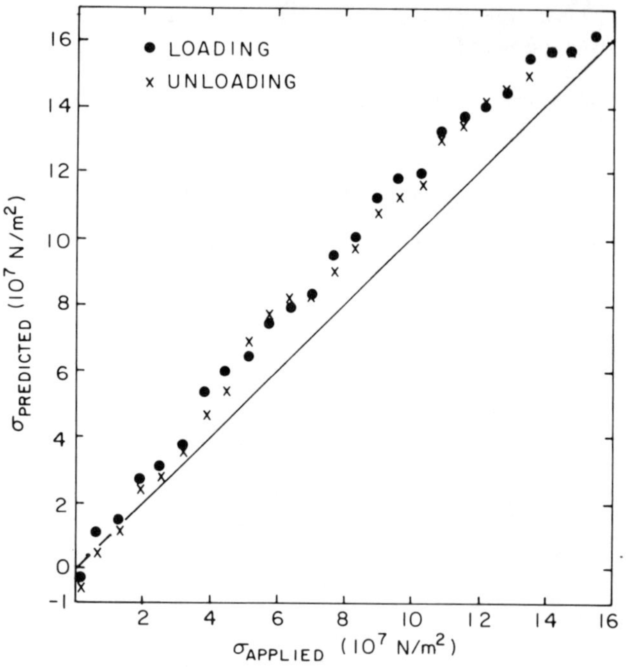

Fig. 5. Comparison of predicted and applied stress in 304 stainless steel.

Other than the example illustrated in Fig. 4, a detailed experimental study of this proposed procedure for non-parallel principal stress and material symmetry axes has not yet been completed. However, when Ω = 0, a number of samples have been studied.[6,8-10] Figure 5 presents, as an example, the results in a 304 stainless steel sample having an unstressed SH_o velocity anisotropy of 1.5%. Agreement of predicted and applied stress is excellent. Similar results have been obtained on 6061 aluminum, commercially pure copper, and 1100 aluminum plate. In the latter case, tensile plastic deformation of 0.6% did not destroy the agreement. In an additional sample of titanium,[10] errors occurred due to the absence of the assumed orthorhombic symmetry in the unstressed state.

DETERMINATION OF TEXTURE

In the measurement of stress, the effects of texture were suppressed by comparing the SH_o velocities measured before and after rotating the propagation direction by 90°. However, those suppressed terms may be used to directly determine important parameters of the preferred orientation. The possibility of inferring preferred orientation from measurements of the anisotropies of the elastic constants or sound velocities has been demonstrated by several authors.[11-13] The present discussion differs from the prior work in two regards. First, the nondestructive measurement of the velocity anisotropy can be made with couplant free EMAT transducers.[14] Hence operation in an on-line, process control mode at room or elevated temperatures is possible.[15] Second, using relationships recently reported by Sayers and Allen[4] the velocity anisotropies can be directly related to coefficients of the expansion of the crystallite orientation distribution function (CODF) in terms of generalized Legendre functions.[15,16]

For a material with unknown stress orientations, the procedure would be as follows. First the techniques described in the above section should be employed to determine the principal stress directions and differences. This allows the angular variation of the term T to be removed from Eqs. (1) and (2). If one restricts attention to the velocities measured at 0°, 45°, and 90°, one finds

$$\bar{C}_T + (\sigma_a+\sigma_b)/2 = \rho V_T^2(0°) - [(\sigma_a-\sigma_b)/2] \cos 2\Omega \qquad (6a)$$

$$\bar{C}_T + (\sigma_a+\sigma_b)/2 + \bar{\beta}\,\bar{C}_T - \frac{\bar{\alpha}^2}{16} \frac{\bar{C}_L^2}{(\bar{C}_L - \bar{C}_T)} = \rho V_T^2(45°) - [(\sigma_a-\sigma_b)/2] \sin 2\Omega \qquad (6b)$$

$$\bar{C}_T + (\sigma_a+\sigma_b)/2 = \rho V_T^2(90°) + [(\sigma_a-\sigma_b)/2] \cos 2\Omega \qquad (6c)$$

$$\bar{C}_L + (\sigma_a+\sigma_b)/2 + \frac{\bar{\alpha}}{2}\bar{C}_L = \rho V_L^2\ (0°) - [(\sigma_a-\sigma_b)/2] \cos 2\Omega \qquad (6d)$$

$$\bar{C}_L + (\sigma_a+\sigma_b)/2 - \bar{\beta}\,\bar{C}_T + \frac{\alpha^2}{16} \frac{\bar{C}_L}{(\bar{C}_L - \bar{C}_T)} + \bar{C}_{16} + \bar{C}_{26} = \rho V_L^2\ (45°) - [(\sigma_a-\sigma_b)/2] \sin 2\Omega \qquad (6e)$$

$$\bar{C}_L + (\sigma_a+\sigma_b)/2 - \frac{\bar{\alpha}}{2}\bar{C}_L = \rho V_L^2\ (90°) + [(\sigma_a-\sigma_b)/2] \cos 2\Omega \qquad (6f)$$

In each of these equations, the experimentally observable velocities, and the inferred principal stress orientations and differences are placed on the right-hand side and can be considered as known quantities.

After noting that Eqs. (6a) and (6c) are not independent because of the relationship between $[V_T^2(0°)-V_T^2(90°)]$ and $[\sigma_a-\sigma_b]$, one concludes that there are five equations in the six unknowns, $\bar{C}_T$, $\bar{C}_L$, $(\sigma_a+\sigma_b)$, and $(\bar{C}_{16} + \bar{C}_{26})$. Hence a rigorous solution is not possible, and some further approximations will be required. Since $\bar{C}_{16}$ and $\bar{C}_{26}$ appear only in Eq. (6e) and are not of primary interest, this equation will be dropped leaving four equations in five unknowns. By elementary manipulations it can be concluded that

$$\bar{C}_L + (\sigma_a+\sigma_b)/2 = \rho[V_L^2(0°) + V_L^2(90°)]/2 \qquad (7a)$$

$$\bar{\alpha}\,\bar{C}_L = \rho[V_L^2(0°) - V_L^2(90°)] - (\sigma_a-\sigma_b) \cos 2\Omega \qquad (7b)$$

$$\bar{C}_T + (\sigma_a+\sigma_b)/2 = \rho[V_T^2(0°) + V_T^2(90°)]/2 \qquad (7c)$$

$$\bar{\beta}\,\bar{C}_T - \frac{\bar{\alpha}^2}{16} \frac{\bar{C}_L^2}{(\bar{C}_L - \bar{C}_T)} = \rho\ [V_T^2(45°) - V_T^2(0°)] + [(\sigma_a-\sigma_b)/2]\ [\cos 2\Omega - \sin 2\Omega] \qquad (7d)$$

The stress terms on the left-hand side of Eqs. (7a) and (7c) will generally be small with respect to $\bar{C}_L$ and $\bar{C}_T$ and can be neglected. However, those on the right-hand side must be retained as they appear in conjunction with the difference of two numbers, which could be of comparable magnitude. It should also be noted that, in general, $\bar{\alpha}$, $\bar{\beta}$, $\bar{C}_T$, and $\bar{C}_L$ are themselves weakly modified by stress through the stress dependence of the $\bar{C}_{ij}$. In general, these are small effects and will be neglected hereafter. In many applications, determination of texture would be done in the unstressed state and the above considerations are unnecessary. Finally, the term proportional to $\bar{\alpha}^2$ in Eq. (7d) can be dropped for modest textures. The result is the simplified system

$$C_L \cong \rho\ [V_L^2(0°) + V_L^2(90°)]/2 \qquad (8a)$$

$$C_T \cong \rho\ [V_T^2(0°) + V_T^2(90°)]/2 \qquad (8b)$$

$$\alpha C_L \cong \rho\ [V_L^2(0°) - V_L^2(90°)] - (\sigma_a-\sigma_b) \cos 2\Omega \qquad (8c)$$

$$\beta C_T \cong \rho\ [V_T^2(45°)-V_T^2(0°)] + [(\sigma_a-\sigma_b)/2]\ [\cos 2\Omega-\sin 2\Omega], \qquad (8d)$$

which can readily be solved for C_L, C_T, α, and β.

The coefficients of the CODF expansion are finally obtained from the relationship between α, β, C_L, and C_T and the coefficients $W_{\ell mn}$ as defined by the CODF expansion[4,16]

$$W(\xi,\psi,\Phi) = \sum_{\ell=0}^{\infty} \sum_{m=-\ell}^{\ell} \sum_{n=-\ell}^{\ell} W_{\ell mn} Z(\xi) e^{-im\psi} e^{-in\Phi} \qquad (9)$$

where W is the CODF expressed in terms of the Euler angles ξ, ψ, Φ and $Z_{\ell mn}$ are generalized Legendre functions. Sayers and Allen have derived the relationship between the ultrasonic velocities and the $W_{\ell mn}$ using the method of Voigt to compute polycrystalline averages.[4] Following Bunge,[17] they conclude that only three independent coefficients enter in the calculation of elastic constants for cubic crystallites, W_{400}, W_{420}, and W_{440}. One additional coefficient, W_{000}, is determined by normalization. In texture analysis, the preferred direction in Eq. (9) is generally taken to be the rolling direction, which leads to a different axis identification than that shown in Fig. 1 and used throughout the rest of this paper. After rotation to such a system, as employed by Sayers and Allen[4], in which the 1, 2, and 3 axes correspond respectively to the normal, transverse, and rolling directions, one concludes

$$\rho[V_L^2(0°) + V_L^2(90°)]/2 = C_{11}° - 2C° \left[\frac{1}{5} - \frac{11}{35}\sqrt{2}\,\pi^2 W_{400} - \frac{4}{35}\sqrt{5}\,\pi^2 W_{420} - \frac{2}{35}\sqrt{35}\,\pi^2 W_{440}\right] \qquad (10a)$$

$$\rho[V_T^2(0°)+V_T^2(90°)]/2 = C_{44}° + C° \left[\frac{1}{5} - \frac{16}{35}\sqrt{2}\,\pi^2 W_{400} - \frac{16}{35}\sqrt{5}\,\pi^2 W_{420}\right] \qquad (10b)$$

$$\rho[V_L^2(0°) - V_L^2(90°)] - (\sigma_a-\sigma_b)\cos 2\Omega = 4C° \left[\frac{5}{35}\sqrt{2}\pi^2 W_{400} - \frac{4}{35}\sqrt{5}\,\pi^2 W_{420} - \frac{2}{35}\sqrt{35}\,\pi^2 W_{440}\right] \qquad (10c)$$

$$\rho[V_T^2(45°)-V_T^2(0°)]+[(\sigma_a-\sigma_b)/2][\cos 2\Omega - \sin 2\Omega] = C°\left[\sqrt{2}\,\pi^2 W_{400} + \frac{28}{35}\sqrt{5}\,\pi^2 W_{420} + \frac{2}{35}\sqrt{35}\,\pi^2 W_{440}\right] \qquad (10d)$$

where $C_{11}°$, $C_{44}°$, and $C_{12}°$ are single crystal elastic constants and $C° = C_{11}° - C_{12}° - 2C_{44}°$. The W_{4mo} are defined in a coordinate system with the 3-axis along the rolling direction, in contrast to the elastic constants appearing in previous equations in this paper. However, the experimental observables on the left-hand side of Eq. (10) are unambiguous.

This constitutes an overdetermined system, since there are four equations in the three unknowns, W_{4mo}. Either experimental error or inaccuracies in the Voigt averaging scheme would render an exact solution of all equations impossible. A detailed analysis of the most stable data reduction scheme has not yet been conducted. However, it can be speculated that Eqs. (10c) and (10d) should be weighed heavily since the experimental quantities are relative velocities, rather than the absolute velocities appearing in Eqs. (10a) and (10b). It is also likely that Eq. (10b) will prove more accurate than Eq. (10a) because no correction of plate to plane wave velocities is required. Whereas such a correction may be quite satisfactory for relative measurements, it may be inadequate for absolute measurements. Thus a first attempt at reduction to practice might involve disregarding Eq. (10a) and solving the three remaining systems simultaneously.

As a check, one should use the computed $W_{\ell mn}$ values to test the accuracy of the isotropic approximation to the correction relating the velocities V_{L_o} and V_{S_o} as given by Eq. (4). For larger textures, an iterative solution might be required in which first estimates of the W's are used to recalculate the velocities $V_{Lo}(0°)$ and $V_{Lo}(90°)$ using Eq. (4) and the polycrystalline averaged elastic constant expressions[4]

$$C_{13} = C_{12}° + C° \left[\frac{1}{5} - \frac{16}{35}\sqrt{2}\,\pi^2 \left(W_{400} - \sqrt{5/2}\,W_{420}\right)\right] \qquad (11a)$$

$$C_{23} = C_{12}° + C° \left[\frac{1}{5} + \frac{4}{35}\sqrt{2}\,\pi^2 \left(W_{400} - \sqrt{70}\,W_{440}\right)\right] \qquad (11b)$$

Again, a mixed notation has been used in this equation. The elastic constants appearing on the left-hand side of Eq. (11) are defined in the coordinated system of Fig. 1. The W_{4mo} are defined after Sayers and Allen[4] as discussed above. As expected, the deviations from the isotropic assumptions will become greater when either the single crystal anisotropy, C°, or the degree of preferred orientation, as indicated by the W_{4mo}, are large.

The above describes one approach to deduce the coefficients of the CODF from the angular dependence of the SH_o and S_o plate mode velocities. Smith et al.[18] have discussed a closely related approach in which the velocities of higher order, horizontally polarized plate modes, SH_n, are measured and used to predict the anisotropic elastic constants. The experimental parameters of that approach could be related to the $W_{\ell mn}$ using arguments similar to the one presented above.

SUMMARY AND CONCLUSION

It has been shown that measurements of the angular dependence of the SH_o and S_o ultrasonic modes of a plate are influenced by both stress and texture. Data reduction schemes are presented for making independent estimates of a) the orientations and difference of in-plane principal stresses and b) the coefficients W_{400}, W_{420}, and W_{440} of the CODF expansion. Since the data can be obtained with couplant free, EMAT probes, there appears to be a high potential for in-line process control applications.[19] Figure 6 sketches the required measurements. Since the use of the CODF as a means of correlating texture with mechanical properties has shown considerable recent promise,[20] such a technique might be of considerable value in process control.

ACKNOWLEDGEMENT

This work was performed for the U.S. Department of Energy, Office of Basic Energy Sciences, Division of Materials Sciences, under contract No. W-7405-Eng-82.

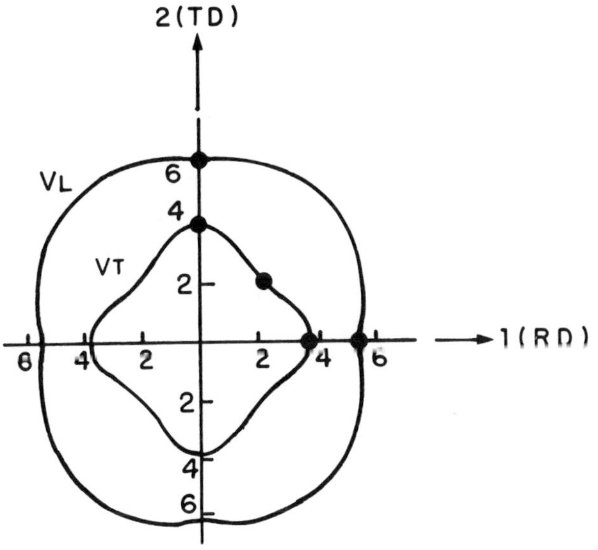

Fig. 6. Measurements required for nondestructive determination of stress and texture parameters. The solid lines schematically indicate the angular dependence of the velocities and the dots indicate the points at which data is required.

REFERENCES

1. Kappmeyer, Keith, "Process Control Needs for Future Steelmaking", this proceedings
2. Thompson, R. Bruce, unpublished results.
3. J. F. Nye, "Physical Properties of Crystals", Oxford University Press, London (1957)
4. Sayers, C. M. and Allen, D. R. J., Phys. D., 17, 1399 (1984)
5. B. A. Auld, "Elastic Waves and Fields in Solids", Wiley, NY, (1973)
6. Thompson, R. B., Lee, S. S., and Smith, J. F., in "Review of Progress in Quantitative Nondestructive Evaluation 3", D. O. Thompson and D. E. Chimenti, Eds., Plenum Press, NY, 1311 (1984)
7. Thompson, R. B., Smith, J. F., Lee, S. S., in "Review of Progress in Quantitative Nondestructive Evaluation 2", D. O. Thompson and D. E. Chimenti, Eds., Plenum Press, NY, 1337 (1983)
8. Thompson, R. B., Smith, J. F., and Lee, S. S., Appl. Phys. Lett. 44, 296 (1984)
9. Thompson, R. B., Smith, J. F., and Lee, S. S., in "Nondestructive Evaluation: Application to Materials Processing", O. Buck and S. M. Wolf, Eds., Amer. Soc. Metals, Metals Park, Ohio, 137 (1984)
10. Lee, S. S., Smith, J. F., and Thompson, R. B., "Evaluation of the Absolute Acoustoelastic Stress Measurement Technique", in "Review of Progress in Quantitative Nondestructive Evaluation 4", D. O. Thompson and D. E. Chimenti, Eds., Plenum Press, NY, (in press)
11. Liu, Y. C., and Alers, G. A., Trans. Met. Soc. AIME 236, 489 (1966)
12. Papadakis, E. P., Trans. Met. Soc. AIME 236, 1609 (1966)
13. Tittmann, B. R. and Alers, G. A., Met. Trans. 3, 1307 (1972)
14. Thompson, R. B. and Alers, G. A., in "Proc. 9th Symposium on Nondestructive Evaluation", South Texas Section of ASNT and Southwest Research Institute, San Antonio, TX, 6 (1973)
15. Thompson, R. B., in "1977 Ultrasonics Symposium Proceedings", IEEE, NY, 74 (1977)
16. Roe, R. J., J. Appl. Phys. 36, 2024 (1965)
17. Bunge, H. J., Krist. Tech. 3, 431 (1968)
18. Smith, J. F., Alers, G. A., Armstrong, P. E., and Eash, D. T., "Separation and Characterization of Stress Levels and Texture in Metal Sheet and Plate: I. Principles and Initial Test", to be published in J. Nondestructive Eval.
19. Droney, B. E., in "Nondestructive Methods for Material Property Determination", C. O. Ruud and R. E. Green, Eds., Plenum Press, NY, 237 (1984)
20. Sowerby, R., "Deformation Textures in Some Non-Ferrous Metals", to be published in the proceedings of the symposium on texture of non-ferrous metals held as part of the 1984 Fall TMS-AIME Meeting, Sept. 16-20, 1984, Detroit, Michigan

ULTRASONIC CHARACTERIZATION OF POROSITY IN POWDER METALS

S. Howard, J. Tani, H. Arnold, H. Schwetlick, Wolfgang Sachse
Department of Theoretical and Applied Mechanics
Cornell University
Ithaca, NY 14853

ABSTRACT

We present preliminary results of a project aimed at investigating ultrasonic techniques to measure the bulk density, pore size, and spatial variation of porosity in powder metal specimens. Specific techniques considered are ultrasonic wavespeed, attenuation and wave dispersion measurements, and the identification of spectral features in the amplitude spectra. Also discussed is the inverse medium problem for reconstructing the acoustic impedance profile along the direction of wave propagation in a specimen.

The advantages of near net-shape fabrication of complex-shaped parts from powder metals has long been recognized. The strength and moduli of such parts, never as high as an equivalent part machined from a solid metal, is a strong function of their density [1,2]. Controlling the strength is not only the total void fraction but also its shape and distribution [2]. The in situ non-destructive measurement of material density of powder metal parts is most readily achieved by x-ray radiographic techniques, eddy currents and mechanical vibrations and ultrasonic wave propagation measurements. It is the latter which are considered in this paper.

Several techniques for ultrasonic porosity measurements in powder metal specimens have been proposed. Among the first used were direct wavespeed measurements. Using pulse-echo, time-of-flight measurements, Brockelman [3] established a correlation between longitudinal wave velocity and tensile strength in sintered iron powders which was also related to the pressing direction in the specimens. In another test, the ultrasonic wavespeed measurements were shown to correlate with x-ray density measurements made along the central region of a 3.5-inch bar of sintered steel. Later, Papadakis and Petersen [4], using ultrasonic velocity measurements along different directions in briquetted and sintered specimens of steel powders, established a correlation between wavespeed and density. Other investigators have attempted to establish a relationship between ultrasonic attenuation and the porosity of a material. However, up to now, most measurements appear to have been restricted to porous rocks. A recent example is the work by Winkler [5] who measured attenuation and phase velocity spectra in sedimentary rocks under high pressure. Other work appears to have focused on measurements in two-phase materials. The work of Latiff and Fiore [6] was on a model material consisting of Cu spheres imbedded in an elastically isotropic polymethylmethacrylate (PMMA) matrix. When attempts were made to interpret the data with the multiple scattering theory developed by Waterman [7], only qualitative agreement was found.

Most of the recent work may be classified under the heading of ultrasonic spectral feature identification. In this area, two approaches have been pursued. Gubernatis and Domany [8] developed a theoretical model of elastic wave scattering to establish the relationship between the average pore radius and a frequency peak or transition in the spectrum of the scattered signals. They also showed the connection between the magnitude of this amplitude and the pore concentration. In the work of Tittman, et al [9], the focus was on the determination of the size and concentration of pores in IN-100 alloy from the slope of the backscattered ultrasonic amplitude spectrum. In two samples, the average pore size recovered from ultrasonic measurements was in agreement with the values determined from quantitative microscopy measurements. In the analytical approach, developed by Sayers and Smith [10], a relationship between porosity and ultrasonic wave dispersion effects is established which can be determined from the amplitude or phase spectra of the ultrasonic echoes.

EXPERIMENTS

Our research was conducted on hot-pressed, porous powder samples of iron, copper and aluminum. After fabrication, each sample was machined flat into the shape of a disk of about 2.54 cm in diameter and polished. The samples ranged in thickness from 3 to 6 mm and were flat to within 10.0 μm (0.0004 inches). There were seven specimens of aluminum, six of iron and nine of copper. The specimen densities were determined from weight and volume measurements. The sintered aluminum powder specimens spanned a density range from less than 1.97 g/cm^3 to 2.69 g/cm^3, corresponding to porosities ranging from 1.4% to 27%. The densities of the copper specimens ranged from 5.44 g/cm^3 to 8.75 g/cm^3, corresponding to porosities ranging from 2.4% to 39%, while the iron powder specimens ranged in density from 4.87 g/cm^3 to 7.72 g/cm^3, corresponding to porosities ranging from 2.0% to 38%.

The ultrasonic measurements were made using a broadband ultrasonic system in which highly-damped, longitudinal and shear piezoelectric transducers of varying center frequencies were shock excited and the detected echoes were amplified and processed. Both time-of-flight (velocity) and amplitude (attenuation) measurements were made. In the first experiments, the time-of-flight measurments were made to a resolution of approximately 20 nsec with a digital oscilloscope system. Preliminary measurements of the spatial variation of velocity were made with an immersion transducer moved in a rectangular grid pattern over the test specimen immersed in water. The time-of-flight was measured using a microprocessor-controlled time-interval counter. In these experiments, each data point represented the average of 10^3 time-interval determinations.

Measurement of the dispersion of ultrasonic pulses in a solid can be conveniently made using the phase spectroscopy technique described by Sachse and Pao [11]. In this technique, a broadband ultrasonic pulse is input into a specimen from a transducer and the signal received after propagating a distance L through a specimen is recorded either with the same transducer acting as a receiver or a second, receiving transducer. Both input and output signals are Fourier-transformed into the frequency-domain and their respective, unwrapped phase functions, ϕ and ϕ_0 are determined. From the difference of these, the dispersion relation can be determined according to

$$k(\omega) = (\phi(\omega) - \phi_0(\omega)) / L \qquad (1)$$

From this the phase and group velocities are found, where

$$v_{phase} = \omega / k = \omega L / (\phi(\omega) - \phi_0(\omega)) \qquad (2a)$$

$$v_{group} = \partial\omega / \partial k = L\,[d\phi/d\omega]^{-1} \qquad (2b)$$

Ultrasonic attenuation measurements were made using the buffer-rod technique described by Papadakis [12]. In this technique, a buffer rod is attached to the test specimen and the broadband, buffer-specimen echo, A(t), and two successive specimen echoes, B(t) and C(t), are analyzed in the frequency-domain. This permits determination of the reflection coefficient of the buffer-specimen interface and the attenuation of the test piece, independent of the excitation amplitude or the attenuation of the buffer rod. The frequency-dependent reflection coefficient and attenuation of the specimen is given by

$$R(f) = [\,\bar{A}\,\bar{C} / (\bar{A}\,\bar{C} - 1)]^{\frac{1}{2}} \qquad (3a)$$

$$\alpha(f) = 1 / L \; \ell n\,[R(f)/\bar{C}] \qquad (3b)$$

where $\bar{A}(f) = A(f)/B(f)$ and $\bar{C}(f) = C(f)/B(f)$. The value obtained from evaluation of Eq. (3a) can be compared with the expected value for the reflection coefficient

$$R = \frac{\rho_s c_s - \rho_b c_b}{\rho_s c_s + \rho_b c_b} \qquad (4)$$

RESULTS

The results of wavespeed measurements using longitudinal and shear waves in sintered aluminum powder specimens is shown in Figs. 1(a) and (b). The longitudinal wavespeed measurements shown were made with two transducers possessing different center frequencies. A linear least-squares fit was made to the wavespeed data. The solid line and open data points in Fig. 1(a) correspond to data taken with a 2.25 MHz transducer generating a pulse with center frequency near 2 MHz. The other data (dashed line, crosses) corresponds to measurements made with a 10 MHz transducer generating a pulse with center frequency near 7 MHz. The average slope of these two data sets was 0.420 cm/μsec/g/cm^3. The shear wave measurements shown in Fig. 1(b) were made with pulses whose center frequency was near 5 MHz. The slope of the least-squares fit line was 0.160 cm/μsec/g/cm^3. The results of similar experiments made in iron powder specimens is shown in Figs. 2(a) and (b). The average of the slopes of the two least-squares linear fits to the longitudinal wavespeed/density data is 0.104 cm/μsec/g/cm^3; for shear waves the slope is 0.0487 cm/μsec/g/cm^3. The results of the corresponding experiments with copper specimens are shown in Figs. 3(a) and (b). The longitudinal and shear wave/density slopes are determined to be 0.0774 and 0.0285 cm/μsec/g/cm^3, respectively. We note that the variation in wavespeed with density is highest for aluminum and lowest for copper.

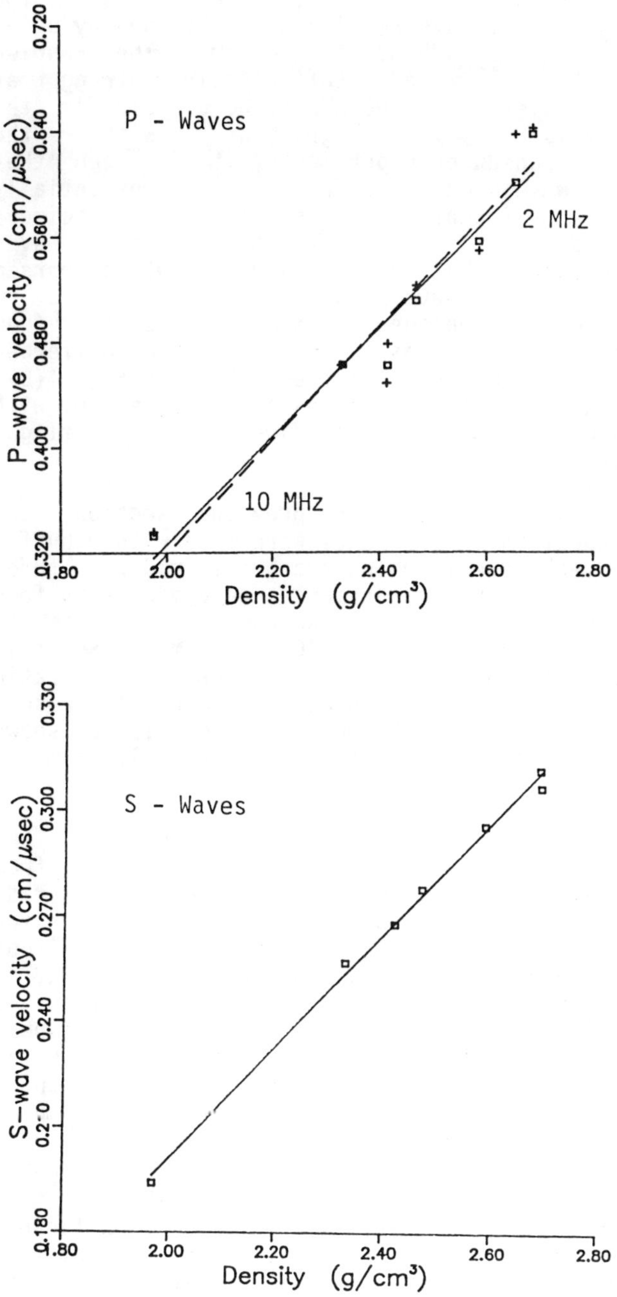

Figure 1(a)-(b) - Hot-pressed Aluminum

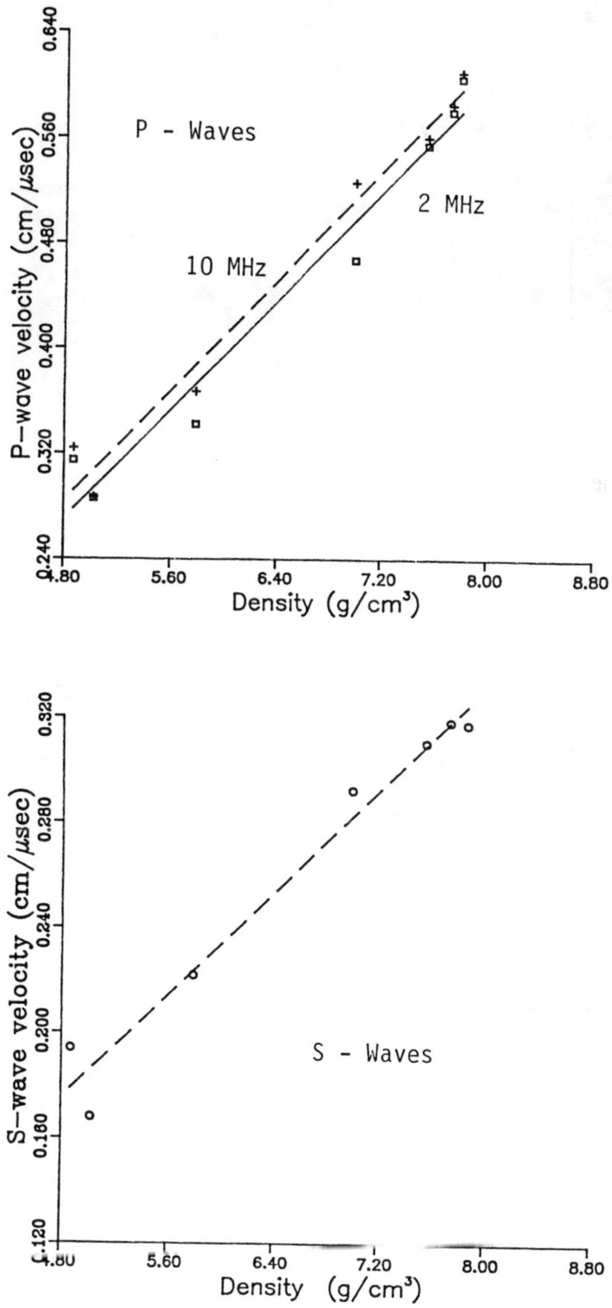

Figure 2(a)-(b) - Hot-pressed Iron

The importance of the results shown in Figs. 1 - 3 is that they provide a means for converting wavespeed measurements to density measurements for a specific material. Because of the measured linear relationship between wavespeed and density, the values of slope and intercept should be sufficient to convert ultrasonic wavespeed data into a density determination. To evaluate the reliability of this procedure, we calculated the error in density expected when using our measured data as input to the conversion algorithm. The root-mean-square deviation in longitudinal

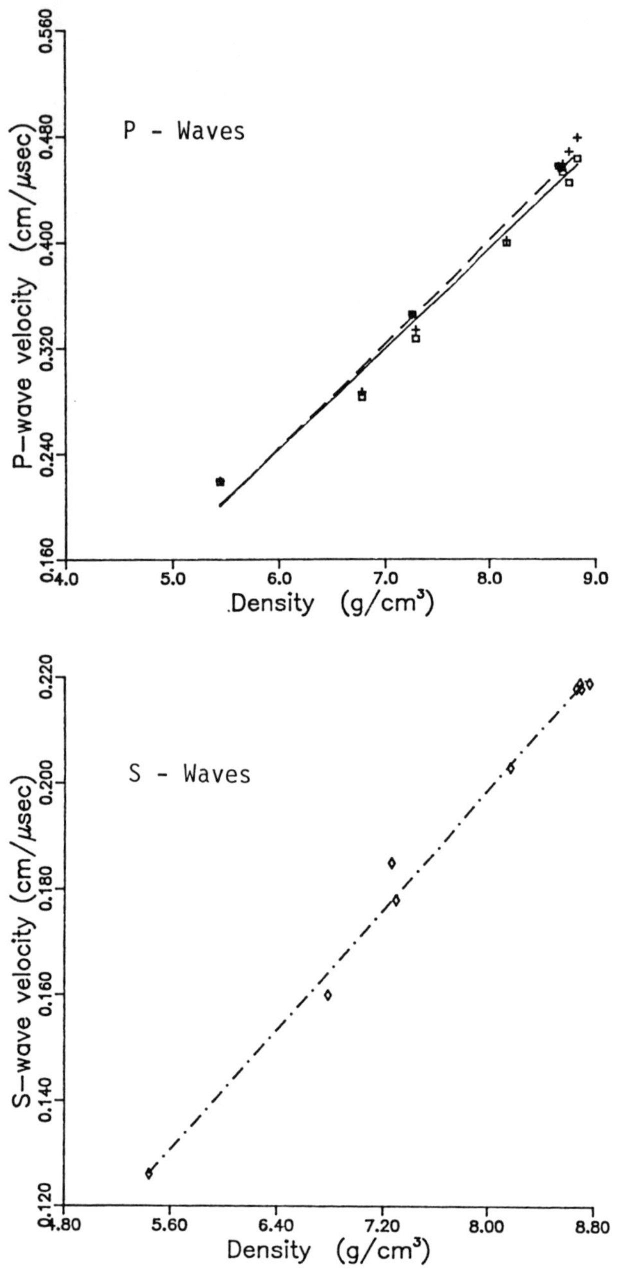

Figure 3(a)-(b) - Hot-pressed Copper

wavespeed, Δc, was evaluated for the wavespeed data shown in Figs. 1(a)-3(a) and the corresponding variation in density, $\Delta\rho = \Delta c/m$ was then determined for each material, where m is the slope of the lines in the figures. Dividing $\Delta\rho$ by the maximum density for each material, we obtained numbers corresponding to the measurement resolution with the 2.25 Mhz and the 10 MHz transducer data sets. Another two sets of data were obtained using a 5 MHz longitudinal wave transducer. The four sets of longitudinal wave data gave resolution values ranging from 1.74-2.37%, 2.16-2.92% and 1.58-1.90% for aluminum, copper, and iron, respectively. It is possible that these values can be improved by better transducer/specimen coupling techniques or by using non-contact transduction techniques.

Two sample results of phase velocities determined with the phase spectroscopy technique are shown in Figs. 4(a) and (b). The longitudinal wave results were obtained by using a 10 MHz broadband transducer. The Fourier amplitudes were sufficiently strong at frequencies up to 50 MHz to be useable for the analysis. In agreement with the multi-transducer experiments, it is seen that the wavespeed is found to be essentially frequency-independent for all the specimens tested. Similar experiments with shear waves also showed that the pulses were not dispersed since the wavespeeds were found to be frequency-independent. Sample results from various specimens, spanning the different materials and porosities are shown in Fig. 4(b). The average veocity values are in agreement with those determined from direct time-of-flight measurements.

As described in the previous section, the attenuation measurements were made with a buffer rod technique in which the buffer rod was fused quartz. The measurements were made with four different broadband transducers of nominal frequencies of 2.25, 5, 10 and 50 MHz. Shown in Fig. 5 is the specimen to buffer signal ratio obtained in an iron-powder specimen. The result obtained from the 50 MHz transducer is not shown since its ratio differed appreciably from the others.

The determination of the frequency-dependent reflection coefficients from the amplitude ratios $\bar{A}(f)$ and $\bar{C}(f)$ was made according to Eq. 3(b). Two sample results for iron specimens attached to quartz buffer rods are shown in Figs. 6(a) and (b). Also shown, as a dashed line, are the expected reflection coefficients computed from density and wavespeed values for the particular specimens shown. Although these results were computed without incorporating any corrections for diffraction or band limitations of the excitation pulse or transducer response, even at the mid-frequency points, near 2 MHz, the difference between expected and measured values was significant. For this reason, no attenuation values were evaluated. This result raises questions regarding the reliability of amplitude measurements which seriously affect any porosity determination procedure based on an attenuation measurement.

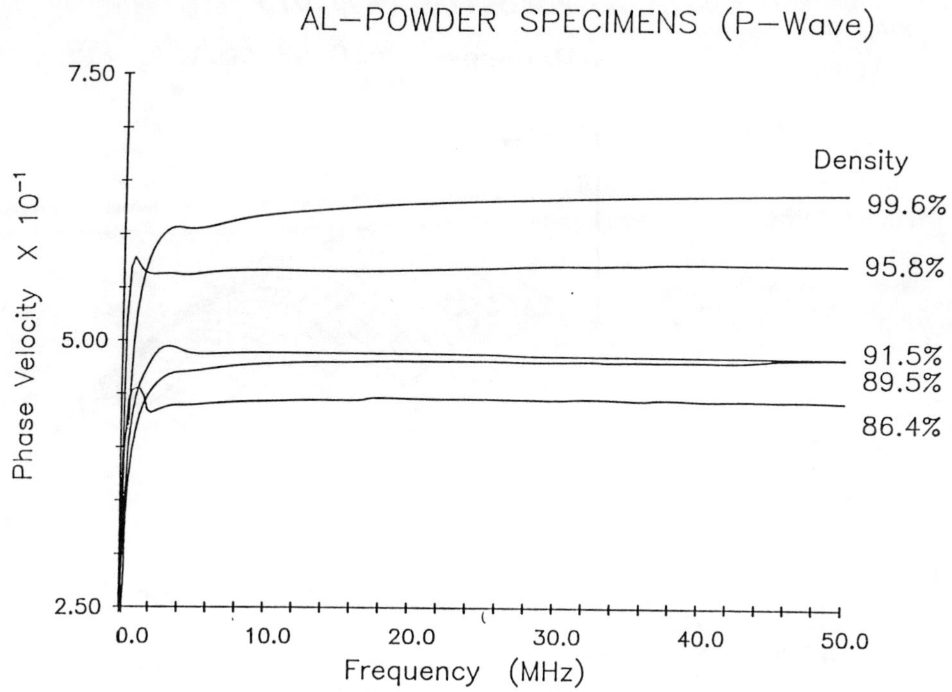

Figure 4(a) - Longitudinal phase velocity in Aluminum.

(Units: cm/μsec)

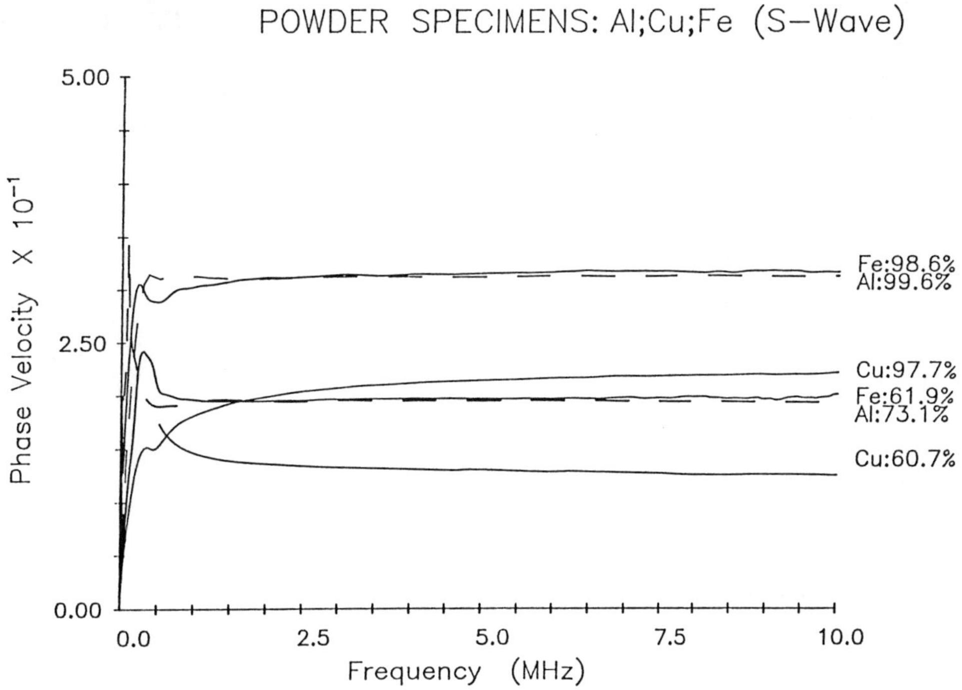

Figure 4(b) - Shear wave phase velocity in Al, Cu and Fe hot-pressed specimens.

(Units: cm/μsec)

Two examples of the determination of two dimensional spatial variations of porosity are shown in Figs. 7(a) and (b). These two specimens were unfired Al-powder briquettes, pressed at 20 ksi and 27 ksi with a resulting porosity of 7.3% and 5.4% respectively. The specimens were polished flat to within 10 μm (0.0004 inches) which corresponds to a possible variation in the travel time of approximately 2.0 nsec. It is seen that there is a significant variation above this value across the cross section of the disks. The variation was almost 20 nsec for the more porous specimen corresponding to a higher wavespeed near the edge of the briquette than at its center. A similar but not as pronounced effect is seen with the less porous sample.

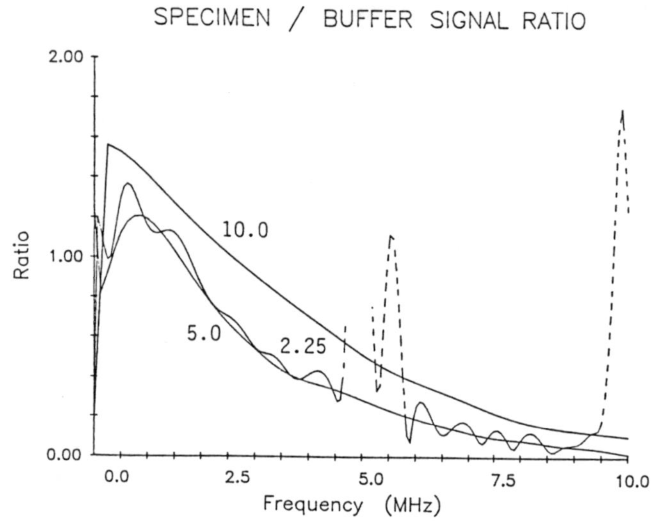

Figure 5 - P-wave Signal in Hot-Pressed Fe, 88%

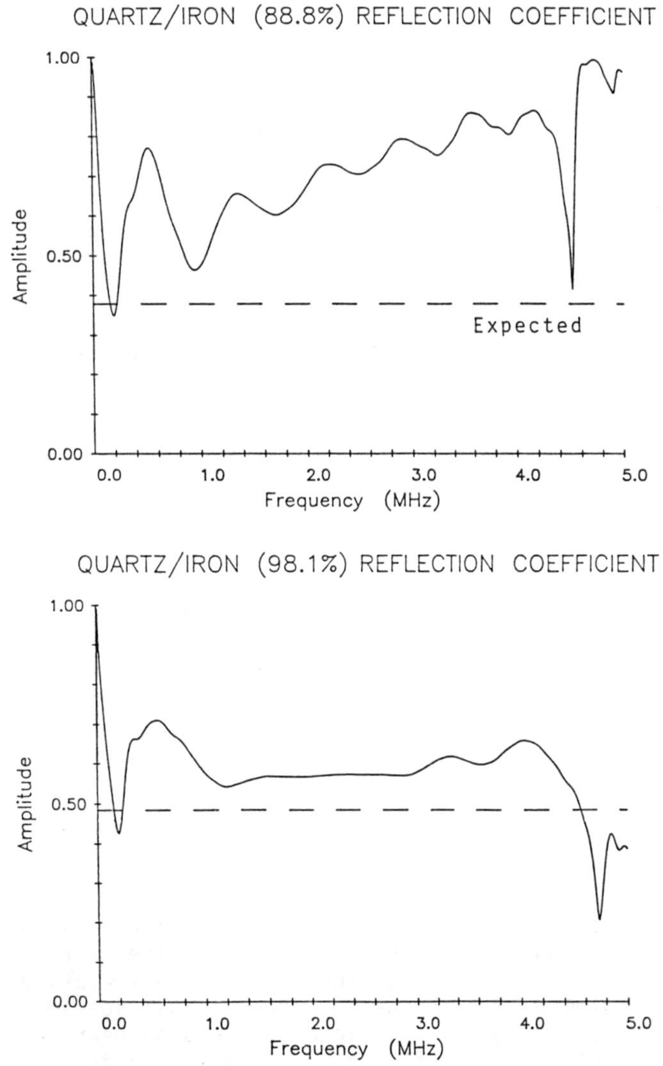

Figure 6(a)-(b) - Measured, Expected Reflection Coefficient; Iron/Quartz Interface

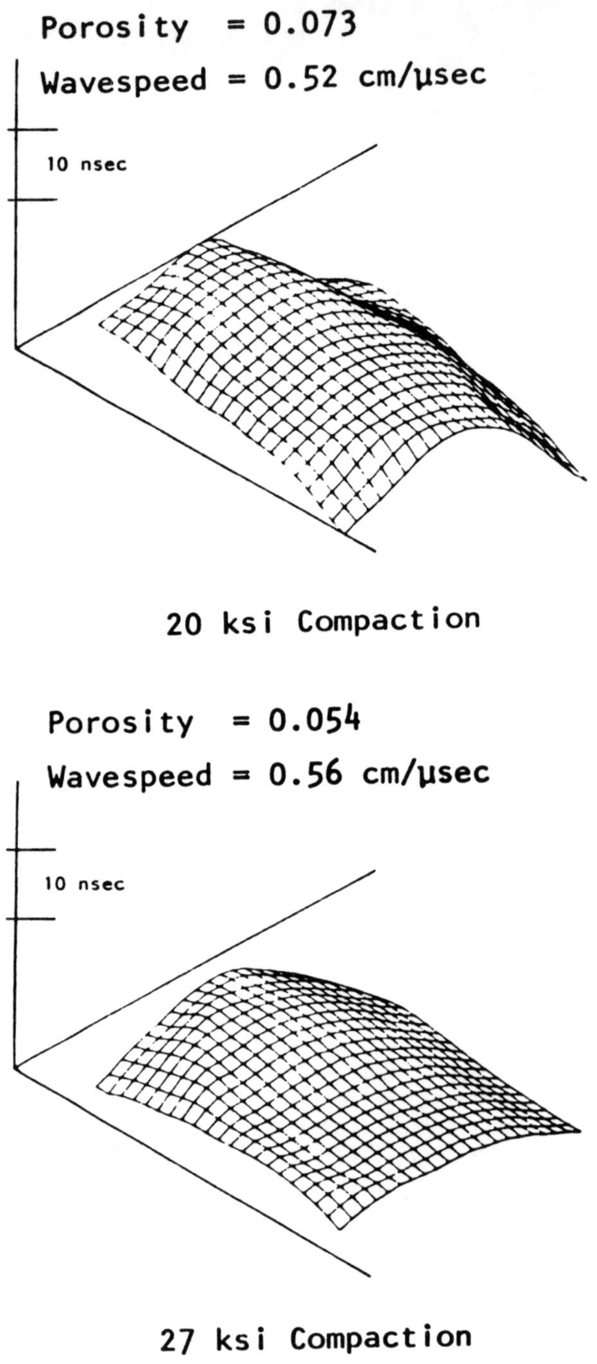

Figure 7(a)-(b) - Time-of-Flight in Al-powder Briquettes: (a) - 20 ksi; (b) - 27 ksi Compaction

The determination of the density and wavespeed variation along the direction of propagation from measurement of input and output signals forms the basis of the inverse medium problem which has been the subject of much research. Algorithms for obtaining the solution of such problems when band-limited excitation signals corresponding to typical ultrasonic pulses are used are still being developed [13]. We show, therefore, the results of a simulation experiment. As shown in Fig. 8(a) a specimen of length L was simulated which has in its interior a region of impedance 10% higher than the surrounding outer regions situated normal to the direction of wave propagation. The synthesized excitation pulse produced by a transducer is shown in Fig. 8(b) and the signal received by the same transducer is in Fig. 8(c). To make this simulation more realistic and to verify the stability of the inversion scheme which is based on the method of characteristics [14], broadband noise whose rms value was 30% of the peak signal amplitude was added to the received signal. The broadband noise signal was the actual signal obtained from a broadband ultrasonic preamplifier. The results shown in Fig. 8(d) are the reconstructed impedance profiles from the noise-free and the 30% noise added waveforms.

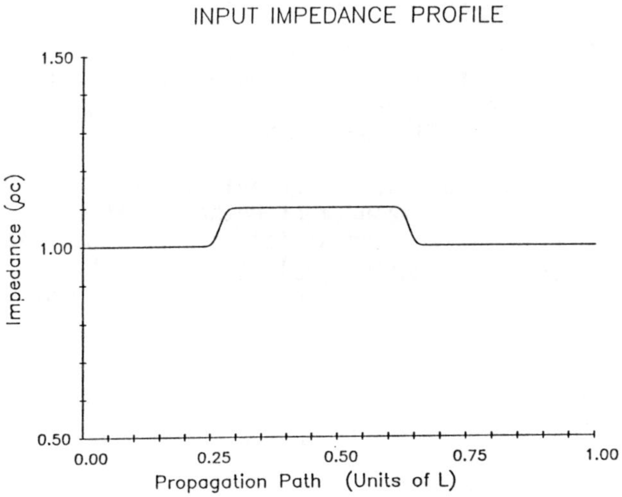

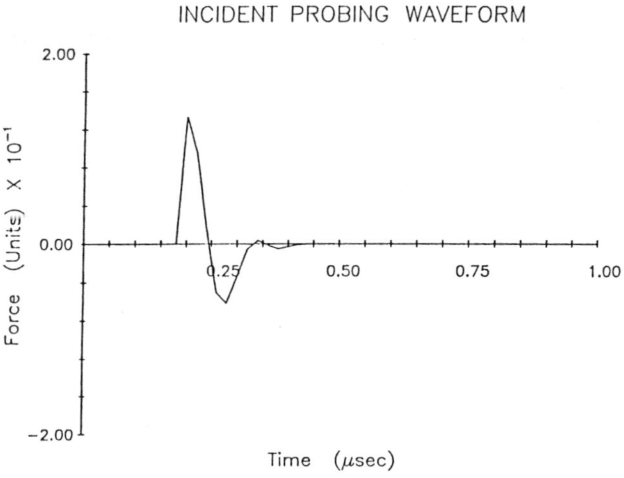

Figure 8 - Impedance Profile Reconstruction Simulation
(a) Input Profile; (b) Probing Waveform

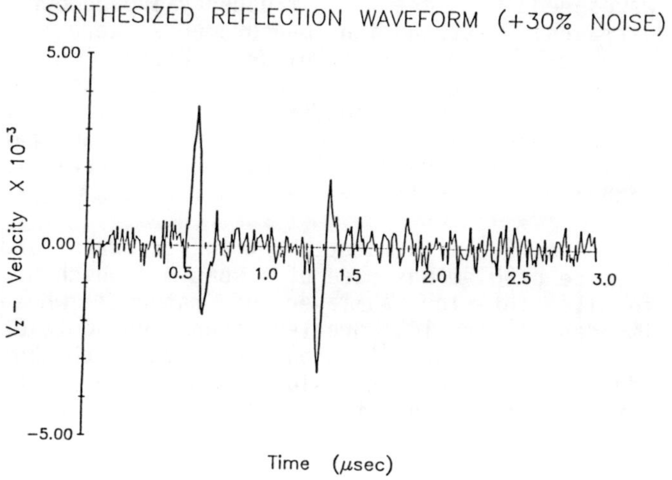

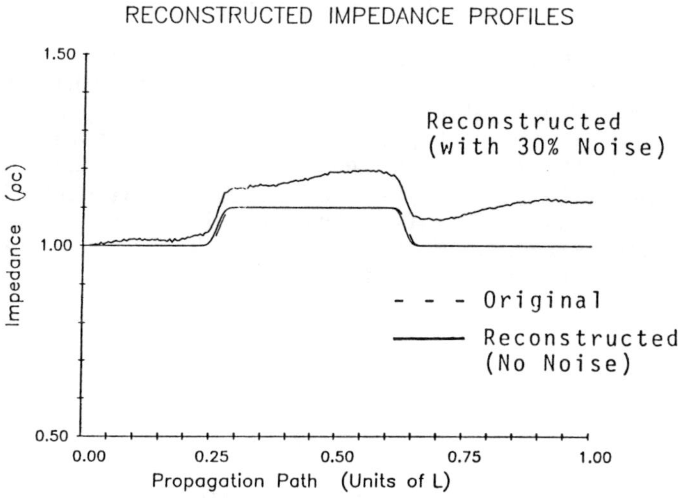

Figure 8 - Continued
(c) - Synthesized Reflection Waveform

CONCLUSIONS

We have reported results of ultrasonic longitudinal and shear wave measurements made in powder metal specimens of aluminum, iron and copper ranging in porosity from 1% to 39%. Ultrasonic time-of-flight (velocity) data were used to determine a relationship between density and longitudinal and shear wavespeeds. This relationship can be inverted to permit determination of the density or porosity of porous metal specimens. Using a simple error model, a density resolution was determined. It was established that errors in the wavespeed measurements lead to density variations of approximately 2% which are not resolvable. A more reproducible measurement technique is needed to yield resolution of smaller density variations.

Ultrasonic phase spectroscopy measurements showed that in the measured frequency range, the wavespeeds are frequency-independent. Ultrasonic attenuation measurements were made with a buffer-rod technique. Based on a comparison between the recovered and expected wave reflection coefficients from the buffer/specimen interface, it was shown that amplitude and hence attenuation measurements cannot be made reliably.

A numerical simulation of the forward and inverse problem was made of a specimen which has in its interior a layer of material whose impedance is 10% greater than the outside layers. The synthesis shows that even with 30% added noise, an inversion scheme based on the method of characteristics can be used to recover the impedance profile of the specimen.

ACKNOWLEDGMENTS

This work was carried out under the Grants-in-Aid for Education Program of Bethlehem Steel Corporation with Dr. B. E. Droney project coordinator. Partial support was also obtained from the National Science Foundation through a grant to the Materials Science Center at Cornell University. This support is gratefully acknowledged. We thank also Mr. B. F. Addis for his care in preparing the specimens.

REFERENCES

1. W. D. Jones, Fundamental Principles of Powder Metallurgy, Edward Arnold Publ., Ltd., London (1960), pp. 839-844.

2. J. S. Hirschhorn, Introduction to Powder Metallurgy, Am. Powder Metall. Inst., New York (1969), pp. 253-261.

3. R. H. Brockelman, "Dynamic Elastic Determination of the Properties of Sintered Powder Metals", in: Advanced Experimental Techniques in Powder Metallurgy, Plenum Press, New York (1970), pp. 201-224.

4. E. P. Papadakis and B. W. Petersen, "Ultrasonic Velocity as a Predictor of Density in Sintered Powder Metal Parts", Materials Evaluation, 37, 76-80 (1979).

5. K. W. Winkler, "Attenuation and Phase Velocity Spectra in Sedimentary Rocks under High Pressure", 1982 Ultrasonics Symposium Proceedings, Vol. 2, Inst. Electr. Electr. Eng., New York (1982), pp. 1054-1058.

6. R. H. Latiff and W. F. Fiore, "Ultrasonic Attenuation and Velocity in Two Phase Microstructures", J. Acoust. Soc. Am., 57, 1441-1446 (1975).

7. P. C. Waterman and R. Truell, "Multiple Scattering of Waves", J. Math. Phys., 2, 512-537 (1961).

8. J. E. Gubernatis and E. Domany, "Effects of Microstructure on the Speed and Attenuation of Elastic Waves: Some Results for Porous Materials", Wave Motion, 16, 580-589 (1984).

9. B. R. Tittman, J. Ahlberg, K. W. Fertig, "Ultrasonic Microstructureal Noise Parameters in a Powder Metal Alloy", in Review of Progress in Quantitative NDE, Vol. 3A, D. O. Thompson and D. E. Chimenti, Eds., Plenum Press, New York (1984), pp. 57-64.

10. C. M. Sayers and R. L. Smith, "The Propagation of Sound in Porous Media", Ultrasonics, 20, 201-205 (1982).

11. W. Sachse and Y.-H. Pao, "On the Determination of Phase and Group Velocities of Dispersive Waves in Solids", J. Appl. Phys., 49, 4320-4327 (1978).

12. E. P. Papadakis, "Ultrasonic Velocity and Attenuation Measurement Methods with Scientific and Industrial Applications", in Physical Acoustics, Vol. XII, W. P. Mason and R. N. Thurston, Eds., Academic Press, New York (1979), pp. 277-374.

13. Y.-H. Pao, F. Santosa and W. Symes, "Inverse Problems of Acoustic and Elastic Waves", in Inverse Problems of Acoustic and Elastic Waves, F. Santosa, Y.-H. Pao, W. Symes and C. Holland, Eds., SIAM, Philadelphia (1984), pp. 274-302.

14. F. Santosa and H. Schwetlick, "The Inversion of Acoustical Impedance Profile by the Method of Characteristics", Wave Motion, 4, 99-110 (1982).

ULTRASONIC ATTENUATION IN POWDER-METAL ALLOY

Bernhard R. Tittmann
Lloyd A. Ahlberg
Kenneth W. Fertig
Monem Abdel-Gawad
Rockwell International Science Center
Thousand Oaks, California

ABSTRACT

THE ULTRASONIC ATTENUATION of longitudinal waves was measured in IN-100, a powder metallurgy alloy used for aircraft engine components. This material was selected as a model system for testing the feasibility of characterizing the microstructure of a variety of inhomogeneous media including powder metals, ceramics, castings and composites. The data were obtained with a variety of different transducers placed either in a waterbath or in sample contact for a frequency range 50-100 MHz. Micrographical examination provided size and number distributions for grain and pore structure. The results showed that the predominant source for the ultrasonic attenuation and backscatter was a dense (~ 100/mm^3) distribution of small micropores (~ 10 μm diameter). Two samples with different micropore densities were studied in detail to test the feasibility of calculating from observed microstructural parameters the frequency dependence of the ultrasonic attenuation in the regime for which the wavelength is larger than or nearly equal to the size of the individual scattering centers. Excellent agreement was found between predicted and observed values so as to demonstrate the feasibility of solving the forward problem. The results suggest a way toward the nondestructive detection and characterization of anomalous distributions of micropores when conventional ultrasonic imaging is difficult. The findings are potentially significant toward the application of the early detection of porosity during the materials fabrication process and after manufacturing of potential sites for stress-induced void coalescence leading to crack initiation and subsequent failure.

THE ACQUISITION AND INTERPRETATION OF ULTRASONIC DATA for the inference of microstructural information is not new and has been carried out by several researchers in the recent past [1-7]. The objective of this work is to apply these techniques to the types of powder metallurgy alloys used in the manufacture of aircraft engine disc components. The particular application motivating this work is the development of an approach to the nondestructive detection and characterization of anomalous microstructure when conventional ultrasonic imaging is difficult. Such an approach could be important in tracking porosity during the materials process as well as during service as a finished part in the early locating of potential sites for crack initiation.

The material chosen for these tests is IN-100, a Ni-base alloy prepared by Pratt & Whitney Aircraft Co. This material is used in the F-15 and F-16 fighter aircraft engine at a rate of 1000 lb of billet material per engine to manufacture three compressor discs, four turbine discs and two spacers. The following sections discuss the experimental approach, the analytical models and the comparison between theory and experiment.

EXPERIMENTAL APPROACH

This section treats the micrographical and ultrasonic characterization of the samples, the determination of bulk microstructure parameters from the 2D optical micrographs and the ultrasonic measurement technique [8].

MICROGRAPHICAL CHARACTERIZATION - Suitable samples were prepared from a number of different parent samples supplied by Pratt & Whitney Co. Selected sample surfaces were polished and examined under the optical microscope. Figure 1 shows an optical micrograph of the grain structure. Figure 2 gives histogram data displaying the grain size distribution for two samples. These data show a uniform grain size of about 2-4 μm.

One of the samples tested (sample C) is a section of a used F100 engine 2nd disc which was pulled out for spin pit tests. The section is shown in cross section in Fig. 3 together with lines drawn to indicate traverses with the microscope to study surface features. Figure 4

SC82-18879

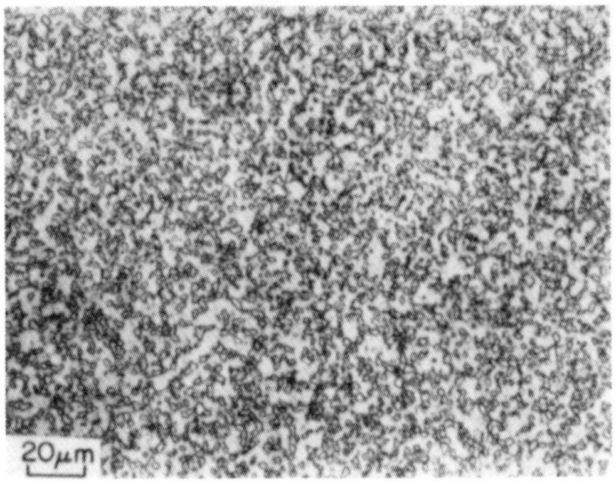

Fig. 1 - Photomicrograph of IN-100 grains

SC82-18888

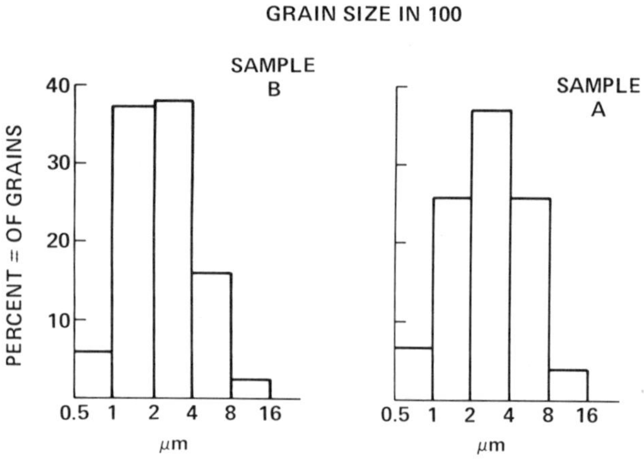

Fig. 2 - Grain size distribution in IN-100

SC82-19519

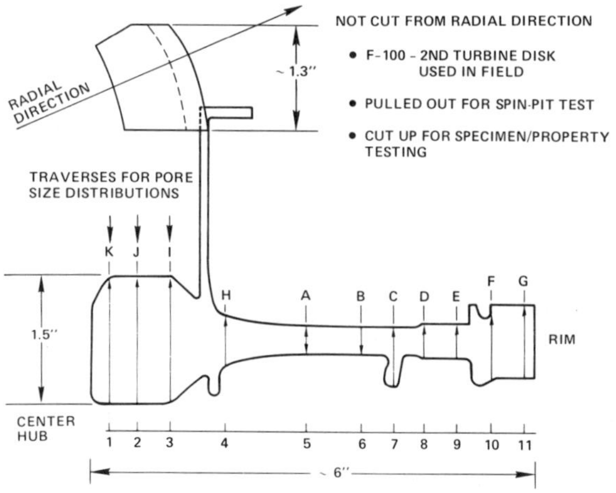

Fig. 3 - Section of F100 engine disc

shows sample microphotos taken along some of these traverses indicating the presence of micropores. The microporosity is a by-product of the powder metallurgy process and has its origin in hollow powder particles with argon gas trapped during the atomization process. Although a sparse, uniform distribution of micropores is not thought to be the limiting factor in low cycle fatigue strength, the micropores do contribute significantly to the ultrasonic noise background and must be taken into account. Table 1 presents a summary of the micropore populations found along the various traverses. Statistical analysis of the micrograph data indicate that the pore distribution is not random but shows a tendency towards clustering. Figure 5 presents the information of Table 1 in graphical manner. Figures 6a and 6b give pore size distribution histograms for two of the samples tested. Sample A was found to have an approximate micropore density of 330/mm^3 whereas sample C had 90/mm^3. For sample C majority of micropores fall in the range of between 5 and 10 μm in diameter whereas in sample A the diameters range from ~ 1 μm to 50 μm.

In order to employ the histogram data obtained from the optical examinations to estimate volume microstructural information, the expressions for the mean volume porosity ρ_v and the expectation value for the sixth power of the pore radius $E(r^6)$ were derived:

$$\rho_v = \frac{1}{\pi} \int_o^\infty \frac{n_2(r')}{r'} dr' = \frac{1}{\pi} \sum_j \frac{n_j}{r_j} \Delta r_j$$

$$E(r^6) = \frac{1}{\rho_v} \int_o^\infty r^6 n_3(r)\, dr \doteq \frac{8}{\rho_v} \sum_j (\Delta r_j)\, r_j^5\, n_j \qquad (1)$$

where $n_2(r')dr$ is the number of circles per unit area (with radii between r' and $r' + dr'$) as measured from micrographs. Each circle represents a cut through a pore by the polished sample surface. $n_3(r')$ is the number of micropores per unit volume element, Δr_j is the bin width of the pore radius histogram, r_j is the average pore radius for each bin and n_j is the number of pores in each bin.

These formulae were used for calculations in the detailed examination of two demonstrative samples discussed in later sections.

ULTRASONIC CHARACTERIZATION - In the ultrasonic attenuation measurements the data were obtained in the 10-100 MHz range of frequencies in several smaller subbands. A low attenuation buffer (Al or fused silica) was typically used to obtain the transducer response and the reflection coefficient of the sample interface. The technique and calculations are illustrated in Fig. 7 for the case of longitudinal waves. For the case of shear waves, an additional technique was employed for rapid data acquisition and is illustrated in Fig. 8. Here the buffer is water and the angle of incidence is such that the longitudinal waves in the water produce shear waves in the sample. Because of the possible generation of leaky waves generated at the interfaces and the noncircular beam spreading associated with nonnormal incidence, the measurements were augmented with calibration

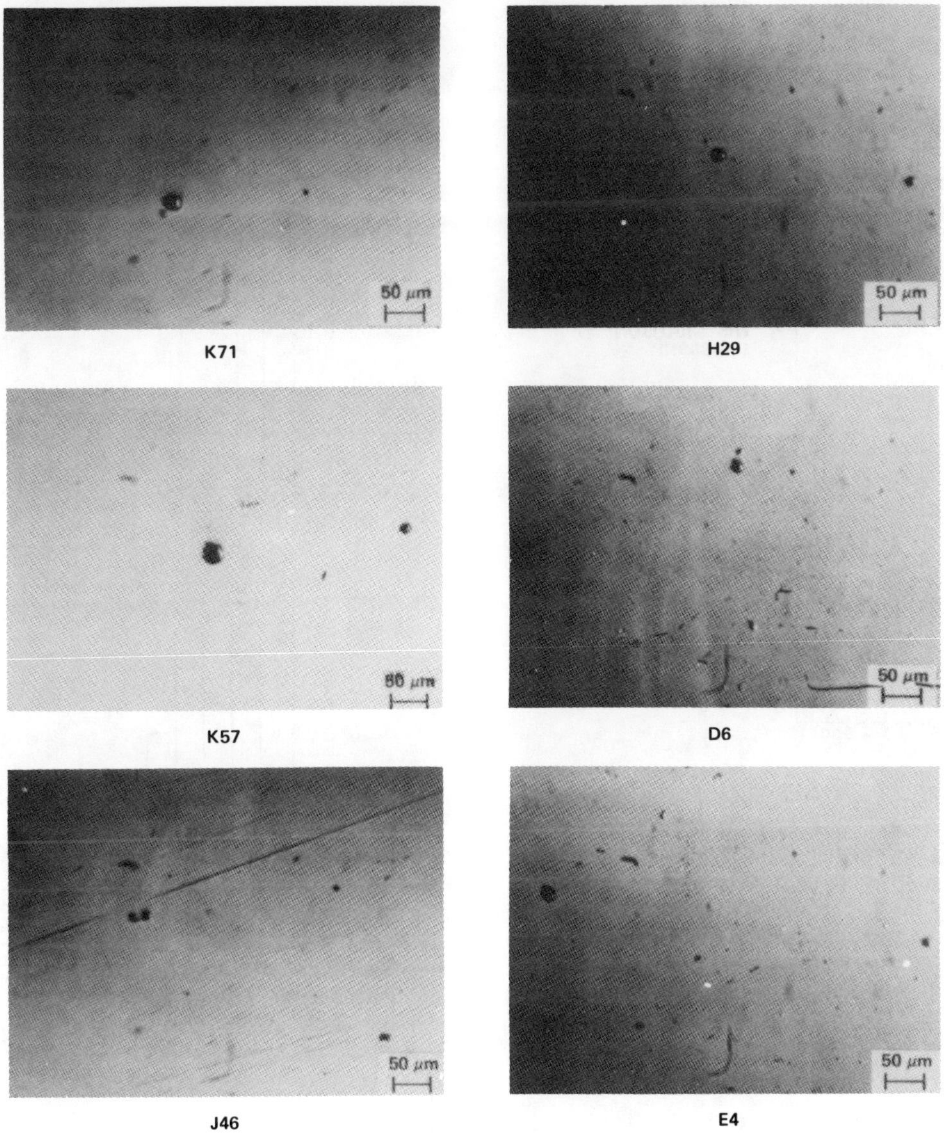

Fig. 4 - Microphotos of area elements (along tranverses of Fig. 3) showing microporosity

Table 1. Summary of Observations to Spin pit Test Sample

Traverse Number	Traverse Symbol	Traverse Length mm	Sample Length mm	Number of Frames	Sample Area (mm²)	Number of Pores	Number of Pores/ Linear mm	Number of Pores mm²	Pore Spacing (Av) mm
1	K	40	36.9	81	21.9	29	0.79	1.3	1.27
2	J	40	39.6	87	23.5	28	0.7	1.2	1.4
3	I	40	40.6	89	24	54	1.3	2.3	0.75
4	H	16	15	33	8.9	27	1.8	3.0	0.55
5	A	8	7.3	16	4.3	10	1.4	2.3	0.7
6	B	8	8.2	18	4.9	3	0.36	0.6	2.7
7	C	19	18.2	40	10.8	35	1.9	3.2	0.5
8	D	10	10.5	23	6.2	15	1.4	2.4	0.7
9	E	10	10.5	23	6.2	14	1.3	2.3	0.75
10	F	21	16	31	9.5	34	2.1	3.6	0.47
11	G	21	15.5	34	9.2	22	1.4	2.4	0.7
Total Sample			218		129	271	1.27	0.87	

	Pore Size Distribution (μm)						
	5-10	10-15	15-20	20-25	25-30	30-35	35-40
Number of Pores	247	18	5	0	0	0	1
Percentage	91.1	6.6	1.8	0	0	0	0.4

measurements carried out with narrow-band quartz shear-wave transducers in direct tightly bonded contact with the sample.

In reference to Fig. 8, T_{01}, T_{10} are the water-sample, sample-water transmission coefficients; α_S is the sample shear wave attenuation in nepers/cm, α_W is the water longitudinal attenuation in nepers/cm, ℓ_{S1}, ℓ_{S2} are the shear wave paths in the sample under test, ℓ_{W1}, ℓ_{W2} are the wave path lengths in water.

The measurements are summarized in Figs. 9 and 10 which display attenuation values as a function of frequency for two different samples and for three different locations for the same sample.

Table 2 summarizes the information collected by given representative values for attenuation and velocity for two powder metal alloys, IN-100 and Ti-6246, compared to a common engine disc material of nonpowder fabrication, i.e., Waspaloy.

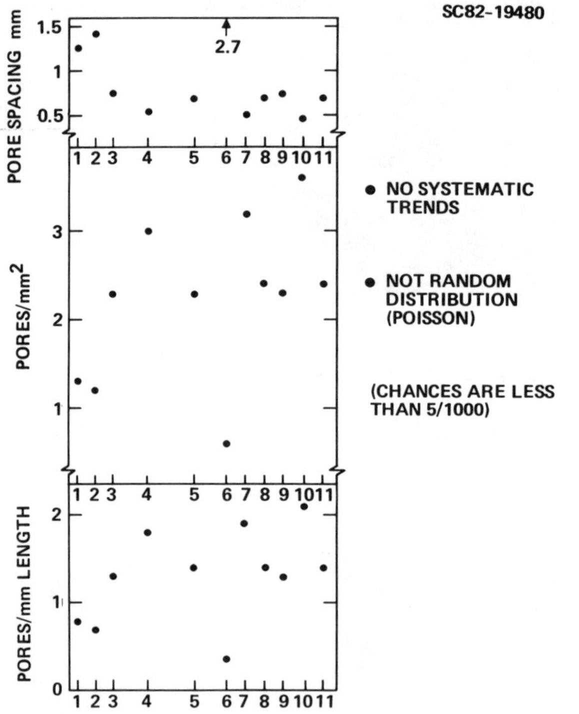

Fig. 5 - Graph of number of pores per unit length, number of pores per unit area and pore separation plotted vs distance from disc center to rim (as indicated by traverse number, see Fig. 3)

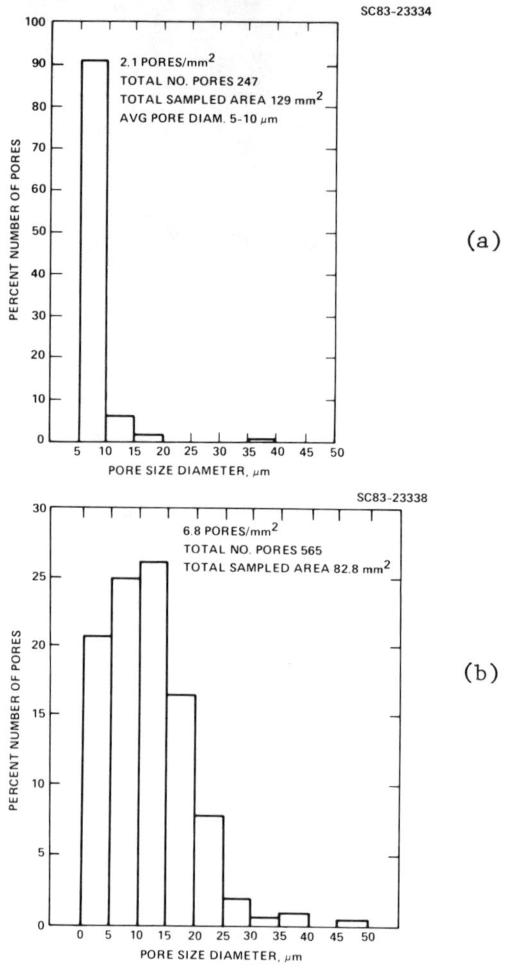

Fig. 6 - Pore size distribution (a) sample C - IN-100, (b) sample A - IN-100

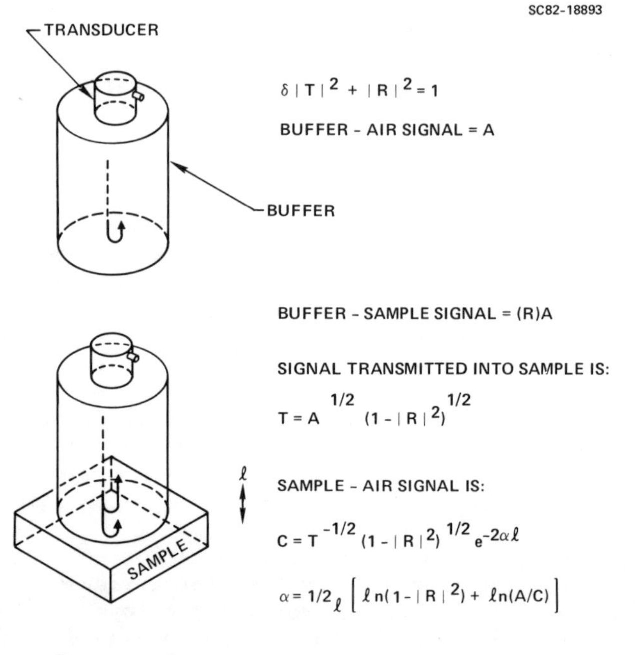

Fig. 7 - Sample configuration for measurement of longitudinal wave attenuation

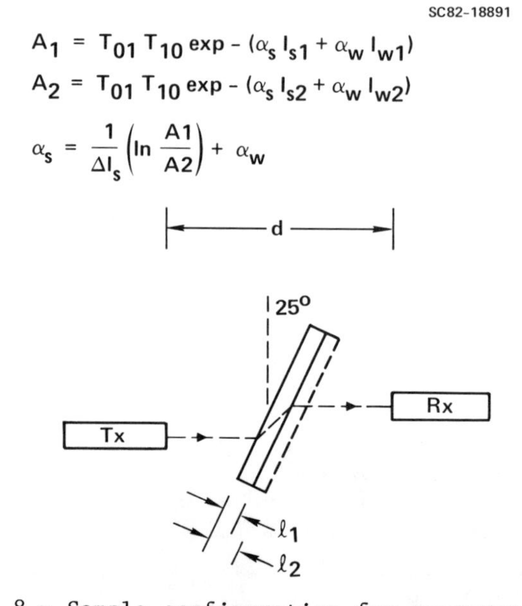

Fig. 8 - Sample configuration for measurement of shear wave attenuation

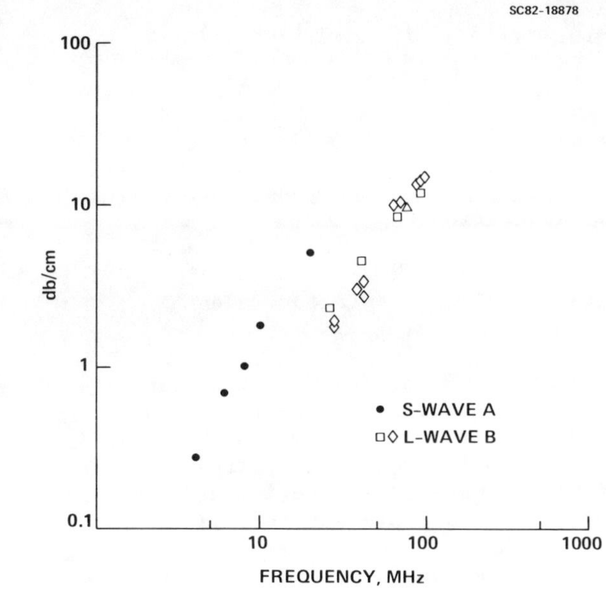

Fig. 9 - Longitudinal (L) and shear (S) wave attenuation for 2 samples of IN-100

Table 2

Sample	Velocity (mm/μm) v_L	v_S	Grain Size (μm)	Atten. (dB/cm) α_L	α_S	Freq. (MHz)	Freq. Dep. α_L	α_S
Waspaloy	6.04	3.23	100	4	11	10	$f^{2.2}$	$f^{2.5}$
IN-100	6.52	3.29	2	4 ± 1	20	50	f^2	$f^{2.2}$
Ti-6246	6.11	3.16	2-4	10	-	100		$f^{3.5}$

ANALYSIS

This section presents the analytical developments leading up to the calculations for a comparison between the experimental observations and the predicted frequency dependency of attenuation.

ULTRASONIC ATTENUATION - Attenuation due to scattering of micropores distributed throughout the medium is modelled as follows.

Consider a volume element δV_j contained within the sample and centered at r_j. Let $\delta V_j N_{3j}(a)$ be the actual number of pores with centers in δV_j and with radii less than or equal to a. Thus, $N_{3j}(a)$ is a random variable representing the number of pores per unit volume having radii less than or equal to a and having centers "near" r_j. Then the observed attenuation per unit length at frequency f due to scattering from pores in δV_j only is

$$\hat{\alpha}_j = \frac{1}{2} \sum_{k=1}^{\delta V_j N_{3j}(\infty)} \Omega(a_k,f)/\delta V_j$$

$$\equiv \frac{1}{2} \int_0^\infty \Omega(a,f)\, dN_{3j}(a) \quad , \qquad (2)$$

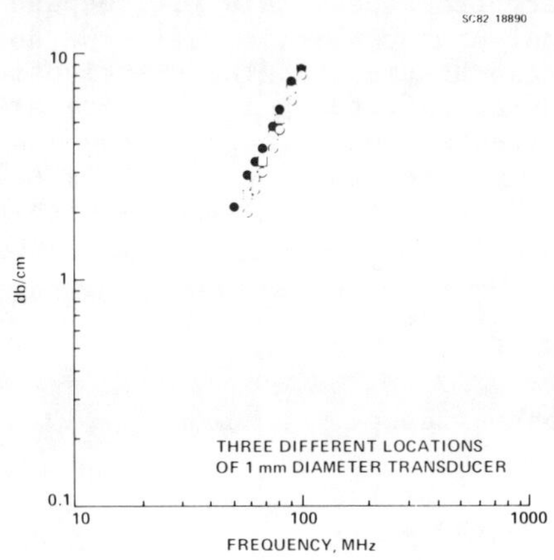

Fig. 10 - Longitudinal wave attenuation at three different locations on sample C

where $\Omega(a,f)$ is the total scattering cross section at frequency f due to a pore of radius a (assumed spherical).

The integral in (2) is defined as an ordinary Stieltjes integral with respect to the random step function $N_{3j}(a)$. The attenuation has units nepers/cm in the case that a has units of cm, δV_j has units of $(cm)^3$, and $\Omega(a,f)$ has units of $(cm)^2$. (The attenuation in dB/cm is found from $\alpha_{dB/cm} = \alpha_{nepers/cm} \cdot 20 \log 10(e)$).

The <u>observed</u> attenuation α, measured in an actual experiment over a volume

$$V = \bigcup_{j=1}^{J} \delta V_j \quad , \quad \text{satisfies}$$

$$\hat{\alpha} = \frac{1}{V} \sum_{j=1}^{J} \delta V_j \hat{\alpha}_j \quad .$$

We can easily compute the expected attenuation, α, as

$$\alpha = \frac{1}{2} \int_0^\infty \Omega(a_1,f) \nu_3(a)\, da \quad ,$$

where, for all a_1, a_2

$$\int_{a_1}^{a_2} \nu_3(a)\, da = E\left(N_{3j}(a_2) - N_{3j}(a_1)\right) \quad .$$

Thus, $\nu_3(a)\Delta a$ is the expected number of pores per unit volume with radii in (a, a + Δa).

The variation in α is more complicated. It depends upon the correlation structure of $(N_{3j}(a + \Delta a) - N_{3j}(a))$ over a j. In particular, one needs to know

$$E\left(N_{3j}(a + \Delta a) - N_{3j}(a)\right)\left(N_{3j'}(a' + \Delta a') - N_{3j'}(a')\right)$$

$$= h(r_j, r_{j'}, a, a', \Delta a) \quad .$$

For stationary processes, this will depend upon r_j and $r_{j'}$ only through $|r_j - r_{j'}|$. We note that if one can assume that the distribution of pores is completely random, i.e., there are no clustering effects and the correlation lengths are essentially zero, then $\delta V_j\,(N_{3j}(a + \Delta a) - N_{3j}(a))$ is a Poisson random variable with mean

$$\delta V_j \int_{a_1}^{a_2} \nu_3(a)da$$

and hence has variance

$$\delta V_j \int_{a_1}^{a_2} \nu_3(a)da \quad .$$

Further, nonoverlapping intervals (a, a + Δa), (a',a' + Δa) produce independent variates. In such a case

$$\mathrm{var}(\hat{\alpha}_j) = \frac{1}{4\delta V_j} \int_0^\infty \Omega^2(a,f)N_{3_j}(a)da$$

In the case where $\hat{\alpha}$ is measured over many volume elements

$$\mathrm{var}(\hat{\alpha}_j) = \frac{1}{4V} \int_0^\infty \Omega^2(a,f)\nu_3(a)da \quad . \qquad (3)$$

We must emphasize that the expression (2) is valid only if the pores are completely randomly distributed throughout the sample. We derive in the appendix a more general expression for the var(α) under the assumption that within each volume element δV_j, $\delta V_j N_{3j}(a)$ can be treated as Poisson random process with intensity function $\nu_3(a|x_j)$ depending on a "hidden" state variable x_j that can change from volume element to volume element. Under the assumption that $N_{3j}(a)$ is independent from $N_{3j'}(a')$, it is seen that (3) above forms a lower bound on the variance of α.

Expression (3) shows that in the case of completely randomly distributed pores, var(α) goes to zero as 1/V. The expression derived in the appendix for a more general distribution also shows that var(α) goes to zero as 1/V. However, the coefficient in front of 1/V can be much larger than that implied by (3). Heuristically, one must sample enough volume elements in the attenuation experiment to account for all possible variation in pore distribution throughout the sample.

We remark, that if we are only interested in the contributions to the attenuation due to the pores in a certain size range, the above equations can be modified by restricting the range of integration to that size range. Thus, the observed attenuation $\alpha(a_1,a_2)$ due to pores in (a_1,a_2), is given by

$$\hat{\alpha}(a_1,a_2) = \frac{1}{2} \int_{a_1}^{a_2} \Omega(a,f)dN_3(a) \quad ,$$

where

$$N_3(a) = \frac{1}{V} \sum_{j=1}^{J} \delta V_j N_{3j}(a) \quad .$$

It has expectation

$$\alpha(a_1,a_2) = \frac{1}{2} \int_{a_1}^{a_2} \Omega(a,f)\nu_3(a)da \quad .$$

SCATTERING CROSS SECTION - The scattering cross section Ω(a,f) for longitudinal waves is computed from Truell, Elbaum and Chick [9] for $(2\pi fa)/c \geqslant 0.5$. Specifically, Truell, et al define the normalized scattering cross section $\gamma_N = \Omega(a,f)/\pi a^2$ and present formula for γ_N as a function of ka.

The computational form for γ_N is numerically sensitive for small ka. In this range we use the Rayleigh limit given by

$$\Omega(a,f) = g_0 f^4 a^6 \quad .$$

The coefficient g_0, which depends upon material properties and the geometry of the scatterers is actually determined by requiring Ω(a,f) to be continuous at ka = 0.5. Thus, since $\Omega(a,f) = \pi a^2 \gamma_N(ka)$, we have

$$g_0 = \pi\gamma_N(ka = 0.5)(4\pi/c)^4 \quad .$$

Similarly, $\gamma_N(ka)$ is computationally difficult for ka ⩾ 10. There, for our materials, it is sufficient to assume γ_N is constant. Thus for ka ⩾ 10,

$$\Omega(a,f) = \gamma_N(ka = 10)\pi a^2$$

$$= g_1 a^2 \quad ,$$

where

$$g_1 = \gamma_N(10)\pi \quad .$$

PORE SIZE DISTRIBUTION - Let $\Lambda_3(a)$ be the expected number of pores per unit volume with radii greater than a. Writing $\nu_3(a)da$ as the expected number of pores per unit volume in the size range a to a + da, we have

$$\Lambda_3(a) = \int_a^\infty \nu_3(a)da \quad .$$

Let $\nu_2(a)da$ be the expected number of circles per unit area observed on a random plane cut through the sample and having radii in the range a to a + da. Then [10]

$$\Lambda_3(a) = \frac{1}{\pi} \int_a^{\infty} \frac{\nu_2(a')da'}{a'^2 - a^2} \tag{4}$$

Expression (4) is valid if either the pores are randomly distributed throughout the volume or the plane used to cut the volume is randomly positioned and $\nu_2(a)$ is defined over all such random planes.

The computational forms given in the Analysis section required knowledge of ν_3. Unfortunately, estimates of ν_3 based on observed distributions of circles on a plane are very unstable. A better approach is to perform the integration in the expression for α by parts so that only $\Lambda_3(a)$ need be known, thus

$$\alpha(a_1,a_2) = \frac{1}{2} \int_{a_1}^{a_2} \Omega(a,f)\nu_3(a)da$$

$$= \Omega(a_1,f)\Lambda_3(a_1) - \Omega(a_2,f)\Lambda_3(a_2)$$

$$+ \int_{a_1}^{a_2} \frac{\partial\Omega(a,f)}{\partial a} \Lambda_3(a)da \quad . \tag{5}$$

Equation (5) shows that the scattering cross section depends on the sixth power of the sphere radius. In the IN-100 each pore acts to scatter energy out of the ultrasonic beam and contributes to the attenuation. It is reasonable to expect that the attenuation values depend strongly on pore density and pore size distribution. In the case of sparse distributions, the large pores would be expected to dominate the attenuation and could influence the values differently depending on the specific sample or even the site for a given sample where the measurements are obtained.

COMPARISON BETWEEN THEORY AND EXPERIMENT

Following the procedure developed in the Analysis section, the histogram data for sample C were used to determine the pore size distribution function, which is plotted in Fig. 11 in terms of the expected number of pores per unit volume with diameters greater than 2a as a function of the diameter (microns). The use of segments of the distribution function, corresponding to smaller ranges of pore size, allows the exploration of how various groups of pores contribute to the overall attenuation. Figure 12 shows such contributions to the longitudinal attenuation from various pore diameter ranges.

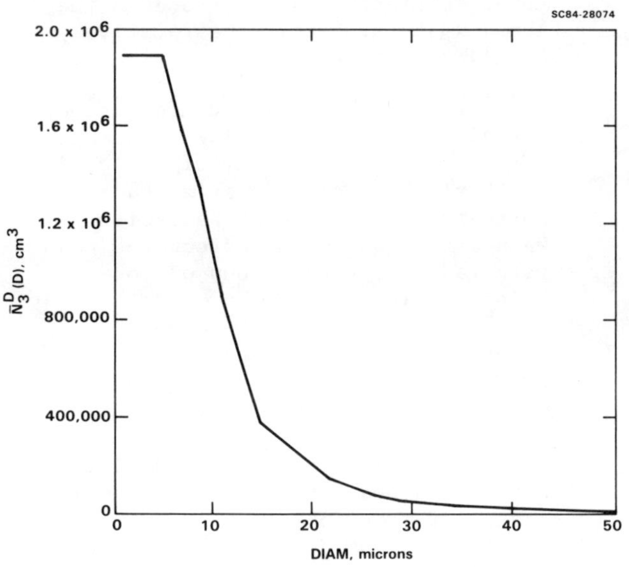

Fig. 11 - Plot of pore size distribution function. N^{-D}_3 is the expected numbers of pores per unit volume greater than 2a vs diameter (2a) for sample C

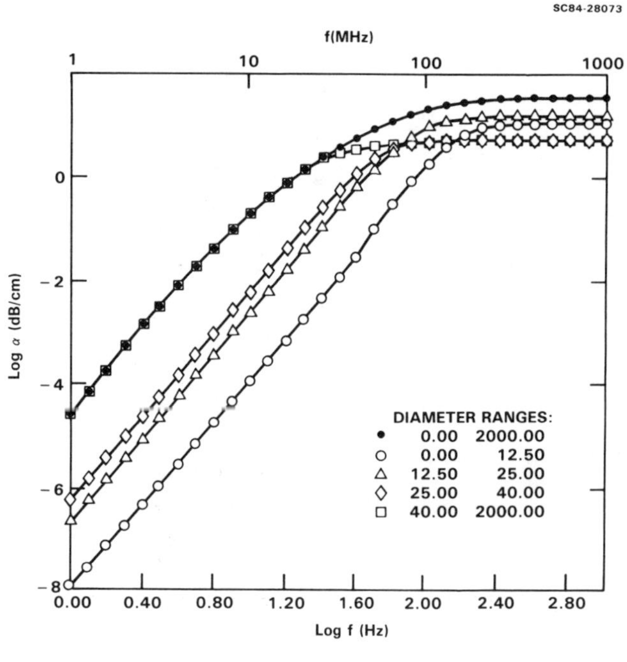

Fig. 12 - Contribution to the frequency dependent longitudinal attenuation from various pore diameter ranges

The calculated data points are plotted on a log-log plot of attenuation vs frequency for 5 diameter ranges. Several important conclusions can be drawn from these results.

1. Below about 50 MHz the total attenuation from the contributions of all the pores is nearly that of those in the 40-50 μm range, i.e., the relatively few large pores.

2. The curves all have in common a steep rise in attenuation at low frequencies corresponding to the Rayleigh regime of scattering and a turnover at ka = 1, corresponding to where the reciprocal wavenumber equals an "effective" mean pore radius for the group of pores being considered.

3. As expected the turnover becomes more gradual and rounded as the total range of pore sizes becomes larger.

To allow a comparison with the experimental data which was obtained between 50 MHz and 100 MHz, this range of frequencies is emphasized in Fig. 13. It is apparent from this figure that there are two parameters which describe the contributions from the various groups of pores, namely the vertical positions of curves, corresponding to the level of attenuation, and the line slopes, corresponding to the frequency dependence. As the "effective" mean pore diameter increases, the level of attenuation rises and the frequency dependence diminishes.

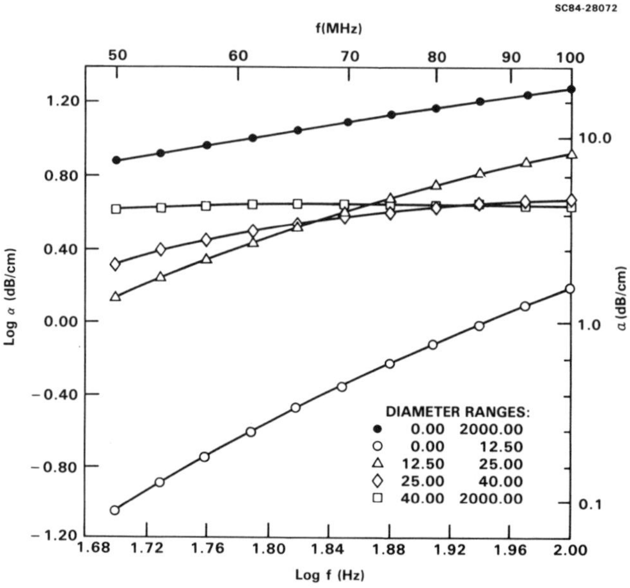

Fig. 13 - Log-log plot of calculated attenuation vs frequency emphasizing 50-100 MHz range

In order to obtain a comparison between theory and experiment for the total attenuation, additional experiments were carried out to obtain the contribution to the attenuation due to scattering by grains alone. The attenuation was measured [8] in another powder metal alloy known to be free of micropores, Ti 6-2-4-6, which has nearly the same grain-size distribution. The use of these data in the context of this discussion in IN-100 makes the assumption that the difference in crystal structure between the IN-100 grains and the Ti grains may be neglected for the purpose of this comparison. An improved procedure would be to obtain a sample of IN-100 that has undergone the same heat treatments as Sample A but has no pores. Such a sample could not be obtained. An alternative procedure is to carry out calculations for the correction necessary to take into account the difference in crystal structure but the theory required for such a treatment is not currently available in the regime near ka = 1.

Figure 14 shows the comparison between experiment and theory in a log-log plot. Here, the dashed line represents the total attenuation predicted on the basis of the pore size distribution function. The solid line was obtained by adding the observed attenuation due to grain scattering. The data points are the observations based on three sets of measurements carried out on three different locations on the sample. The spread in the experimental data is attributed to inhomogeneities in the sample and noise in the deconvolution and normalization procedures. In spite of the inaccuracies in the assumptions leading to this comparison, the agreement is rather favorable and demonstrates the robustness of this approach in solving the forward problem.

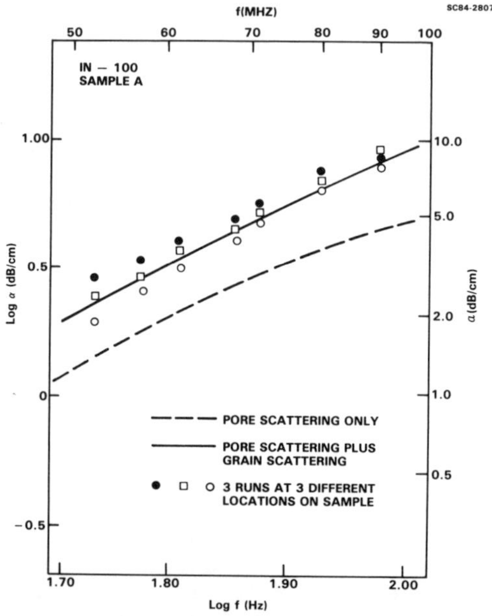

Fig. 14 - Comparison between theory and experiment

It is interesting to determine the contributions to the pore scattering attenuation at one frequency. This is done in Fig. 15 for 100 MHz for 5 μm segments of the pore size distribution function. This plot shows again the relative strength of the few large pores in comparison to that of the many small pores. The plot also shows a peak for the contributions in the 20-25 μm range of pores, representing the optimum combination of pore number and pore size.

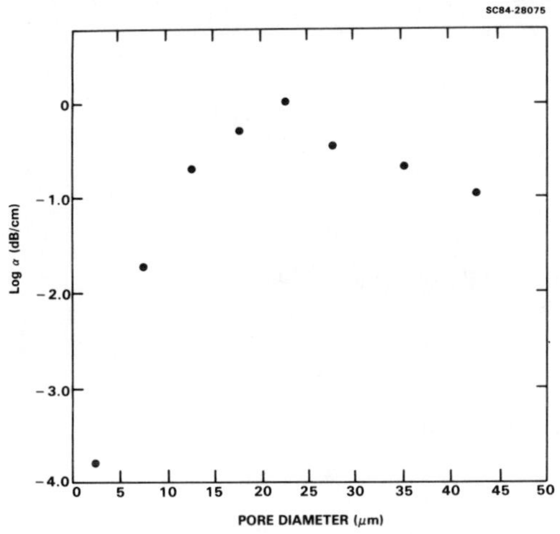

Fig. 15 - Plot of log α vs pore diameter at 100 MHz showing contribution of 5 μm wide pore size ranges

CONCLUSION

The microstructure of a nickel-base powder metal alloy IN-100 has been studied with the aid of micrographical and ultrasonic techniques. The material showed grains of small diameter with a narrow distribution in the range of 2-4 μm and sparsely distributed micropores with diameter in the 5-20 μm range. The micropores dominated the ultrasonic attenuation characteristics which was selected to carry out quantitative measurements and calculations for comparison of theory and experiment. The frequency dependence from 50-100 MHz of the longitudinal wave attenuation was in good agreement with calculations which included the size distribution of scatterers. The results point the way toward the nondestructive detection and characterization of anomalous distributions of micropores where conventional ultrasonic imaging is difficult. Such information may be useful in the early detection of potential failure sites where stress-induced void coalescence leads to crack initiation.

ACKNOWLEDGEMENT

This work was sponsored in part by Rockwell International Research and Development Funds and in part by the Center for Advanced Nondestructive Evaluation, operated by the Ames Laboratory, USDOE, for the Defense Advanced Research Projects Agency and the Air Force Wright Aeronautical Laboratories/Materials Laboratory under Contract No. W-7405-ENG-82.

REFERENCES

1. O'Donnell, M. and J. G. Miller, J. Appl. Phys. 52, 1056 (1981)

2. Goebbels, K., Patent 4026,157

3. Papadakis, E. M., J. Acoust. Soc. A. 37, 711 (1965)

4. Mason, W. P. and H. J. McSkimin, J. Appl. Phys. 19, 940 (1948)

5. Lifshitz, I. M. and G. D. Parkhomovskii, J. Expt. Theoret. Phys. (USSR) 20, 175 (1950)

6. Evans, A. G., B. R. Tittmann, L. Ahlberg, B. T. Khun-Yakub, G. S. Kino, J. Appl. Phys. 49, 1669 (1978)

7. Adler, L., Proceedings of the Review of Progress in Quantitative NDE (1984)

8. Tittmann, B. R., L. A. Ahlberg and K. Fertig, Proceedings of the Review of Progress in Quantitative NDE (edited by D.O. Thompson and D. Chimenti) 1983 and 1982

9. Truell, R., C. Elbaum and B. Chick, "Ultrasonic Methods in Solid State Physics," Academic Press, New York (1969) p. 162, also R. L. Roderick and Rohn Truell, J. Appl. Phys. 23, 267 (1952)

10. Cahn, J. W., and R. L. Fullman (1966)

APPENDIX

Let $\delta V_j N_{3j}(a)$ be the number of pores having centers found in a volume element δV_j and having radii less than or equal to a. The volume elements are assumed chosen so that $N_{3j}(a)$ and $N_{3j'}(a')$ are independent for all, a, a' and all $j \neq j'$.

Let $M3(a1,a2;x_j) = \delta V_j\,(N_{3j}(a2) - N_{3j}(a1))$ be the actual number of pores with centers in δV_j and having radii in (a1,a2). The variable x_j is introduced as a hidden "state" variable that once specified allows $\delta V_j N_{3j}(a)$ to be treated as a Poisson random counting process with independent increments. Thus

$$E(M_3(a_1,a_2;x_j)|x_j) = \delta V_j \int_{a_1}^{a_2} \nu_3(a|x_j)da$$

and

$$\mathrm{cov}\left(M_3(a_1,a_2,x_j)\ ,\ M_3(a_3,a_4,x_j)\right) = \begin{cases} \delta V_j \int_{a_1}^{a_2} \nu_3(a|x_j)da\ , & a_1 = a_3\ ,\ a_2 = a_4 \\ 0 & , (a_1,a_2)\ (a_3,a_4) = \phi \end{cases}$$

where ϕ is the empty set.

If we think of x_j as a random variable which has different realizations for differing volume elements, then $\nu 3(a|x_j)$, which depends upon x_j, is a random process. Thus, though $M3(a1,a2;x_j)$ is a Poisson variate conditioned on x_j, it is generally not a Poisson variate unconditionally.

We may write however

$$E\left(M_3(a_1,a_2;x_j)\right) = E_{x_j}\,\delta V_j \int_{a_1}^{a_2} \nu_3(a|x_j)da$$

$$= \delta V_j \int_{a_1}^{a_2} \nu_3(a)da\ ,$$

where

$$\nu_3(a) = E_x[\nu_3(a|x)]\ .$$

We write $\hat{\alpha}_j$ as the attenuation per unit length due to pores is δV_j. Thus

$$\hat{\alpha}_j = \frac{1}{2}\int_0^\infty \Omega(a,f)dN_{3j}(a)\ .$$

The measured attenuation, $\hat{\alpha}$, measured over

$$V = \bigcup_{j=1}^{J} \delta V_j\ ,\quad \text{is}$$

$$\hat{\alpha} = \frac{1}{V}\sum_{j=1}^{J} \delta V_j \hat{\alpha}_j\ .$$

We let $\alpha(x)$ be the expected value of $\hat{\alpha}_j$ when $x_j = x$. Thus,

$$\alpha(x) = E(\hat{\alpha}_j|x)$$

$$= \lim_{\Delta a\to 0} \frac{1}{2}\sum_k \Omega(a_k,f)E(M_3(a_k,a_k + \Delta a,x)/\delta V_j)$$

$$= \lim_{\Delta a\to 0} \frac{1}{2}\sum_k \Omega(a_k,f)\nu_3(a_k|x)\Delta a$$

$$= \frac{1}{2}\int_0^\infty \Omega(a,f)\nu_3(a|x)da\ .$$

Further,

$$\mathrm{var}(\hat{\alpha}_j|x) = E(\hat{\alpha}_j - \alpha(x))^2$$

$$= \frac{1}{4\delta V_j}\int_0^\infty \Omega^2(a,f)\nu_3(a|x)da\ .$$

From these expressions, we may compute the unconditional expectation over x as

$$\alpha = E\left(\alpha(x)\right) = \frac{1}{2}\int_0^\infty \Omega(a,f)\nu_3(a)da$$

and

$$\mathrm{var}(\hat{\alpha}_j) = E\ (\mathrm{var}(\hat{\alpha}_j|x)) + E(\alpha(x) - \alpha)^2$$

$$= \frac{1}{4\delta V_j}\int_0^\infty \Omega^2(a,f)\nu_3(a)da$$

$$+ \frac{1}{4}\int_0^\infty \Omega^2(a,f)\sigma^2(a)da$$

$$+ \frac{1}{4}\int_0^\infty\int_0^\infty \rho(a,a')\Omega(a,f)\Omega(a',f)dada'\ ,$$

where

$$\nu_3(a) = E(\nu_3(a|x))$$

$$\sigma^2(a) = E\big(\nu_3(a|x) - \nu_3(a)\big)^2$$

$\rho(a,a')$

$$= E\big(\nu_3(a|x) - \nu_3(a)\big)\big(\nu_3(a'|x) - \nu_3(a')\big) \; ,$$

$$a \neq a' \quad .$$

For the observed attenuation, $\hat{\alpha}$, measured over

$$V = \bigcup_{j=1}^{J} \delta V_j \quad ,$$

we have that the mean is

$$E(\hat{\alpha}) = E\Big(\frac{1}{V} \sum_{j=1}^{J} \delta V_j \hat{\alpha}_j\Big)$$

$$= \frac{1}{2} \int_0^\infty \Omega(a,f)\nu_3(a)da \quad ,$$

and the variance is

$$\mathrm{var}(\hat{\alpha}) = \frac{1}{4V} \int_0^\infty \Omega^2(a,f)\nu_3(a)da$$

$$+ \frac{1}{4} \; \frac{\mathrm{var}(\delta V)}{V^2} + \frac{1}{J} \; \int_0^\infty \Omega^2(a,f)\sigma^2(a)da$$

$$+ \int_0^\infty \int \rho(a,a')\Omega(a,f)\Omega(a';f)da\,da' \quad ,$$

(A-1)

where var(δV) is the variance in the "size" of the δV_j used to make up V and J is the number of δV_j. We recall that δV_j has to be chosen so that $N_{3j}(a)$ and $N_{3j}'(a')$ are independent for $j \neq j'$. This will typically mean that δV_j has to be sufficiently large so that $|r_j - r_{j'}|$ is large compared to the correlation lengths of distribution of pore locations. (For example, this will be on the order of a cluster size if clustering of pores is present.) On the other hand, δV_j has to be small enough so that $M_3(a_1,a_2;x_j)$ can be treated as a Poisson process; that is, the pores within δV_j are randomly distributed throughout δV_j. With these conditions in mind, if we set $\delta V_j = \delta V_{min}$ and consider (A-1) as a function of V, we see that since $J = V/\delta V_{min}$, (A-1) approaches zero as 1/V. However, the coefficient of the 1/V term can be significantly larger than that shown in (3) if there are significant correlation lengths in the pore location distribution.

NONDESTRUCTIVE CHARACTERIZATION OF RAPIDLY SOLIDIFIED ALLOYS BY ULTRASONIC TECHNIQUES

John J. Smith, Moshe Rosen
Johns Hopkins University
Baltimore, MD 21218

Haydn N. G. Wadley
National Bureau of Standards
Gaithersburg, MD 20899

C. Lee Friant
Martin Marietta Laboratories
Baltimore, MD 21227

ABSTRACT

The enhancement of certain material properties by rapid solidification is closely related to the novel microstructures that can result from high cooling rates. Since small variations in processing greatly affect these microstructures, NDE sensors are required to detect and monitor these microstructural features during processing. An ultrasonic technique, utilizing laser generated ultrasound has been developed to detect changes of sound velocity associated with microstructural transformations occurring in rapidly solidified melt-spun ribbon specimens. Elastic moduli, calculated from the ultrasonic measurements, were used to characterize the decomposition of extended solid solutions of aluminum-manganese alloys and the crystallization of amorphous metal alloys.

RAPIDLY SOLIDIFIED METALLIC ALLOYS have been the subject of an extensive number of scientific studies in the last 10 years. By virtue of their different and often enhanced physical properties (as compared to conventionally solidified alloys) both scientific and commercial interest are warranted. These enhancements are achieved by more homogeneous alloy element distribution, reduced grain size, nonequilibrium microstructural constituents, and, sometimes, a noncrystalline (amorphous) microstructure. In order to achieve high cooling rates the methods of rapid solidification result in fine powder or thin ribbon samples. Thus, it is difficult to characterize rapidly solidified materials in their initial state with traditional nondestructive techniques designed for bulk geometries. However, it is precisely in the initial powder or ribbon form, that it is essential to ensure that the desired metastable microstructures have formed. Ideally, these characterizations should be in-process, making possible closed loop control of the production processes.

In this paper, ultrasonic velocity measurements are used to characterize (both before and after thermal processing designed to decompose metastable microstructure constituents) rapidly solidified microcrystalline and amorphous alloys that were prepared by melt spinning. The experimental advantages and disadvantages of some ultrasonic methods will be discussed. Both a microcrystalline alloy system (aluminum-manganese) and amorphous systems are examined.

The amorphous structure that results when some alloys are correctly melt-spun is caused by the large quench rates (up to 10^6 K/sec) that are attainable by melt spinning. A most important need in the case of the amorphous alloys is a nondestructive method of determining whether any crystallization has occurred during solidification or during subsequent processing. Conversely, in other alloys such as some in the Al-Mn system, the intent is not to create an amorphous structure but rather to extend the solubility of a solute (Mn) in the matrix (Al) beyond that achievable under conventional (equilibrium) processing conditions. The resulting product is crystalline with a grain size of 10 to 40 microns. It is essential to develop a means of ensuring that a sufficient concentration of alloy elements is retained in solid solution both during solidification and during subsequent thermo-mechanical treatments.

EXPERIMENTAL PROCEDURES

In the thin narrow (e.g., 2 mm x 50 μm) ribbons of melt spun material, only guided ultrasonic waves are propagated. There are several ways of generating and receiving these guided ultrasonic waves. Two will be discussed here: piezoelectrically generated and detected signals in a pitch - catch mode

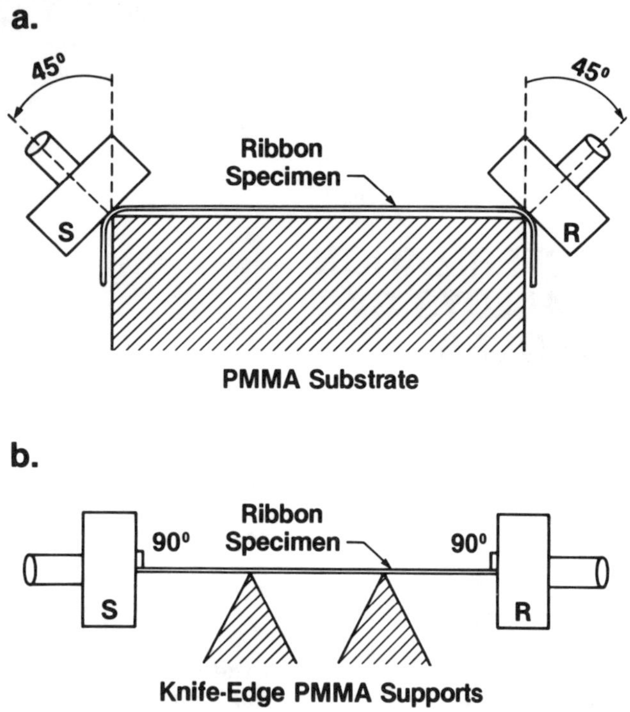

Fig. 1 - Sending (S) and receiving (R) transducers-specimen configurations for longitudinal (symmetric) Lamb wave generation and detection for a. substrate supported and b. self-supported samples

and laser generated signals in a pitch -piezoelectric catch mode. The first represents only a slight modification to conventional ultrasonic techniques. The second method is not traditional and was especially developed for measurements on specimens with ribbon geometries, and may ultimately be capable of noncontact application by replacing the detecting piezoelectric transducer with an EMAT or laser interferometer.

PIEZOELECTRIC METHOD - For acoustically two-dimensional samples the group velocity of the fundamental longitudinal Lamb wave and the fundamental horizontally polarized shear wave (SH(0) were measured using piezoelectrically generated waves. Commercial longitudinal transducers with nominal frequencies of 1, 2.25, 5, and 10 MHz were used for generation and detection of the fundamental longitudinal Lamb wave. AC-cut quartz or Panametrics (V155) normal beam (0.5 in.) shear wave transducers operating at 5 MHz were used to generate and receive the SH(0) wave. The quartz crystal transducers required impedance matching networks whereas the commercial transducers did not. For both wave types a viscous silicone fluid was used as an acoustic couplant between the transducer and the ribbon specimen. A two transducer pitch-catch arrangement was used for the time-of-flight measurements instead of the single transducers pulse-echo method due to the

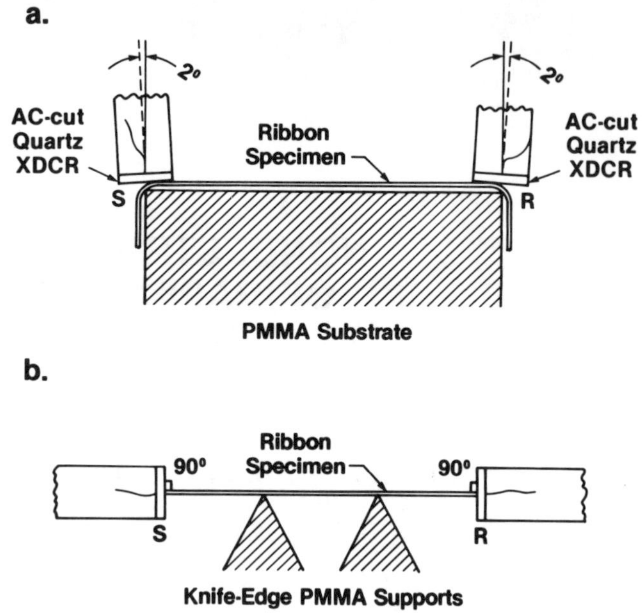

Fig. 2 - Sending (S) and receiving (R) transducer-specimen configurations for SH-type shear wave generation and detection for a. substrate supported and b. self-supported samples

absence of at least two echoes of appreciable amplitude.

The orientations of the generating and receiving transducers relative to the ribbon specimen proved to be an extremely important consideration. The transducer-specimen configurations used in this study are based on those reported by Gelles (1, 2) in his work on films, foils, thin plates, whiskers, fibers and fine wires. In particular, two transducer positions for each type of wave were found to yield the largest received signal amplitudes. The longitudinal Lamb wave transducer-specimen configuration is shown in Figure 1 and the SH(0) wave transducer-specimen configuration is shown in Figure 2. In both cases the "a" configuration was used for samples that could not support their own weight but could be easily bent around a sharp corner. The "b" configuration was used for stiffer samples that could support their own weight but could not be easily bent around a sharp corner. For generation and detection of the SH(0) type wave the transducers were positioned in such a manner as to orient the shear particle displacements parallel to the width direction in the plane of the ribbon or sheet specimen.

A schematic diagram of the experimental apparatus used for velocity measurements is shown in Figure 3. The apparatus is essentially equivalent to that used for the pulse-echo overlap method (3).

The generation of longitudinal sound waves in specimens with ribbon geometries by conventional techniques (i.e. with piezoelec-

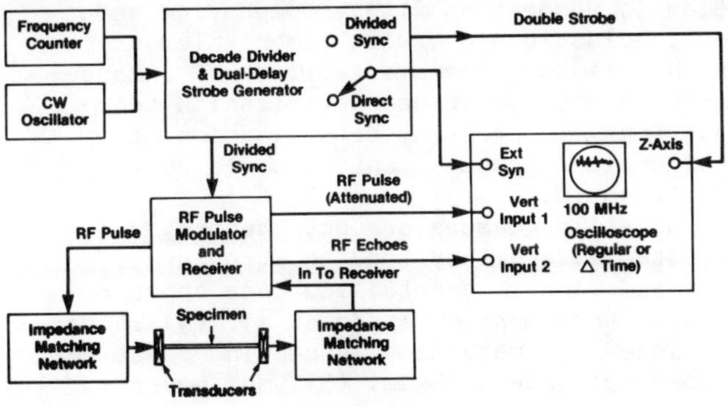

Fig. 3 - Schematic diagram of conventional through-transmission time-of-flight measuring system using piezoelectric generation and detection

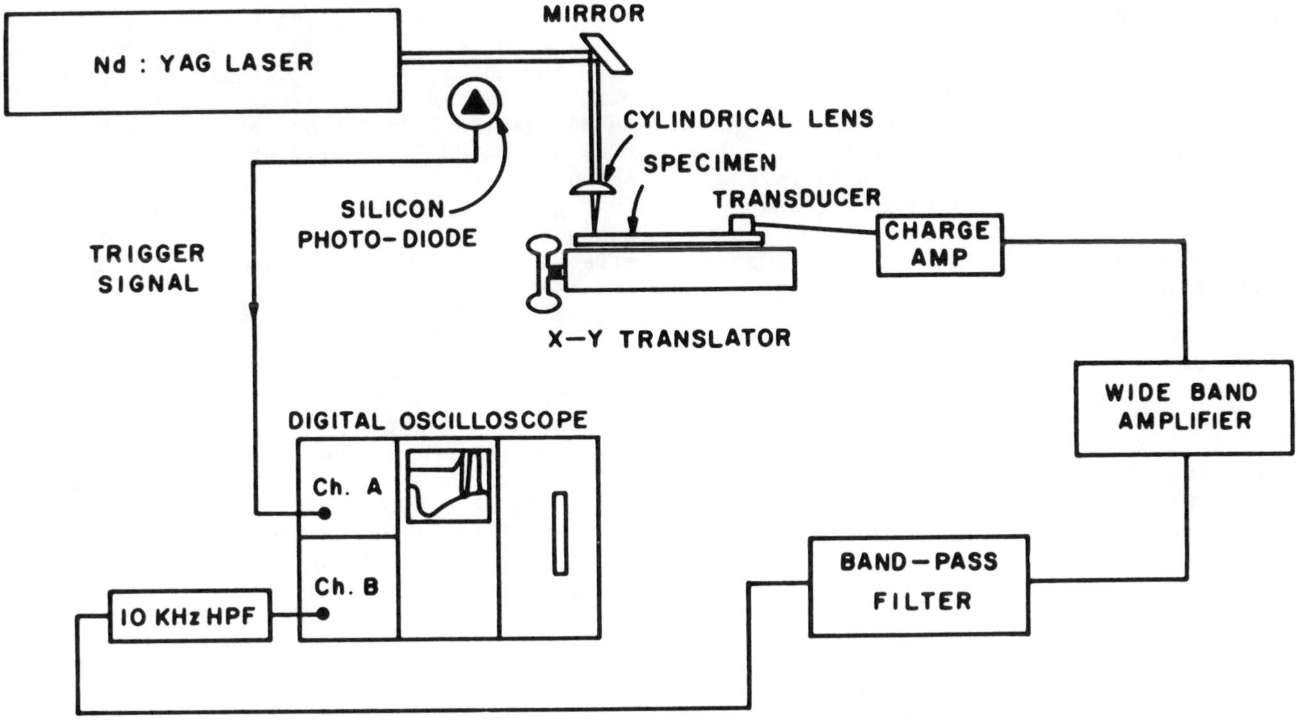

Fig. 4 - Schematic diagram of time-of-flight measurement apparatus using pulsed laser generation and piezoelectric detection

tric transducers) is difficult. The efficiency of energy transfer from transducer to specimen and back is low because the small cross-sectional area that can be exposed to the transducer allows only a small fraction of the sound energy radiated from the transducer to be passed into the ribbon. The ribbon specimens examined in this study typically had a thickness of only 40 microns. Furthermore, it would be extremely difficult to apply this technique in-process.

LASER ULTRASOUND GENERATION - The ultrasonic velocity measuring system is illustrated in Figure 4. A single shot from a Q-switched Nd:YAG laser (wavelength = 1.064 μm) was directed at the ribbon surface. A fraction of the electromagnetic pulse was absorbed, heating the ribbon and thereby thermoelastically generating an ultrasonic pulse at one end of the ribbon. The pulse propagated the length of the ribbon and was detected by a 3 mm diameter acoustic emission transducer. The frequency of the detected waves was below 300 kHz, corresponding to an extensional wavelength of ≈ 1.7 cm in aluminum (11 to 17 times the specimen ribbon width), satisfying the dimensional requirement for extensional wave propagation. The amplified transducer signal was digitally recorded at a 50 ns sampling rate with a 2 channel digital storage oscilloscope. A silicon photodiode, activated by the scattering of the light pulses from the sample was used to trigger the oscilloscope. The trigger and transducer signals were recorded simultaneously and displayed on the oscilloscope screen. Transit time was determined by reading the time difference between the trigger signal and the arrival of the ultrasonic pulse at the transducer. A typical oscilloscope display is shown in Figure 5.

A number of procedures were followed to ensure accurate measurements: the laser pulse that launched the ultrasonic wave was shaped into a line by a 0.5 mm wide slit; the ribbon was held flat on a PMMA (Plexiglass) sheet; and measurements of transit time were taken at two different propagation distances for each specimen. This differential tech-

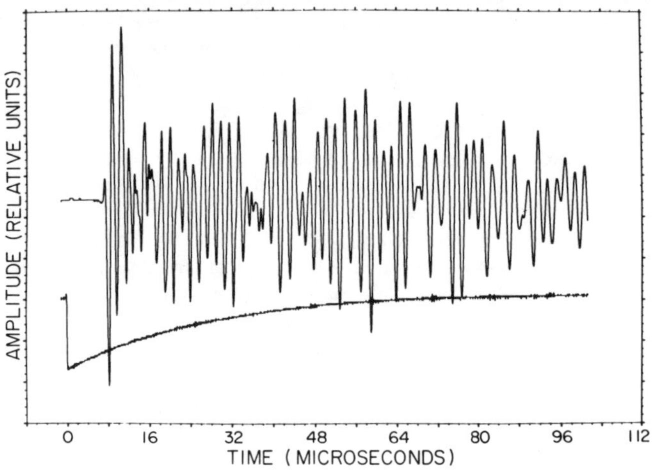

Fig. 5 - Typical ultrasonic signal (top) and trigger signal (bottom) recorded and displayed

nique eliminated ambiguities about distance (i.e. to what part of the transducer should distance be measured) and about time (i.e., where the leading edge of the sound wave is). The ribbon was mounted on a calibrated translation stage that allowed distances to be measured to a precision of 0.001 inches. Reproducible time measurements were made by choosing a distinctive part of the initial wave arrival to find the time of flight. The experimental error of the ultrasonic velocity measurements was estimated to be approximately 1%.

NDE EXAMINATION OF THE AMORPHOUS TO CRYSTALLINE TRANSITION

Amorphous alloys, or glassy metals, are solids with frozen-in liquid structures. The absence of translational periodicity in the amorphous state along with the macroscopic compositional homogeneity are the main reasons for their improved properties, e.g., high mechanical strength, good corrosion resistance, and excellent magnetic behavior. Their unusual mechanical, chemical, and physical properties have stimulated extensive scientific and technological interest. One serious problem in the processing and utilization of amorphous alloys that may limit their future technological applications is their low thermal stability. When thermomechanical conditions are appropriate, metallic glasses relax structurally and ultimately crystallize into more stable structures resulting in drastic variation in properties. The factors governing the thermal stability of these alloys and their effect on properties are not well understood. For example, upon crystallization, amorphous alloys undergo very large changes in the elastic and anelastic properties with accompanying reduction in plastic properties (embrittlement). For this reason, the availability of a nondestructive ultrasonic characterization technique for both property determination and metallurgical process control could be extremely useful.

Metallic glasses are not thermodynamically stable, and they tend to structurally relax and finally crystallize upon appropriate heat treatment. Waseda et al. (4) showed that annealing metallic glasses increased the short-range order. Egami (5) used energy-dispersive methods to confirm the log-time kinetics of the relaxation process. His work indicated that the activation energy for relaxation continuously increased with time, which may correspond to the removal of quenched-in defects. Although changes in density and electronic structure associated with the crystallization process are rather minute (6), appreciable variations were observed in several physical properties, including the elastic moduli (7-9). The crystallization behavior of metallic glasses was examined by several workers (10-13). It was found that small additions of noble metals (Cu, Ag) to Pd-Si greatly stabilizes the amorphous phase, and particularly in $Pd_{0.775} Cu_{0.06} Si_{0.165}$ glass (14) which became the subject of numerous investigations. The effect of isothermal annealing on the crystallization kinetics of $Pd_{0.775}Cu_{0.06}Si_{0.165}$ was determined by differential scanning calorimetry (14). Associated with relaxation and crystallization of metallic glasses are variations in the elastic and mechanical properties. Young's modulus E and shear stiffness generally increase by 20-40% but the bulk modulus K increases by only about 7% upon crystallization (8). Kursomovic and Scott (15) found that E for $Cu_{60} Zr_{40}$ increased by about 10% due to structural relaxation and another 15% upon crystallization. In contrast, the density of glassy metals generally changes by only 0.3-1.5%.

The main objectives of the present investigation were as follows:

- o Determination of the isothermal transformation kinetics from the amorphous to the crystalline state by means of monitoring changes in sound wave velocity (elastic modulus).
- o Study of the relaxational changes occurring in the amorphous state, prior to crystallization.

Metallic glass ribbons of PdCuSi and CuZr alloys were prepared by the melt spinning technique. A small charge of a casting was melt spun from a fused quartz tube. The alloys were melted inductively and the molten metal was ejected from the fused quartz tube onto the outside of a rotating copper wheel. The samples were approximately 40 µm thick and 1.25 mm wide. Ultrasonic, x-ray, and metallographic examination prior to the crys-

Table 1 - Physical properties of ribbon samples

ALLOY		$V_{Ext.}$ (m/sec)	V_{Lamb} (m/sec)	V_{Shear} (m/sec)	$V_{long.}$ (m/sec)	E (GPa)	G (GPa)	k (Gpa)	ν
Pd-Cu-Si		2970	(3290)	1787	4495	94.5	33.6	167.8	0.41
Cu - Zr	(1)	3300	(3585)	1980	(4670)	79.8	28.7	121.5	0.39
	(2)	3270	(3560)	1950	(4777)	78.1	27.8	154.2	0.40
Metglas	(1)	4640	4900	2680	(5560)	157.2	59.7	145.6	0.32
2605-SC		4460	4720	2750	(5345)	145.6	55.4	134.8	0.33
Metglas 2605-CO		4490	4730	2770	(5279)	152.4	58.8	133.7	0.31

tallization treatment revealed no crystallinity. Crystallization was performed in evacuated borosilicate glass capsules at predetermined isothermal holding temperatures. Optical metallography indicated that the crystallization process in PuCuSi ribbons is not homogeneous in the bulk. Two coexisting growth processes were observed: frontal growth from the free surfaces of the ribbon towards the interior, and homogeneous nucleation and growth in the bulk.

Using the piezoelectric pitch-catch technique in conjunction with equations listed in the Appendix, values for the Young's, shear, and bulk moduli and Poisson's ratio were calculated for Pd-Cu-Si, Cu-Zr, and Metglas in the as-spun state, before they were crystallized. In addition, values for the velocity of the bulk longitudinal wave, V_ℓ, and the longitudinal type extensional wave, V_E, were calculated for Metglas from V_L and V_S. For Pd-Cu-Si and aluminum, V_ℓ, V_S. and V_E were obtained from direct measurements. Finally, for Cu-Zr, V_E and V_S were determined by direct measurements and V_ℓ was calculated. Table 1 summarizes the results obtained for elastic moduli, Poisson's ratios and sound wave velocities.

The variation of extensional velocity of $Pd_{77.5}Cu_{.06}Si_{.165}$ with annealing times was determined with the laser-generation ultrasound technique. Figure 6 shows the variation of the extensional wave velocity, V_E as a function of crystallization time at three isothermal holding temperatures, 380, 390, and 400 °C. The initial sound-wave velocity, in the amorphous state, was found to be 2970 ms^{-1}, whereas in the fully crystallized state it reached an asymptotic value of 3490 ms^{-1}. Taking 10.52 and 10.69 g cm^{-3} as the density of the amorphous and crystalline $Pd_{77.5}Cu_6Si_{16.5}$ respectively, the Young moduli could be calculated. They were found to be 9.46 x 10^3 kg-mm^{-2} for the amorphous state and 13.27 x 10^3 kg-mm^2 for the crystalline

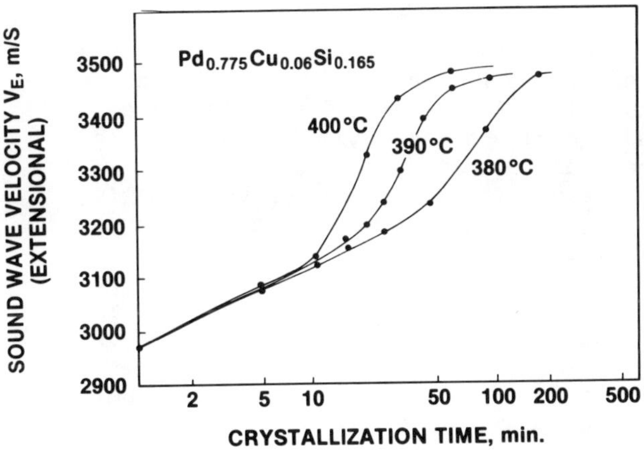

Fig. 6 - Extensional velocity change in $Pd_{77.5}Cu_6Si_{16.5}$ with time at 380 °C, 390 °C, and 400 °C

state, i.e., a difference of about 40%. Figure 6 exhibits a relatively sharp increase in the sound velocity during the first five minutes of the isothermal holding. The variation of the extensional wave velocity with crystallization time manifests the dramatic changes in the elastic properties with an increase of about 40% in Young's modulus. The sigmoid curves represent the crystallization kinetics which are typical of a thermally activated process. Thus, the kinetic parameters, i.e., activation energy and the growth regime parameters can be determined from this nondestructively gathered data (16, 17). Furthermore, the time dependence of the crystallized volume fraction was found to be compatible with a diffusion controlled growth mechanism (18, 19) for a plane boundary growth that is preceded by a rapid nucleation process (20).

The effect of copper content on the Young's modulus of ZrCu alloys in the crystalline and amorphous state indicates that the lowest value of the Young's modulus is

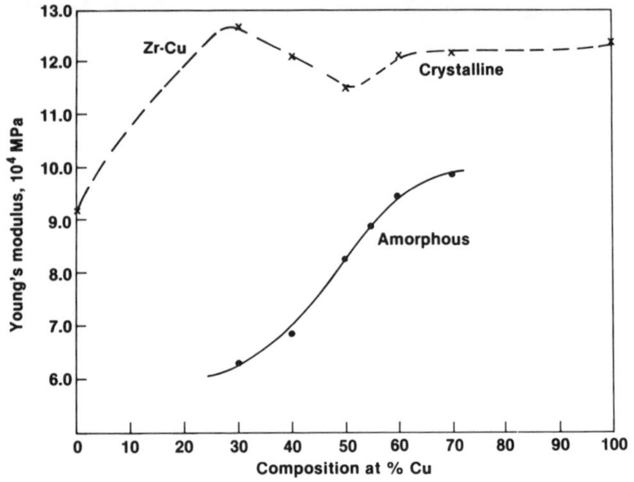

Fig. 7 - Dependence of Young's Modulus on copper content in amorphous and crystalline Zr-Cu alloys

for pure crystalline zirconium, whereas the highest value is for pure crystalline copper (Figure 7). A maximum and a minimum are found for alloys and compositions near the intermetallic compounds Zr_2Cu and ZrCu, respectively. The variations of the Young's modulus with composition of the amorphous alloys, between 30 and 70% copper, is a smooth S-curve with an inflection point at 50% copper. It is conjectured that the amorphous state can be interpreted in terms of short range order structure where the mixing pairs Zr-Zr, Cu-Cu, and Cu-Zr are of equal probability at the composition of 50% Cu. The inflection point on the Young's modulus versus composition curve will then represent the point for maximal configurational entropy of the metallic glass. Analysis of the temperature dependence of the sound wave velocity of Zr-Cu alloys in the composition range between 30 and 70% Cu allowed determination of the activation energies of the diffusional processes governing the crystallization kinetics. The activation energy increased from 90 to 110 kcal/mole as copper content exceeds 50%. The change in crystallization mechanism as copper content exceeds 50 % may be related to the configurational entropy change as the population of Cu-Cu pairs increases. The relationship is of basic importance in the understanding of the transformation of the amorphous state into coexisting compounds upon isothermal crystallization (21).

The initial increase in the extensional sound velocity, of about 3.7% (i.e., 7.4% in the Young's modulus) of PdCuSi and a similar increase in the Zr-Cu amorphous alloys, appears to be related to a structural relaxation that precedes the crystallization process in the amorphous state (15, 19). Structural relaxation was found to be particularly enhanced in glasses obtained at high quenching rates, where greater structural disorder is to be expected (27). The lower elastic stiffness in the glassy state has been attributed to the interatomic displacements inherent in a disordered structure (23). The elastic stiffness should, therefore, increase with increased local ordering as a consequence of the structural relaxation. Initially, the amorphous alloy has a disordered packing of the constituent atoms. Heat treatment in the amorphous region at temperatures below the glass transition temperature, Tg, or for relatively short periods of time at temperatures above Tg causes structural and compositional relaxation. The free volumes are redistributed by short range shuffling of the atoms, thus a more rigid close-packed atomic distribution of high coordination is achieved.

EXTENDED SOLID SOLUTIONS: Al-Mn

The aluminum-manganese system was chosen for investigation because significant extension of the equilibrium solubility limit is possible upon rapid solidification (24-32). Thus, the properties of aluminum-manganese solid solutions could be studied over a wide range of manganese concentrations. The equilibrium solubility of manganese in aluminum is 2 wt.% at the eutectic temperature of 657 °C (930 °K) as shown shown in Figure 8. At higher Mn concentrations an intermetallic compound (Al_6Mn) is formed by a peritectic transformation. Under conditions of rapid solidification this reaction is bypassed and microcrystalline substitutional solid solutions containing up to 15 wt.% Mn reportedly have been formed (30). Increases of hardness (26, 28, 29, 31) and ultimate tensile strength (32) which are proportional to manganese content have been observed for these nonequilibrium solid solutions. To explore the physical basis of the ultrasonic phenomena observed x-ray diffraction, electrical resistivity, and optical microscopy measurements were made.

Aluminum-manganese alloys were prepared by melting 5N purity aluminum and 4N purity manganese and casting under 0.5 atmospheres (380 torr) or argon. Rapidly solidified ribbons were then prepared from these alloys by melt spinning. Thin ribbon specimens, 1 to 1.5 mm wide and 30 to 50 µm thick with continuous lengths of > 1 meter were produced. For thermal stability experiments the as-spun specimens were annealed isothermally at 450 °C (723 °K) for varying lengths of time. Each ribbon was encapsulated in a borosilicate glass tube, in the same manner as the amorphous alloys during annealing.

The effect of composition and heat treatment on the lattice parameter is illustrated in Figure 9. In the as-spun specimens containing less than 5 wt.% Mn the lattice parameter decreased linearly with increasing manganese concentration. Above 5 wt.% Mn

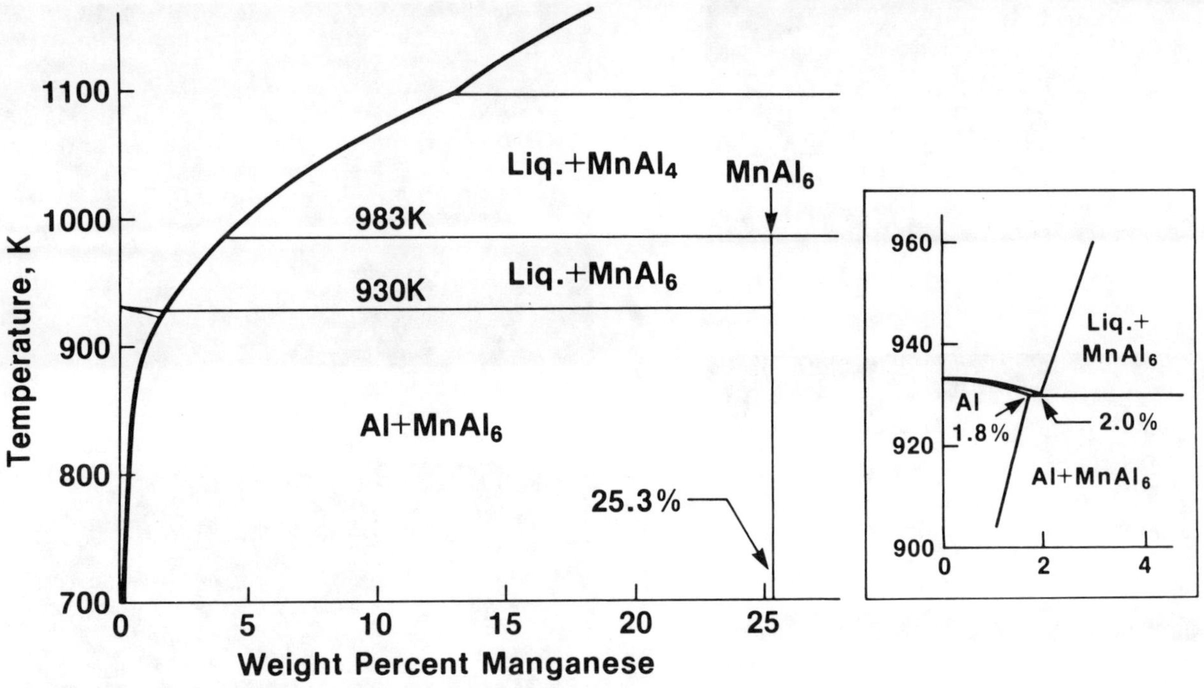

Fig. 8 - Partial phase diagram of the aluminum-manganese system with the region around the eutectic point enlarged

the slope of the relation changed. This deviation from linearity suggested that the solubility of manganese in aluminum may not have been complete in the higher Mn content (9 and 12 wt.% Mn) alloys. The as-spun lattice parameters are consistently above the values Jones (32) derived by averaging data reported in the literature. From 0 to 5 wt.% Mn, the difference is less than .003Å. Beyond 5 wt.% Mn, the deviation between the values of Jones and Figure 9 increases with manganese content.

After annealing for 1 hour at 450 °C the lattice parameter increased from that of the as-spun condition for all compositions above 1 wt.% Mn and peaks, identified as the Al_6Mn phase, were seen on the diffraction patterns of the alloys containing > 5 wt.% Mn. For alloys containing less than 2 wt.% Mn only a small change in the lattice parameter was seen, indicating a more thermally stable solid solution. Annealing for a further 2 hours resulted in a small additional lattice parameter increase in the alloys with less than 3 wt.% Mn.

Optical micrographs of ribbons with 1, 5, and 12 wt.% Mn are shown in Figures 10, 11, and 12 respectively. In the micrographs the straight edge of each sample is the side of the ribbon that contacted the spinning wheel during solidification.

A columnar microstructure, indicative of through thickness directional solidification, is seen in the 1 wt.% Mn sample. In the as-spun state, grain boundaries are slightly visible. Darkish spots seen in the structure are small pores, which may have caused the

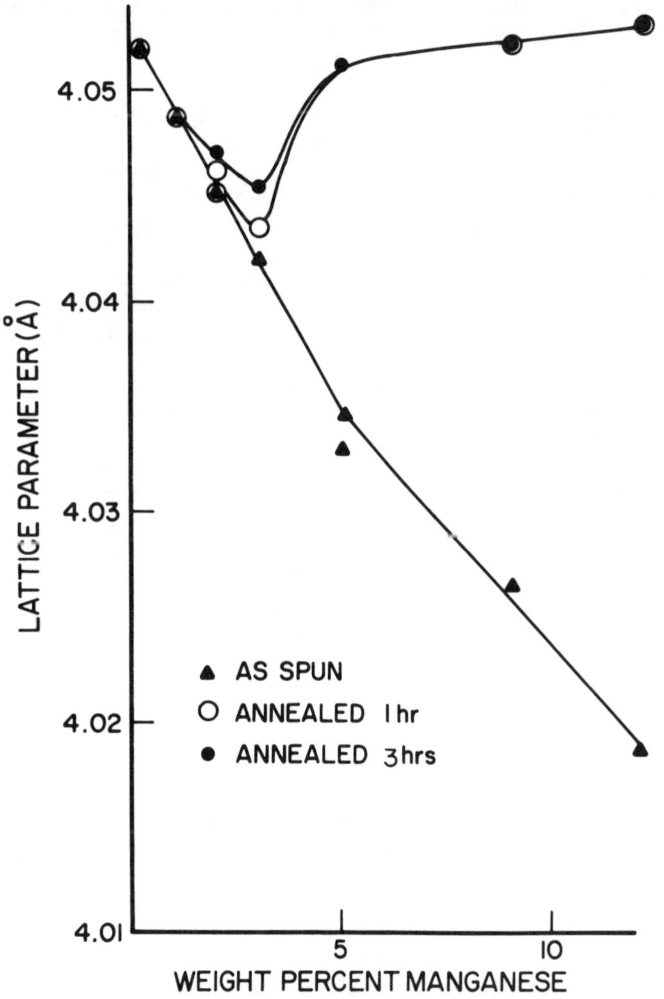

Fig. 9 - Lattice parameter variation of as-spun and annealed Al-Mn alloys

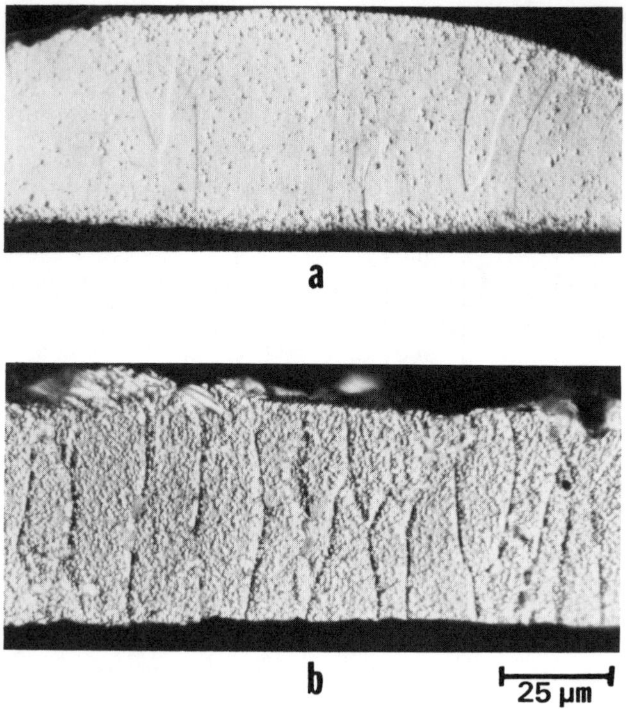

Fig. 10 - Al-1 wt.% Mn ribbon edge as quenched (a) and annealed 3 hours at 450 °C (b)

calculated densities to be somewhat overestimated. The grain boundaries are accentuated after annealing for 3 hours at 450 °C. In addition, a roughened surface is observed, which could be due to precipitates.

The as-spun 5 wt.% Mn specimen appears inhomogeneous only in the center, where the presence of a cellular solidification microstructure indicates a decreased temperature gradient (33). Substantial precipitation exists in the specimen annealed 3 hours at 450 °C (Fig. 11), consistent with the large resistivity decrease and lattice parameter increase reported above.

Figure 12 reveals the microstructure of the as-spun 12 wt.% Mn alloy. Some primary precipitation exists along the grain boundaries. Directional solidification is indicated by the columnar grain structure and by the solute segregation at the edge of the ribbon opposite the wheel side. Evidently, the quench rate was not sufficient to retain all 12 wt.% Mn in solution. A large amount of Al_6Mn precipitated after annealing at 450 °C for 3 hours (Fig. 12).

In Figures 13 and 14 the ultrasonic velocity and Young's modulus as a function of manganese content are shown. The velocity and modulus curves have essentially the same shape, although the slope of the modulus curve is increased because Young's modulus is proportional to the square of the velocity, and the density increases linearly with manganese content (0.02 gm-cm^{-3}/wt.% Mn). It can be seen that manganese concentration had a much larger effect on velocity and Young's

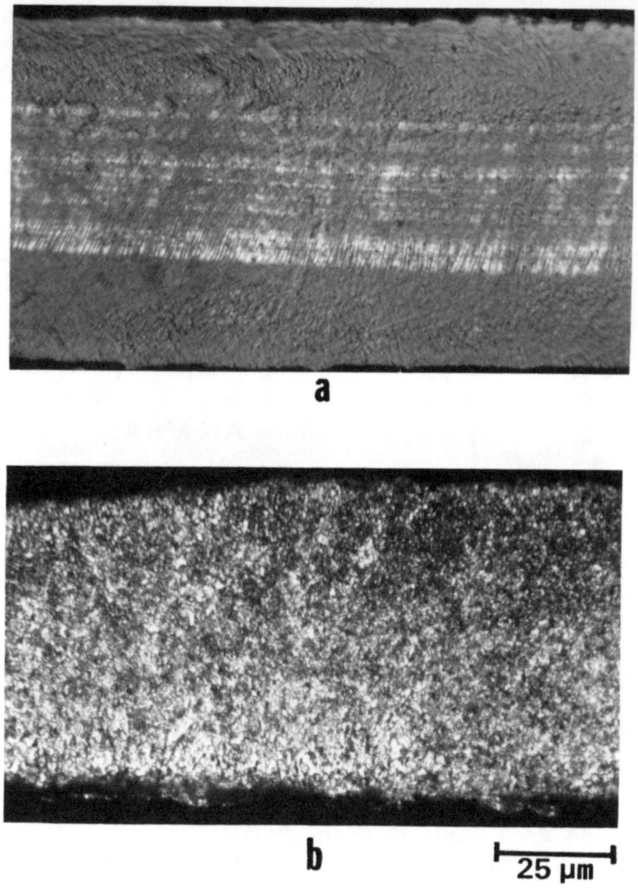

Fig. 11 - Al-5 wt.% Mn ribbon edge as quenched (a) and annealed 3 hours at 450 °C (b)

modulus than did the heat treatment of a particular alloy composition.

Pure aluminum ribbons had a modulus of 6.93 x 10^{11} dynes/cm^2 (69.3 GPa). Upon the addition of 0.1 wt.% Mn the modulus decreased sharply to 6.35 x 10^{11}dynes/cm^2 (63.5 GPa). The modules decreased more slowly with increasing manganese content up to ≃2 wt.% Mn. Above ≃2 wt.% Mn the modulus increased. Above 9 wt.% Mn the rate of increase abated. The Young's modulus of the heat treated alloys was always higher than in the as-spun state. The difference due to annealing was not a simple function of concentration.

The statically determined Young's modulus of conventionally cast Al-Mn alloys has been reported to increase linearly for compositions between 1.7 and 14.5 wt.% Mn (34). This suggests that the Al_6Mn phase has a substantially higher modulus than aluminum. Between manganese concentrations of 3 and 9 wt.% Mn the elastic modulus results of the present study confirm this finding, even though the rapidly solidified ribbons are expected to contain much higher concentrations of Mn in solution and, thus, a smaller volume fraction of Al_6Mn. The leveling of Young's modulus above 9 wt.% Mn could be a manifestation of manganese segregation. Due to the high concentration of manganese near

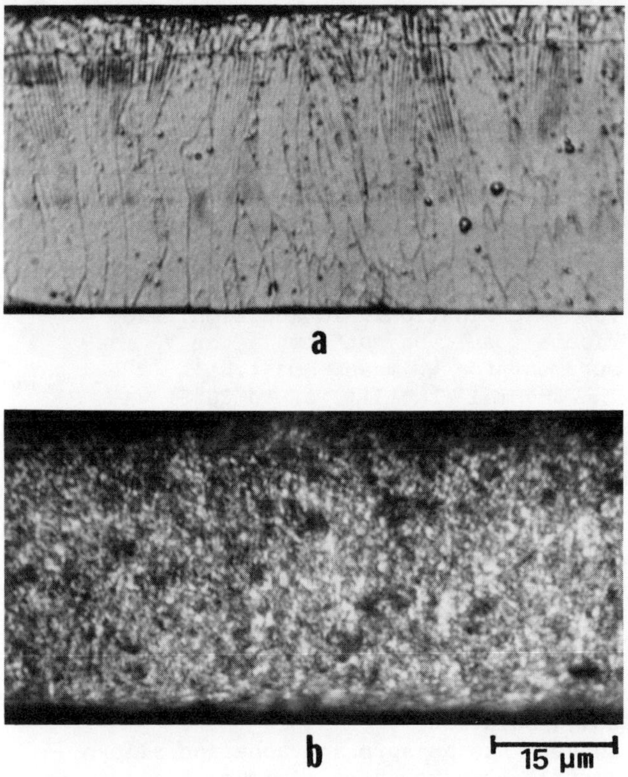

Fig. 12 - Al-12 wt.% Mn ribbon edge as quenched (a) and annealed 3 hours at 450 °C (b)

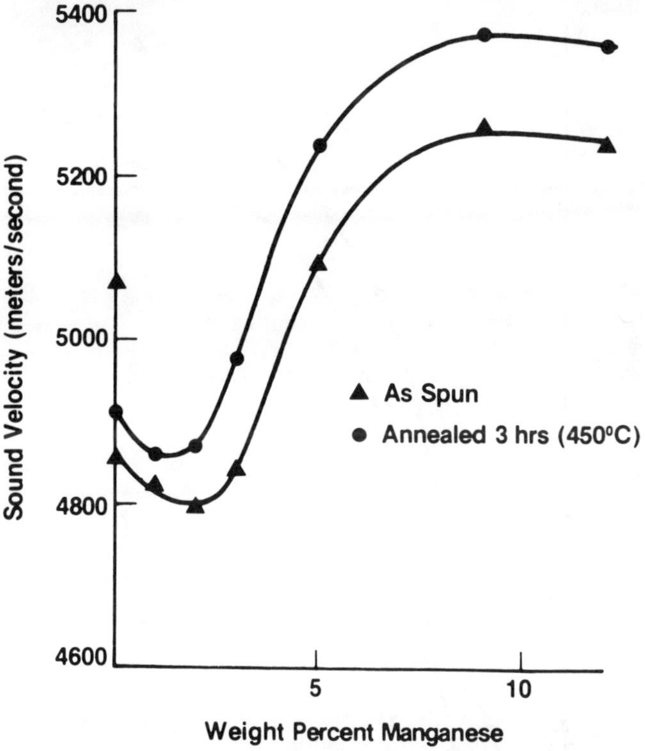

Fig. 13 - Extensional sound wave velocity of as-spun and annealed Al-Mn alloys

the ribbon side not exposed to the melt spinning wheel, a substantial fraction of the specimen volume was depleted of manganese. Therefore, the bulk of the ultrasonic pulse may have propagated in material containing much less than 12 wt.% Mn.

The cause of the abrupt decrease of Young's modulus upon the addition of 0.1 wt.% Mn to pure aluminum is at this time unclear and is still under investigation. Several comments can be made, however about the continued decrease of Young's modulus upon the addition of more Mn to aluminum. A similar decrease in Young's modulus with increasing solute content has been observed in substitutional solid solutions of some other alloy systems, notably ones based on copper and silver (35). Zener proposed a model that predicted, from thermoelastic considerations, an elastic modulus decrease in substitutional solid solutions because of solute induced lattice strain (36). Support for the notion that this can occur in Al-Mn alloys comes from a slight dip of longitudinal sound velocity that was observed in hot pressed Al-Mn alloys at a very low Mn concentration corresponding to a solid solution (37). The extreme conditions of rapid solidification extend the solubility limit and, thus, would be expected to magnify any effect seen in equilibrium solid solutions.

According to Zener's model, Young's modulus would be expected to continue to decrease monotonically with increasing Mn solute in the as-spun ribbons. This was not observed here. Instead, the modulus rose sharply with the addition of manganese above 3 wt.% Mn. Although some precipitation was seen in as-spun alloys containing more than 3 wt.% Mn, the lattice parameter and resistivity results indicate that most of the manganese was quenched in solution. It seems surprising, therefore, that the small volume fraction of Al_6Mn that may have existed in the as-spun state could so strongly increase the modulus. Moreover, the precipitation of much more Al_6Mn during annealing treatments caused only a relatively small modulus increase. This leads one to speculate the possibility that the nature of the solid solution itself undergoes a transition at around 2 wt.% Mn--Interestingly, this is the eutectic composition, at which the equilibrium solubility of manganese in aluminum is maximum.

The rise of Young's modulus caused by annealing at 450 °C (Figure 14) could be due to the elimination of disorder from the lattice by precipitation or clustering, thus annihilating the modulus-decreasing effect of lattice strain which Zener's theory addresses. Another effect of annealing could be precipitation hardening, which is known to increase slightly the Young's modulus of some Al-Cu alloys (38). The contribution of precipitation to the elastic modulus increase is most likely minor, however, because the increase of Young's modulus over the as-spun

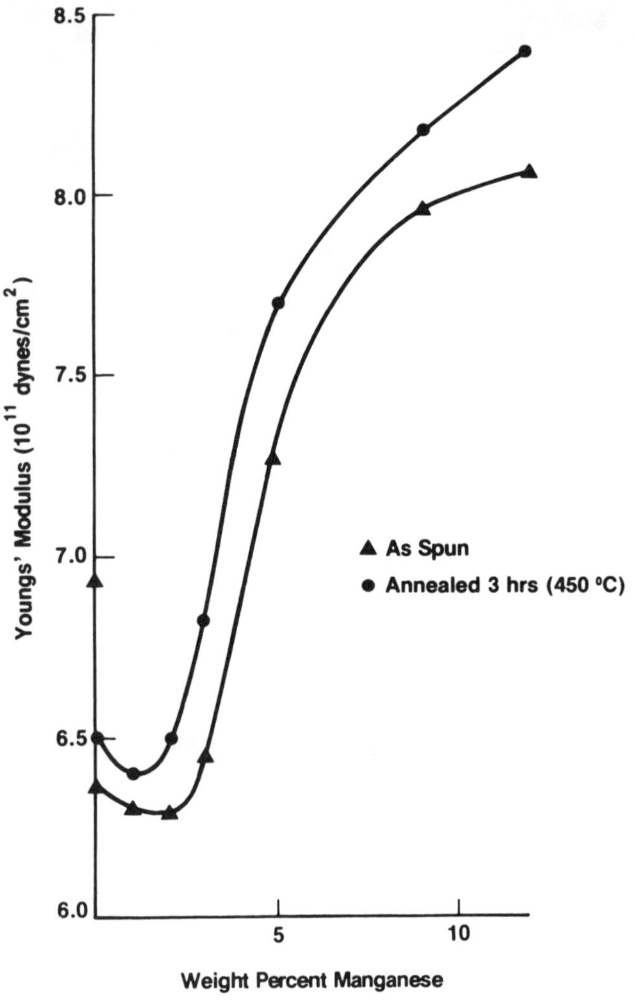

Fig. 14 - Young's modulus of as-spun and annealed Al-Mn alloys calculated from extensional sound velocity and x-ray density

value is relatively independent of manganese content.

CONCLUSIONS

Both the piezoelectric pitch-catch and the laser generation method are useful for NDE of melt-spun specimen.

1. The present investigation showed that it is possible to determine all of the elastic moduli of melt spun metallic ribbons 50 µm or less in thickness. This was achieved by determination of the wave velocities V_E, V_L, and $V_{SH(0)}$ along with the mass density of the materials and insertion of these values into the proper equations.
2. Two methods for obtaining the velocity of the fundamental longitudinal Lamb wave, V_L, were successfully implemented for ribbons and foils which were many acoustic wavelength in width. Laser generated waves could be used to measure V_L directly, whereas, piezoelectrically generated wave velocity measurements had to be made at several frequencies and the data fitted to the proper dispersion curve corresponding to a fixed Poisson's ratio.
3. The low frequency Lamb wave velocity, V_L, could not be determined for narrow ribbons whose width was on the same order as the acoustic wavelength. Instead, the longitudinal extensional wave velocity, V_E, was obtained by the laser generation method and it was used for elastic modulus determination. In addition, wide ribbons could be cut down after V_L measurement for V_E measurement.
4. The thermal stability of Pd-Cu-Si was evaluated. The initial increase in the extensional velocity appears to be related to a structural relaxation that precedes the crystallization process in the amorphous state. The crystallization process obeys conventional kinetics for a thermally activated process.
5. Young's modulus and x-ray diffraction lattice parameter measurements have been made nondestructively on rapidly solidified ribbons of Al-Mn supersaturated solid solutions.
6. In both the as-spun and annealed alloys, Young's modulus has a minimum at a composition around 2 wt.% Mn.
   The percentage of manganese present in Al-Mn alloys appears to have a greater effect on Young's modulus than the state in which manganese is distributed (as solid solution or as part of a precipitate).
   The physical processes that are responsible for the softening and stiffening of the crystal lattice as the Mn concentration increases are at present not fully understood.

ACKNOWLEDGMENTS

Support from the Defense Advanced Research Projects Agency under DARPA Order No. 4275 is gratefully acknowledged, Technical Monitors of this contract were Lt. Col. L. A. Jacobsen and Capt. S. Wax.

REFERENCES

1. I. L. Gelles, J. Acoust. Soc. Amer. 40, 138, (1966).
2. I. L. Gelles and C. H. Bomhard, Rev. Sci. Instrum. 37, 1345 (1966).
3. J. E. Lee May, IRE Nat'l. Conv. Rec. 6, 134, (1958).
4. Y. Waseda, H. Ikazaki, and T. Masumoto, J. Mat. Sci. 12, 1927 (1977).
5. T. Egami and T. Ichikawa, Mat. Sci. Eng. 32, 293 (1978).
6. H. S. Chen and J. T. Krause, Scripta Met. 11, 761 (1977).
7. H. S. Chen, M. J. Leamy, and M. Barmatz, J. Non-Crystalline Solids 5, 444 (1971).

8. B. Golding, B. C. Bagley, and F. S. L. Hsu, Phys. Rev. Letters 29, 68 (1972).
9. H. S. Chen, J. Appl Phys. 49, 3289 (1978).
10. D. Duwez, R. H. Willens, and R. L. Crewdson, J. Appl. Phys. 36, 2267 (1965).
11. J. J. Burton and R. P. Ray, J. Non-Crystalline Solids 6, 393 (1971).
12. T. Matsumoto and R. Maddin, Acta Met. 19, 725 (1971).
13. B. G. Bagley and E. M. Vogel, J. Non-Crystalline Solids 18, 29 (1975).
14. H. S. Chen and D. Turnbull, Acta Met. 17, 1021 (1969).
15. A. Kursumovic and M. G. Scott, Appl. Phys. Letters 37, 620 (1980).
16. M. Rosen, H. N. G. Wadley, and R. Mehrabian, Scripta Metall. 15, 1231 (1981).
17. J. Chang, F. Nadeau, M. Rosen, and R. Mehrabian, Scripta Metall. 16, 1073 (1982).
18. W. A. Johnson and R. F. Mehl, Trans. AIME 135, 416 (1939).
19. J. W. Christian, Physical Metallurgy. R. W. Cahn, ed., North Holland Press, Amsterdam, London (1970), 471.
20. J. W. Cahn, Acta Met. 4, 449, (1956).
21. F. N. Madeau, M. Rosen, and R. Mehrabian, Effects of Composition on the Elastic Properties of Amorphous and Crystalline Cu-Zr Alloys. Johns Hopkins University Report CMR-NDE-7 (1982).
22. H. S. Chen, Rep. Prog. Phys. 43, 355 (1980).
23. D. Weaire, M. F. Ashby, J. Logan, and M. S. Weins, Acta Met. 19, 799 (1979).
24. L. P. Limina, Russian Metallurgy, Metally (English Translation) 2, 95 (1968).
25. E. H. Dix and W. D. Kieth, Proceedings of the Institute of Metals, AIME, 315 (1927).
26. G. Falkenhagen and W. Hoffmann, Zeitschrift fur Metallkunde 43, 69 (1952).
27. I. V. Salli and G. V. Kudryumov, Daklady Akademii Nauk SSR (English Translation) 132 (6), 557 (1960).
28. L. M. Burov and A. A. Yakunin, Russian Journal of Physical Chemistry (English Translation) 39 (8), 1022 (1965).
29. V. I. Dobatkin, V. I. Elagin, and V. M. Fedorov, Russian Metallurgy, Metally (English Translation), 122 (2) (1970).
30. I. N. Fridlyander, A. A. Kolpachev, T. V. Vukolova, and V. D. Sharabkova, Russian Metallurgy, Metally (English Translation) 132 (4) (1979).
31. E. Babic, E. Girt, R. Krsnik, B. Leontic, M. Ocko, Z. Vucic, and I. Zoric, Physica Status Solidi A 16, K21 (1973).
32. H. Jones, Journal of Material Science Letters 1, 405 (1982).
33. M. Cohen, B. H. Kear, and R. Mehrabian, Rapid Solidification Processing: Principles and Technologies II. Ed. R. Mehrabian, B. H. Bear, and M. Cohen, 1, (1980), Claitors, Baton Rouge, LA.
34. N. Dudzinski, J. R. Murray, B. W. Mott, and B. Chalmers, Journal of the Institute of Metals 74, 291 (1948).
35. W. Koster, Zeitschrift fur Metallkunde 39, 111 and 145 (1948).
36. C. Zener, Acta Crystallographica 2, 163 (1949).
37. L. L. Rokhlin, Physics of Metals and Metallography (English Translation 28 (3), 206 (1969).
38. C. L. Friant, M. Rosen, and R. E. Green, Jr., Proceedings of the 13th Symposium on Nondestructive Evaluation, April 21-23, 1981, San Antonio, Texas, edited by B. E. Leonard, Southwest Research Institute, San Antonio, Texas, pp. 123-34 (1981).

APPENDIX

The equations that relate the isotropic elastic moduli to the measured guided wave velocities and to each other are listed below.

$$G = \rho V_S^2$$

$$E = 4G\left[1 - \left(\frac{V_S}{V_L}\right)^2\right] = \rho V_E^2 = \rho V_L^2 (1-\nu^2)$$

$$\nu = 1 - 2\left(\frac{V_S}{V_L}\right)^2 = \frac{V_E^2}{2V_S^2} - 1 = \left[1 - \left(\frac{V_E}{V_L}\right)^2\right]^{\frac{1}{2}}$$

$$V_E = \left[\frac{E}{\rho}\right]^{\frac{1}{2}} \qquad V_L = \left[\frac{E}{\rho(1-\nu^2)}\right]^{\frac{1}{2}} \qquad V_S = \left[\frac{G}{\rho}\right]^{\frac{1}{2}}$$

$$\nu = \frac{E}{2G} - 1 \qquad k = \frac{E}{3(1-2\nu)} \qquad G = \frac{E}{2(1+\nu)}$$

Where:

G = Shear Modulus
ρ = Mass Density
E = Young's Modulus
ν = Poisson's Ratio
k = Bulk Modulus
$V_S = V_{SH(0)}$ = Velocity of fundamental horizontally polarized shear wave
V_L = Velocity of fundamental longitudinal Lamb wave at zero frequency
V_E = Velocity of fundamental extensional wave at zero frequency

EFFECTIVE CASE DEPTH MEASUREMENT BY AN ULTRASONIC BACKSCATTER TECHNIQUE

Morris S. Good
Sigma Research, Inc.
3200 George Washington Way
Richland, WA 99352

ABSTRACT

Ultrasonic shear wave backscatter was evaluated as a nondestructive technique for measuring effective case depth. A focused piezoelectric transducer was used to reduce the masking effect of the front surface scatter. Seventeen specimens ranging in effective case depth from 1.1 to 2.9 mm were analyzed and a linear relation was demonstrated between the values measured ultrasonically and metallurgically.

THE ULTRASONIC SHEAR WAVE TECHNIQUE (illustrated in Figure 1) localizes area measurements by a lateral resolution (approximately 1 mm) and may be effective for case depths of 0.5 mm and greater with an accuracy of 0.3 mm. The lateral resolution of this technique should permit effective case depth measurements to be performed on a variety of geometries such as gears, where the ultrasonic beam can be directed into the area of interest without the formation of spurious signals from part geometry. The range of assured case depth is an extension of methods previously reported in literature (Emerson, 1976; Perennes et al., 1979).

Inclining the transducer to produce a shear wave by mode conversion serves two purposes. First, by directing the front wall surface echo away from the transducer, only the scattered response due to surface roughness is observed and the resulting echoes are several orders of magnitude smaller than the reflected plane wave front. This reduces the masking effect of the front wall echo on superficial scatterers. The masking effect can be further reduced by using a focused transducer to isolate the front surface scatter from the core grain structure scatter response. Second, the incident angle is such that a shear wave is formed in the steel sample of interest. Because the backscatter response of a shear wave incident on the core grain structure is significantly higher in amplitude than for a longitudinal wave, this results in a higher signal-to-noise ratio.

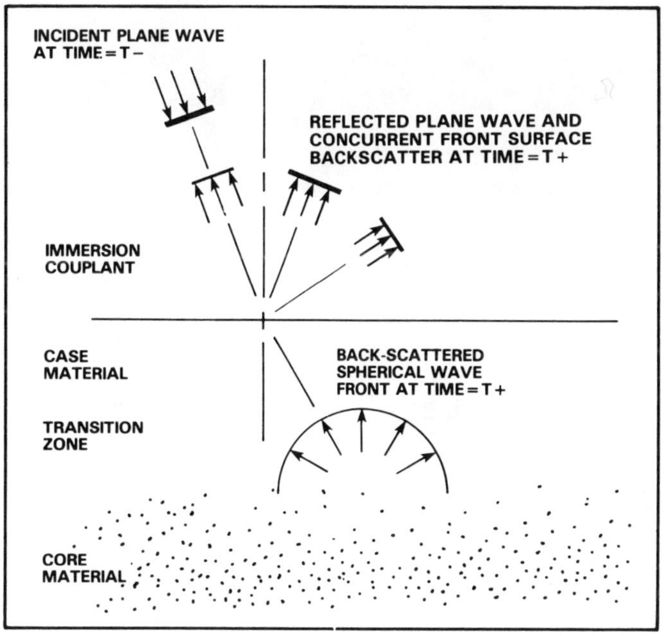

Fig. 1 - Incident, reflected, refracted, and back-scattered wave phenomena resulting from wave-/case-hardened specimen interaction.

SPECIMEN PREPARATION AND MATERIAL CHARACTERISTICS

Seven sets of steel shafts were induction-hardened and identified by a Rc 60 effective case depth. The shafts were fed through the induction unit with a fixed feed rate and quench flow. Changes in case depth were made by setting the unit wattage. The shafts were tempered and ground to the manufacturer's standard production tolerances for pump shafts (Figure 2). Characteristics of the specimens are listed in Table 1. All specimen case structures were martensite, all core structures were open pearlite, and all core region ASTM grain-size numbers were 4.

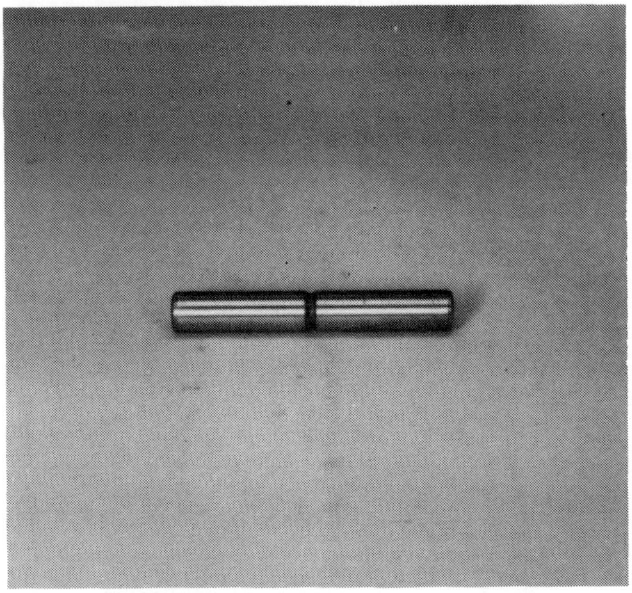

Fig. 2 - Induction-hardened pump shaft.

Table 1 - Pump Shaft Characteristics

Set No.	Rc 60 Effective Case Depth (mm/in.)	Specimens
1	0.41/0.016	3
2	1.09/0.043	2
3	1.57/0.062	3
4	1.83/0.072	3
5	2.34/0.092	3
6	2.72/0.107	3
7	2.87/0.113	3

DATA ACQUISITION SYSTEM

The data acquisition system diagrammed in Figure 3 was used to acquire ultrasonic shear wave backscatter signatures and to perform analog video enveloping of the RF signal and spatial averaging over the specimen. The transducer was aligned and positioned in relation to the specimen by the fixture shown in Figure 4. This device ensured consistent measurement readings by maintaining an 18° inclination between the transducer and specimen cross-sectional plane and a focal length distance between the transducer and the specimen. The transducer was 13 mm in diameter and had a 51-mm focal length and 10-MHz center frequency.

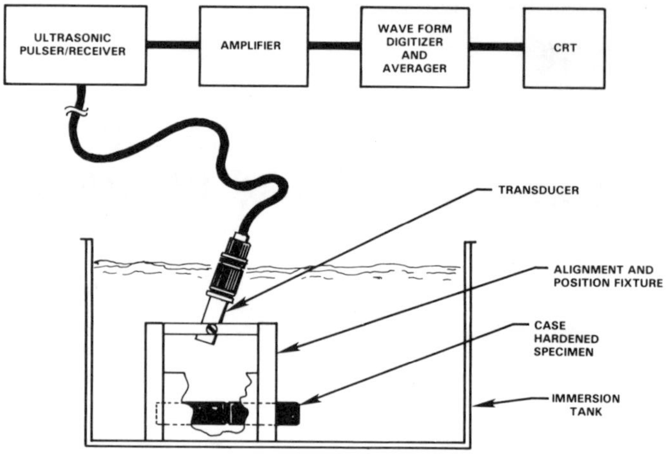

Fig. 3 - Data acquisition system.

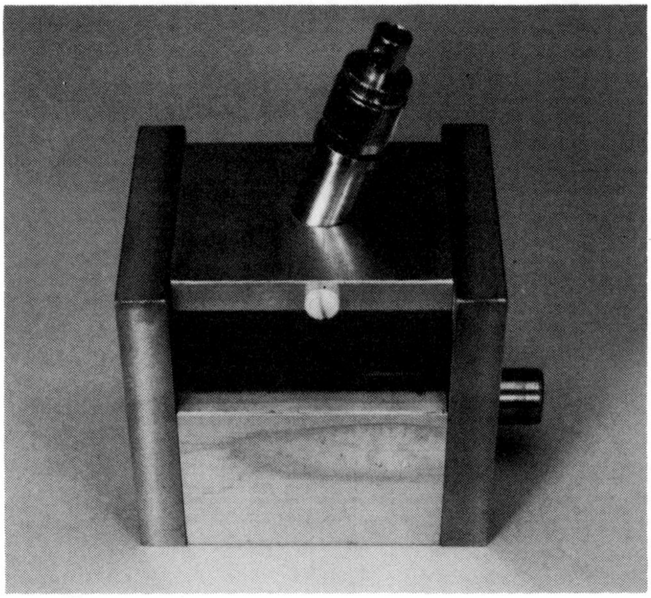

Fig. 4 - Transducer specimen alignment and position fixture.

Backscatter signatures were acquired with a standard ultrasonic pulser-receiver and displayed on a cathode-ray-tube (CRT). Because the core material was a continuum of scatterers formed by individual grains, the phase shift of the respective scatterers caused the ultrasonic signature to be extremely sensitive to position over the specimen (see Figure 5). To minimize this effect, spatial averaging was implemented. To acquire a stable and consistent profile upon which to base the determination of the onset of backscatter from the core grain structure, 256 signal envelopes were averaged as the specimen was rotated.

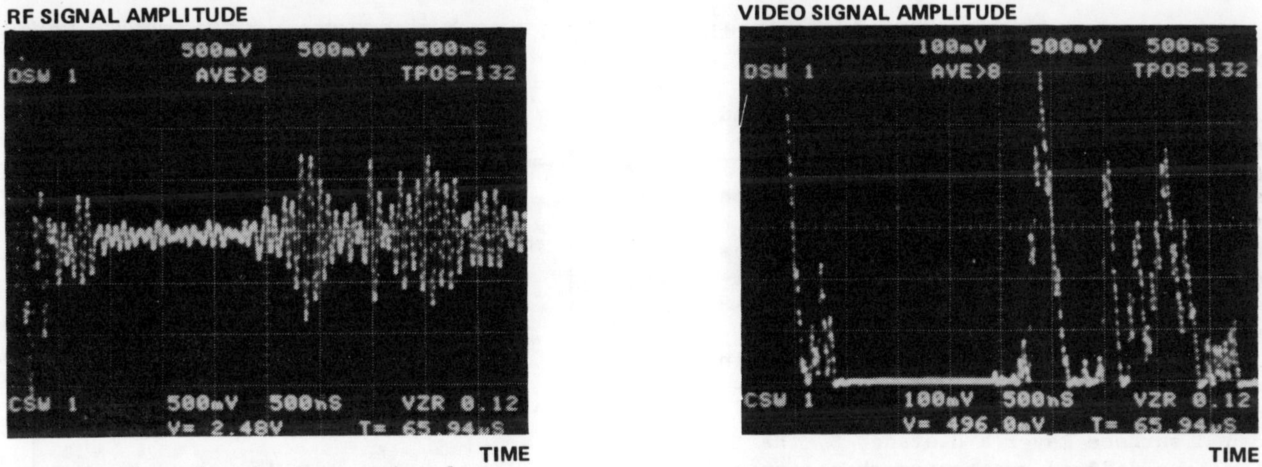

A. RF and video signals from point 1.

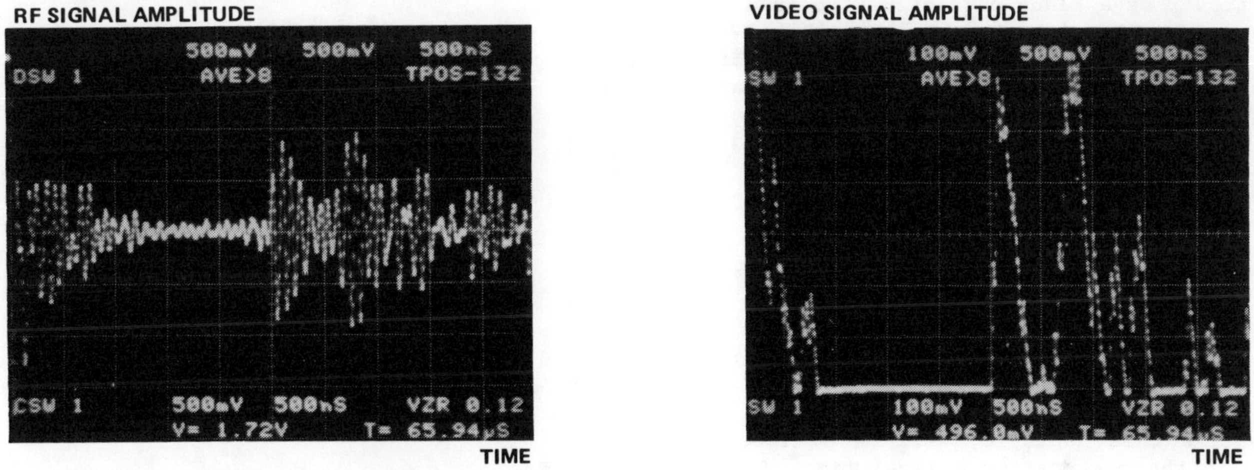

B. RF and video signals from point 2.

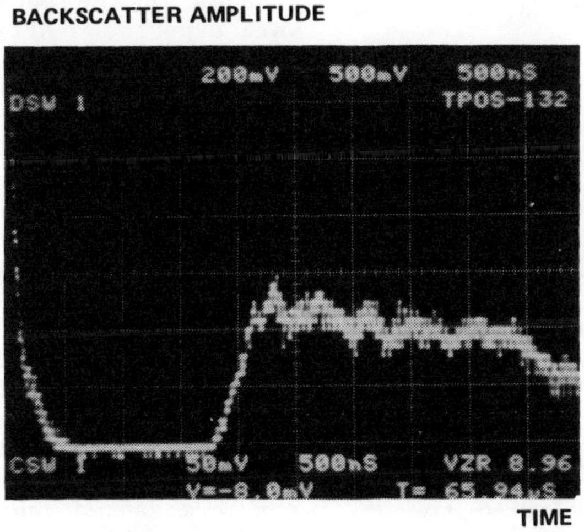

C. Spatially averaged video signal (256 signals).

Fig. 5 - Ultrasonic signature recordings from data acquisition system.

MODEL ANALYSIS

Efforts have been made to measure case depth by an extension of pulse-echo techniques (i.e., depth was determined from the interim between the arrival times of the front surface echo and the detected onset of increased backscatter). This approach, however, has several inherent weaknesses: the accurate spatial determination of the front surface and the assumption that the leading edge of increased backscatter coincides with the boundary layer of the specified effective case hardness value.

To relate effective case depth with the leading edge of increased backscatter, we begin with the refracted shear wave entering a case-hardened specimen. After propagating through the hardened surface layer a distance S, the wave encounters diffuse scatterers characteristic of the core area (Figure 6). In addition, as the wave propagates through the steel specimen, it traverses a plane characteristic of a given hardness value, which segregates the effective case layer from all other material. If we measure the distance along the propagational path between the discriminating hardness plane and the point where significant backscattering is initiated and label this offset U, we can determine the effective case depth by the following equation:

$$d = (S - U) \cos \theta \quad , \text{ or} \qquad (1)$$

$$d = \left(\frac{V_{ss}}{2} \cos \theta\right) T_{bs} - U \cos \theta \quad . \qquad (2)$$

Effective case depth, d, will be a linear function of the backscatter arrival time, T_{bs}, if the offset distance, U, is also a linear function of d in the range of interest. This relation will be examined in the results.

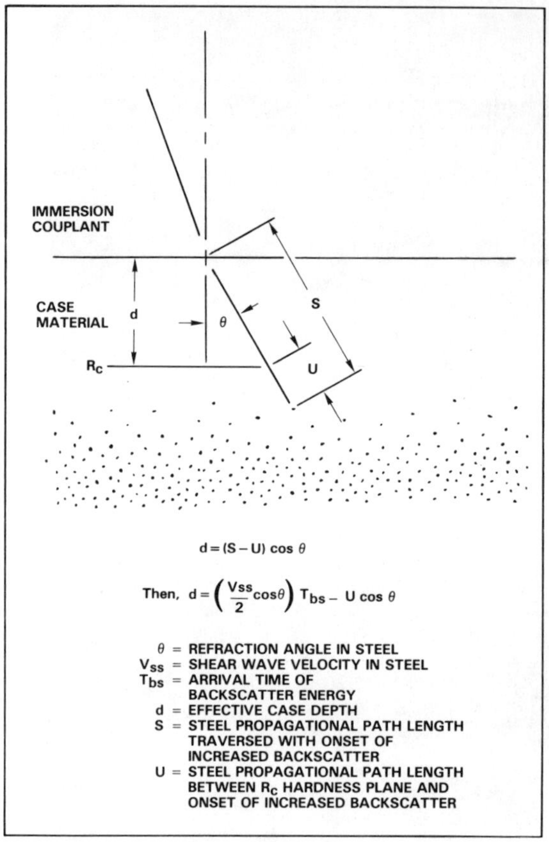

Fig. 6 - Model representation of case-hardened specimen for ultrasonic shear wave backscatter determination of effective case depth.

RESULTS

Four data sets were collected for each of the 20 case-hardened specimens, and a calibration curve was established. To minimize any bias, the 20 specimens were randomly numbered and data were recorded sequentially beginning with sample 1 and ending with sample 20. This cycle was repeated four times.

Leading edge detection of the core structure backscatter was also implemented without knowledge of the metallurgically derived values of case depth. A threshold criteria normalized to the maximum backscatter amplitude was used to determine the leading edge of the core structure backscatter. This approach was selected to compensate for observed amplitude differences between specimens where the scattering coefficient was assumed to vary slightly. The technique was derived in part from the work of K. Goebbles, 1980, who compared backscatter amplitude profiles of two specimens with ASTM grain size numbers of 4 and 6. Because all 20 samples had a grain size number of 4, only subtle changes in the backscatter amplitude profile were expected. After determining the time values associated with each of the 80 amplitude backscatter profiles, a comparison was made with the effective case depth values provided by the manufacturer.

Typical amplitude backscatter profiles are shown in Figure 7. The arrival time used in detection of the leading edge was referenced to the transducer pulsing; therefore, the recorded values include the time the ultrasonic signal was in the coupling medium. Figure 8 is a graph of leading edge arrival time versus metallurgically measured effective case depth. Quantitative measurements were taken on 17 of the 20 specimens. The remaining three specimens, which had an effective case depth of 0.41 mm, could not be measured quantitatively because the front surface response interfered with the core structure backscatter.

The values plotted in Figure 8 indicate a linear relation between backscatter arrival time and effective case depth. As discussed in the model analysis, effective case depth, d, is a linear function of backscatter arrival time if

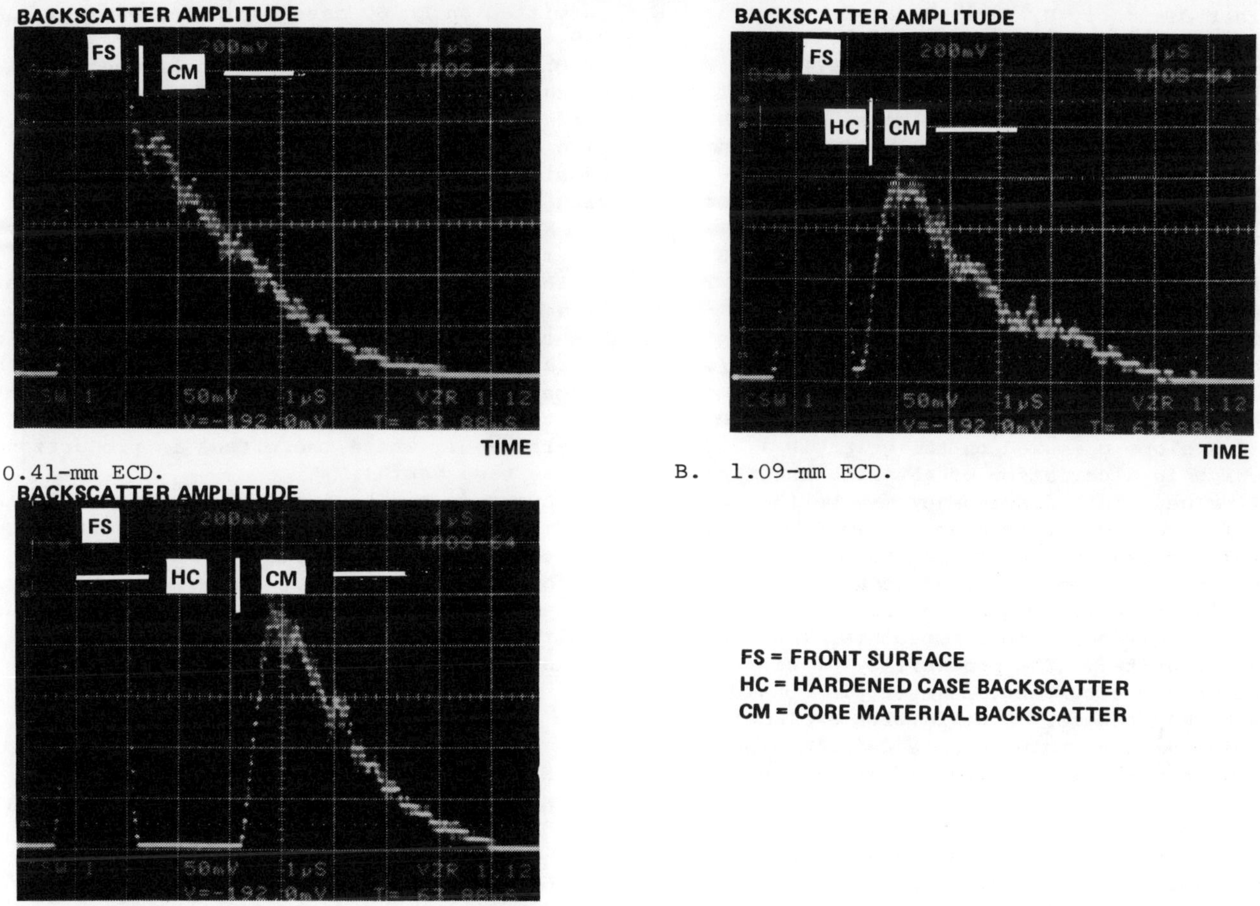

A. 0.41-mm ECD.

B. 1.09-mm ECD.

C. 2.72-mm ECD.

Fig. 7 - Spatial average ultrasonic signatures of the pump shaft hardened to different effective case depths (ECD's).

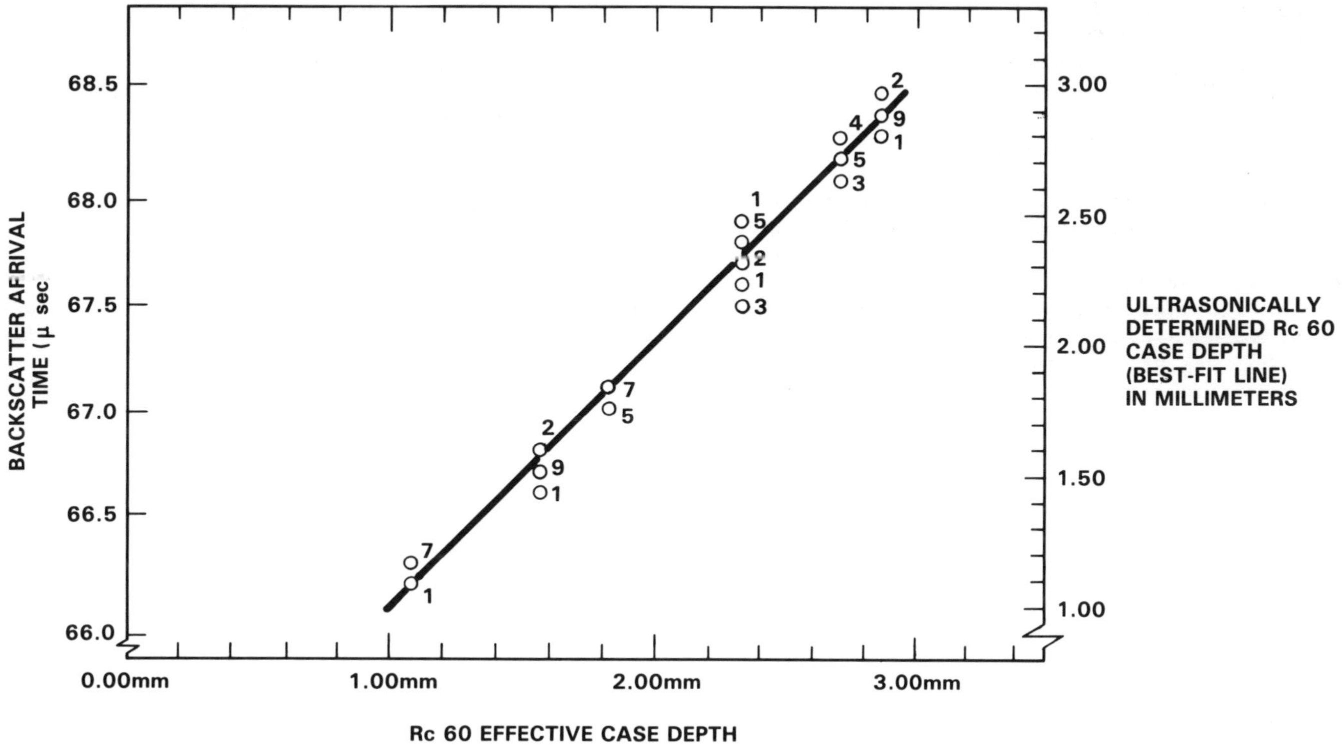

Fig. 8 - Leading edge arrival time versus metallurgically measured effective case depth.

the offset distance, U, is a linear function of effective case depth. Perennes et al., 1979, performed a similar study in which they concluded that U was a constant over the range of 2 to 5 mm. Because the method of case-hardening was not identified, however, the assumption that U is a constant in the case depth range of interest for induction-hardened specimens may not be applicable.

The slope of the best-fit line in Figure 8 can be determined theoretically from Equation 2 if U is considered a constant. The hardened steel ultrasonic wave velocity was not determined but is estimated to be equal to 3,200 m/s and, in accordance with Snell's law, would equal 42°; therefore, theoretically the slope is 840 μm/s. The slope measured on the graph is 1,220 μm/s, which is a deviation of 45% from the theoretical value. This discrepancy may be the result of incorrect transducer inclination, which would cause the sonic beam to be off-axis in relation to the specimen, or of a proportional error in the metallurgically measured case depth values given by the manufacturer (such as a change of units). The probability of either of these errors occurring, however, is considered extremely remote. A definitive explanation of the discrepancy would require detailed information on the behavior of the offset distance, U.

Using the best-fit line in Figure 8 as a calibration curve, a standard deviation of 0.1 mm (0.003 in.) was obtained. This deviation from a mean value could be the result of human error in reading the time values from the amplitude backscatter profiles or the high-frequency noise carried through the analog video envelope circuitry. An increase in sampling frequency and bit resolution during data digitization may have reduced the deviation of points from the mean value.

CONCLUSIONS

A nondestructive ultrasonic shear wave backscatter technique was used to measure effective case depth in induction-hardened pump shafts. Analysis was performed on 20 specimens, and 17 were quantitatively measured by determining the onset of core material backscatter. A best-fit line method was applied to the measured ultrasonic values to form a calibration curve that permitted an Rc 60 case depth between 1.1 and 2.9 mm to be measured with a standard deviation of 0.1 mm. The 3 remaining specimens were not measured due to interference from the front surface scatter response; however, the qualitative amplitude backscatter profiles showed a substantial decrease of case depth when compared with the 17 measured specimens.

When plotting backscatter arrival time versus metallurgically measured effective case depth the slope of the calibration curved formed by a best-fit line through the data points was 70% of the theoretically derived value. This deviation is considered to be an effect of an offset between the effective case depth plane and the plane at which ultrasonic scattering occurs (i.e., where the offset is proportional to the case depth).

Accuracy can be improved by automating the leading edge detection of core backscatter and by reducing the high-frequency noise of the analog video circuitry. It may also be possible to increase accuracy by using a higher frequency digitizer/averager with greater amplitude reolution.

REFERENCES

1. P. J. Emerson, Alternative Methods for Measuring Case Depth in Ferrous Components, British Journal of NDT, 18:2, (1976)

2. C. Perennes, N. Chretien, A. Jessiaume, and D. Gaillot, Nondestructive Testing of Hardening Depth in Steel Cylinders by Two Ultrasonic Methods, "Proceedings of Ninth World Conference on Nondestructive Testing," Session 1B-2, November (1979)

3. K. Goebbels, Structure Analysis by Scattered Ultrasonic Radiation, "Research Techniques in Nondestructive Testing, Vol. IV," Academic Press, New York (1980)

AUTHOR NOTE

This work was performed by Sigma Research, Inc., however, the author's current affiliation is with Battelle Pacific Northwest Laboratories, Battelle Boulevard, Richland, Washington, 99352.

THERMAL WAVE IMAGING FOR NDE OF METALS

R. L. Thomas
Dept. of Physics
Wayne State University
Detroit, MI 48202

L. D. Favro
Dept. of Physics
Wayne State University
Detroit, MI 48202

K. L. Grice
Dept. of Physics
Wayne State University
Detroit, MI 48202
(Present address: Dept. of Physics
U.S. Military Academy
West Point, NY 10996)

L. J. Inglehart
Dept. of Physics
Wayne State University
Detroit, MI 48202

M. J. Lin
Dept. of Physics
Wayne State University
Detroit, MI 48202
(On leave from Dept. of Physics
National Taiwan Normal University)

PRINCIPLES OF THERMAL WAVE IMAGING

SCANNING PHOTOACOUSTIC MICROSCOPY (SPAM), now known generically as thermal wave imaging, is an emerging technique for nondestructive evaluation (NDE) of opaque solids, and is applicable to surface and near surface characterization. Research in this laboratory and elsewhere[1-19] is establishing the experimental and theoretical framework for interpreting thermal wave micrographs for a variety of subsurface defects. These include a simulated subsurface void[2], delaminations in layered structures[3], closed, lateral subsurface cracks[4], and closed slanted cracks.[5] SPAM has also been applied to complex shaped parts such as turbine engine components.[6] In this paper, we give a brief introduction to the principles of thermal wave imaging and discuss its potential for NDE of metals.

Detailed descriptions of thermal wave imaging and its potential for nondestructive evaluation (NDE) applications are given in the literature.[1-19] In thermal wave imaging, an intensity-modulated and focused laser beam (electron beams or ion beams may also be used) establishes a point source of heat which varies in time at the modulating (or "chopping") frequency. The temperature (T) within the sample, beneath the point on the surface being probed, is a function of time (t) and depth (x). The important features of these variations can be seen by solving the one-dimensional thermal diffusion equation for a semi-infinite solid whose temperature at the boundary is a periodic function of time. The thermal wave solution is of the form

$$T(x,t) = T_o \exp[i(qx - \omega t)], \qquad (1)$$

where q and ω are defined below. The wave number in the solid, q, is a complex quantity with equal real and imaginary parts, so that the thermal wave is very highly damped spatially. Significant temperature variations extend only to one thermal wavelength. (The thermal wavelength, λ_t, is defined as $2\pi(2k/\omega\rho c)^{1/2}$, where $\omega/2\pi$ = the heat source modulating frequency, and k, ρ, and c are the thermal conductivity, density and heat capacity of the material). The thermal wave accordingly probes only the region close to the surface, and the depth being probed can be varied experimentally by adjusting the laser (or other heat source) modulation frequency. The thermal wave depth can be varied from 10^{-6} m to 10^{-3} m for the range of frequencies and thermal diffusivities encountered experimentally. Defects between the surface and that depth are detectable due to variation in thermal properties or thermal discontinuities at the boundaries of the defects.

Because of the critically damped form of the thermal waves, in nearly all experimental observations, thermal wave imaging is carried out in the extreme near field, i.e., either the source or the detector (or both) are much closer than a thermal wave length to the subsurface defect (thermal wave scatterer). In this situation, one has the somewhat unusual result that the resolution of the microscope depends only on the depth of the scatterer and not on the wavelength of the thermal waves.[18,19] Thus, it is possible to resolve micron-sized subsurface defects at frequencies for which the thermal wave lengths are hundreds of microns.

THERMAL WAVE SIGNAL DETECTION METHODS - Thermal wave imaging can be performed with any of several thermal wave detection techniques. One method (gas-cell detection)[1-6,9] uses a sensitive microphone in conjunction with a gas cell surrounding the point being probed. As the point is alternately heated and cooled, it in turn heats and cools the air (or other gas) in its vicinity, generating an acoustic signal which the microphone detects. A second acoustic detection technique[7-9] involves the thermal expansion of the heated surface point as an acoustic driving source and a mechanically bonded piezoelectric transducer to detect the plate modes of elastic vibration. This technique may be combined with electron beam heating to carry out thermal wave imaging in a scanning electron microscope (SEM). An optical technique (mirage-effect, or optical beam probe detection)[5,13,14] utilizes a separate laser probe beam directed parallel to the sample surface through the air above the point being probed. The gradient in optical index of refraction of the air, associated with the temperature gradient in the air, causes a deflection of the probe laser beam. The phase coherent ac component of this deflection is detected by a position sensor and forms the thermal wave signal. Another variation of optical technique (thermodisplacement, or thermoreflectance microscopy)[8,15-17] is carried out by reflecting the probe beam from the locally curved surface during thermal expansion and contraction. Finally, the thermal waves may be detected through their influence on the phase coherent ac variation in the infrared emission from the heated surface point (photothermal detection).[12]

Regardless of the type of detection, thermal wave images are generated by recording voltage levels proportional to the amplitude or phase of the signal as a function of position as the relative position between the heating point and the surface of the material under study is scanned. Typically, the scanning and data acquisition are carried out under automatic digital control, and the resulting data are displayed as gray scale images and/or computer-generated perspective plots. When optical heating is employed, corresponding scanned optical reflectance micrographs are obtained using a photodetector to monitor the ac variations in the reflected light, and the same electronic instrumentation to form and process the image.

APPLICATION TO DEFECTS IN METALS

A detailed comparison of gas-cell and mirage-effect techniques for detecting thermal wave scattering from slanted cracks in metals has recently been carried out. The results show that mirage-effect detection is especially useful in detecting vertical cracks, and has the advantage of requiring no contact with the surface being probed. A disadvantage of the mirage-effect technique is that the geometry of the specimen must be such as to allow access by the probe beam, from source to detector, to a line nearly tangent to the surface at all points to be probed.

As an illustration of gas-cell thermal wave detection of slanted cracks, magnitude and phase images of a 45 degree slanted fabricated crack in an aluminum alloy are presented in both perspective plot and gray scale formats in Fig. 1. The important features of these images, as well as images for other crack angles, have been explained theoretically in Reference 5. An important finding from these studies is the fact that when the angle becomes 90 degrees (a perfectly vertical crack), gas-cell detection in principle is incapable of detecting the crack. As a practical matter, fatigue cracks in metals are seldom perfectly vertical, and are thus usually observable by this technique.[10] We have shown, however, both experimentally and theoretically, that the lower detection symmetry of mirage-effect detection can be exploited to advantage in studying vertical cracks. A comparison of gas-cell and mirage-effect images is given in Fig. 2, and the much better detection of the latter technique is evident.
The geometry of mirage-effect detection also makes it quite suitable for examining the inside surfaces of cylindrical holes in metal. To demonstrate this, we have used the technique to locate a fatigue crack on the inside surface of a simulated bolt hole in an aluminum alloy. Although the crack was quite difficult to observe optically or by gas-cell thermal wave detection, it was readily seen in the mirage-effect thermal wave images.[11]

One of the acoustic detection techniques (electron beam excitation and piezoelectric detection via the elastic modes of vibration)[7,8] is noteworthy in its particular sensitivity to detailed metallurgical features, particularly grain boundaries, with a spatial resolution of a few microns. At this time it is uncertain whether this sensitivity is thermal or elastic in origin, or simply related to the higher frequency range of operation of this method (10^6 Hz), but it any case it represents a capability which has good potential for metallurgical diagnostic applications, and is readily adaptable to existing commercial SEM's.

SUMMARY AND CONCLUSIONS

The principles of thermal wave imaging are being established on a good theoretical and experimental base. Thermal wave scattering from several classes of typical subsurface defects is now well understood. This technique is specifically applicable for near surface defects, with a depth range which can be varied experimentally. Several detection techniques are presently under development and have different advantages. The methods have been shown to be useful for studying fatigue cracks in metals and for examination of metallurgical microstructure, with a spatial resolution of the order of 10^{-6}m.

ACKNOWLEDGMENTS

This work was supported in part by ARO under Contract No. DAAG-29-82-K-0152, and in part by the Center for Advanced NDE, operated by the Ames Lab., USDOE, for the Air Force Wright Aeronautical Laboratories/Materials Laboratory under Contract No. W-7405-ENG-82 with Iowa State University. The authors would like to thank Dr. Douglas Rose of the US Army Tank-Automotive Command for supplying images of our data (Figs. 1 and 2).

REFERENCES

1. Wong, Y.H., R.L. Thomas and G.R. Hawkins, Appl. Phys. Lett. 32, 538 (1978).
2. Thomas, R.L., J.J. Pouch, Y.H. Wong, L.D. Favro, P.K. Kuo and A. Rosencwaig, J. Appl. Phys. 51, 1152 (1980)
3. Favro, L.D., P.K. Kuo, J.J. Pouch and R.L. Thomas, Appl. Phys. Lett. 36, 953 (1980).
4. Favro, L.D., P.K. Kuo, L.J. Inglehart, R.L. Thomas, and M. Srinivasan, J. Appl. Phys. 53, 1258 (1982).
5. Grice, K.R., L.J. Inglehart, L.D. Favro, P.K. Kuo and R.L. Thomas, J. Appl. Phys. 54, 6245 (1983).
6. Inglehart, L.J., R.L. Thomas and J. Schuldies, J. Nondes. Eval. 11, 287 (1980).
7. Rosencwaig, A., Science 218, 223 (1982).
8. Rosencwaig, Allan, in VLSI Electronics (to be published).
9. Birnbaum, G. and G.S. White in Research Techniques in Nondestructive Testing, Vol. 7, Academic Press, London (1984), p. 259.
10. Grice, K.R., L.D. Favro, P.K. Kuo and R.L. Thomas, Review of Progress in Quantitative NDE, Vol. 2B, ed. D.O. Thompson and D. Chimenti, Plenum (1983), p. 1019.
11. Grice, K.R., L.J. Inglehart, L.D. Favro, P.K. Kuo and R.L. Thomas, Review of Progress in Quantitative NDE, Vol. 3B, ed. D.O. Thompson and D. Chimenti, Plenum (1984), p. 763.
12. Busse, G., Appl. Opt. 21, 107 (1982).
13. Boccara, A.C., D. Fournier and J. Badoz, Appl. Phys. Lett. 36, 130 (1980).
14. Aamodt, L.C. and J.C. Murphy, J. Appl. Phys. 54, 581 (1983).
15. Jackson, W.B. and N.M. Amer, Phys. Rev. B25, 5559 (1982).
16. Olmstead, M.A. and N.M. Amer, Phys. Rev. Lett. 52, 1148 (1984).
17. Opsal, J., A. Rosencwaig and D.L. Willenborg, Appl. Optics 22, 3169 (1983).
18. Inglehart, L.J., K.R. Grice, L.D. Favro, P.K. Kuo and R.L. Thomas, Appl. Phys. Lett. 43, 446 (1983).
19. Inglehart, L.J., M.J. Lin, L.D. Favro, P.K. Kuo and R.L. Thomas, Proc. 1983 IEEE Ultrasonics Symposium, edited by B.R. McAvoy (IEEE, New York, 1983), p.668.

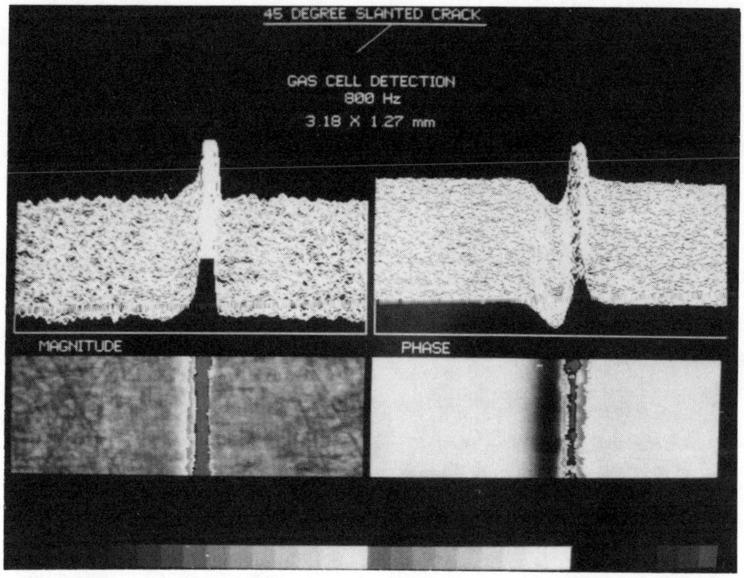

Fig. 1 Gas-cell detection thermal wave magnitude and phase perspective plots (top) and gray scale images (bottom) of a 45° slanted fabricated crack in an aluminum alloy. A detailed description is given in Ref. 5.

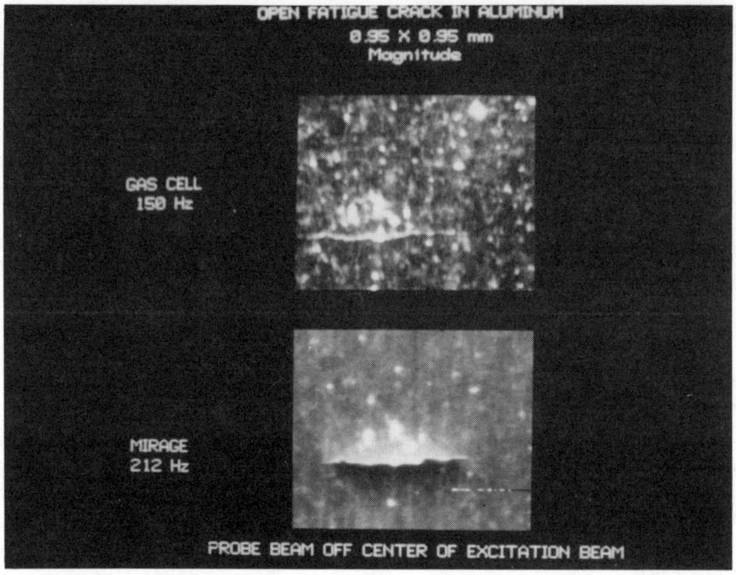

Fig. 2 Gas-cell detection magnitude (top) and mirage effect detection (phased) thermal wave images of an open, nearly vertical fatigue crack in an aluminum alloy. A detailed description is given in Ref. 10.

MICROSTRUCTURAL CHARACTERIZATION BY SMALL ANGLE NEUTRON SCATTERING

R. J. Fields
R. C. Dobbyn
C. I. Glinka
Center for Materials Science
National Bureau of Standards
Gaithersburg, MD 20899

ABSTRACT

A brief review of the NBS small angle neutron scattering (SANS) facility is given followed by examples of its application. The examples cover precipitation of copper in steel, volume changes during deformation of high strength alloys and creep cavitation. Some potential contributions of SANS to NDE are discussed.

INTRODUCTION

When a beam of thermal neutrons is transmitted through a material, the neutrons are scattered by microstructural features that are in the size range of 1 to 1000 nanometers. Scattering by features in this size range occurs in the forward direction within a few degrees of the beam and is referred to as small angle neutron scattering or SANS. Examples of features that scatter thermal neutrons are density fluctuations, phase domains, precipitates, magnetic domains, cavities or microcracks and, to a small extent, dislocations. The scattering can be quantified by measuring the angular dispersion of an initially collimated, monochromatic beam that occurs as the beam goes through the material. Using established theories of diffraction and refraction, the volume fraction and size distribution of the microstructural features can be determined. The size resolution is several orders of magnitude greater than that obtained by x-ray microradiography. If portable neutron sources become available, SANS may become a valuable NDE technique.

Three examples are presented here to indicate how this technique is used at NBS to study materials. In the first example, the existing precipitate microstructure in an age-hardening, HSLA steel is characterized and correlated with yield properties. In the second example, small changes in precipitate size caused by plastic deformation are measured in 300 grade maraging steel and a 7039 aluminum alloy. Finally, creep damage in Type 304 stainless steel is quantified during the earliest stage of creep. This is the stage about which least is known and only SANS can provide information about the nucleation of the cavities that will eventually cause failure.

These three examples demonstrate some of the applications of SANS. They are presented in order of increasing complexity to provide an initial understanding of how this powerful technique can be applied to the microstructural characterization of materials.

SANS Facility at NBS

The application of the SANS technique to the study of structural and magnetic inhomogeneities in the 1 to 1000 nm range has increased dramatically in the last ten years. This growth has been due in part to the development of large area position-sensitive detectors. Interest at NBS in SANS for research in polymers, metallurgy, biology, critical scattering and for nondestructive evaluation has led to the construction of a dedicated instrument for SANS at the NBS Reactor.

A drawing of the layout of the SANS instrument at NBS is shown in Fig. 1. The neutron source is a direct thermal beam from the reactor which passes through a liquid-nitrogen cooled filter, consisting of 20 cm of beryllium and 20 cm of single-crystal bismuth, to remove core gamma rays and fast neutrons

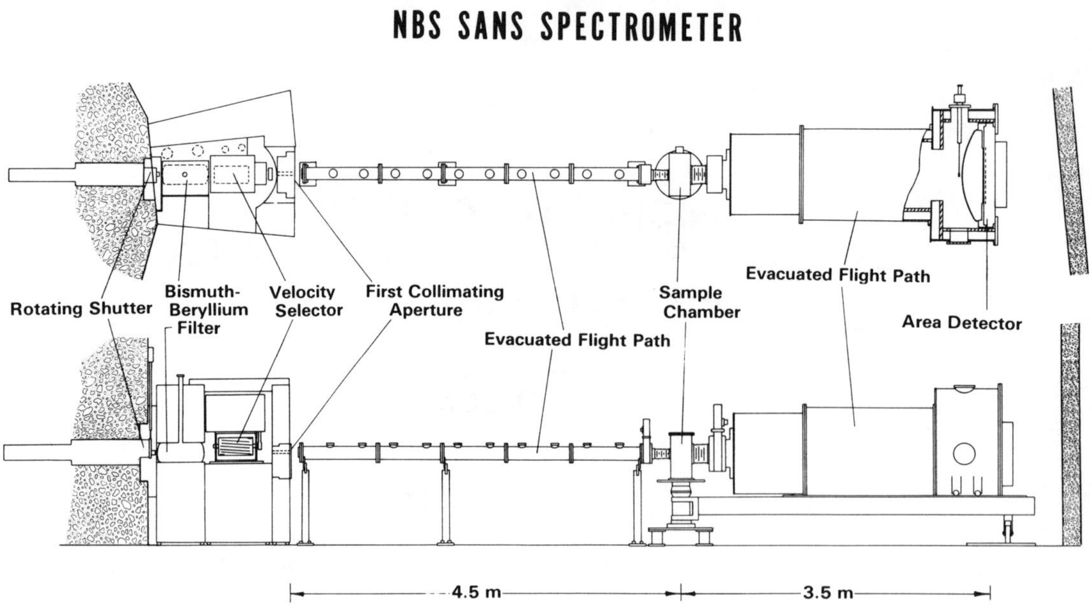

Figure 1. Schematic drawing of SANS facility at NBS.

from the beam. The beam is then monochromatized by a rotating, helical-channel velocity selector. By tuning the frequency of the velocity selector, the mean wavelength of the beam can be varied from 4 to 10 A with a constant $\Delta\lambda/\lambda$ (FWHM) of about 0.25. Following the velocity selector, the beam is transported to the sample (specimen) position through a 4.5 m long evacuated flight tube. The beam is collimated by pinhole apertures at both ends of the flight tube or by a novel high-resolution, focussing collimation system,[1] shown schematically in Fig. 2, for use with larger samples (~1.5 x 1.5 cm^2). A series of twelve masks, with seven apertures each, are aligned along the flight tube so as to produce seven independent beams which congverge to a point at the center of the detector. In this way, the total intensity on the sample (n/sec) is increased sevenfold over what would be obtained with a single pair of apertures with the same angular divergence.

The evacuated post-sample flight path consists of three shielded cylindrical sections which are mounted on a frame which rotates about the sample position. The section which houses the detector rolls on rails in the frame to allow the sample-to-detector distance to be changed from 2.0 m to 3.5 m. Built into the detector shielding is an externally adjustable positioning device for locating a beam stop in front of the detector.

Two area detectors of the Borkowski-Kopp type[2] have been purchased for the instrument. The primary detector has an active area of 65 x 65 cm^2 with 5 mm spatial resolution. The smaller, backup detector has a 25 x 25 cm^2 area and 4 mm resolution. A dedicated minicomputer processes the signals from the detector and provides a real-time display of the data. Complete data sets are then transferred to a larger computer for reduction and analysis as well as for display on an interactive color graphics terminal.

SANS for Microstructural Characterization

An elementary understanding of how SANS works and how it is applied to microstructural characterization can be obtained by considering Bragg's Law:*

$$\lambda = 2d \sin \theta$$

or

$$\frac{\sin \theta}{\lambda} = \frac{1}{2d}$$

where λ is the wavelength, d is a linear dimension of the scattering feature and 2θ is the scattering angle (Fig. 3). In Bragg scattering d is the lattice spacing. In the case of SANS, however, d can be thought of as characteristic dimensions of density fluctuations, phase domains, precipitates, etc. Considering the above equation, larger features scatter at smaller angles. A

*While Bragg scattering and small angle scattering are somewhat different, Bragg scattering may be more familiar to the reader and it correctly predicts the inverse behavior between dimensions and angles.

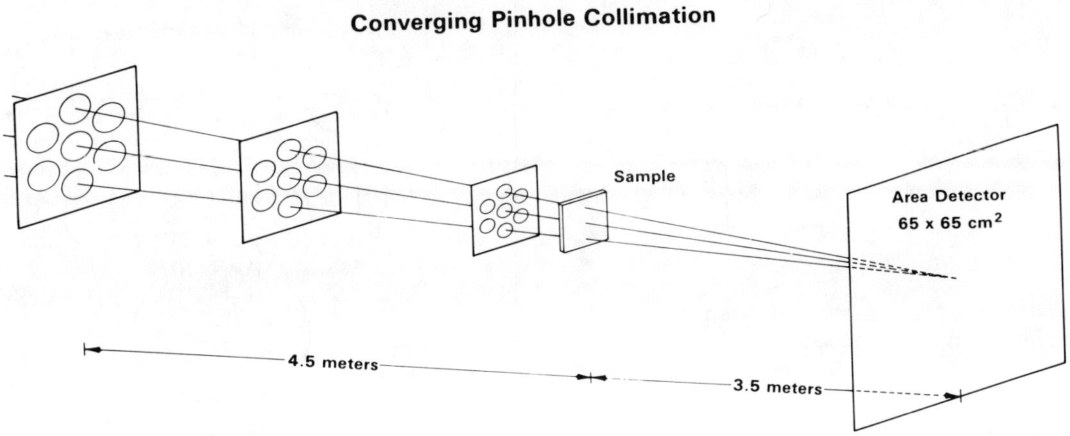

Figure 2. Schematic drawing of focussing collimation system.

quantitative relation between the angular dispersion of a collimated beam and the size of the scatterer has been worked out by Guinier. Furthermore, the intensity of the scattering is related to the concentration of scatterers: the higher the concentration, the more intense the scattering. A detailed treatment of the theory of SANS has been presented by Kostorz[3]. It is based on established theories of diffraction and refraction.

MICROSTRUCTURAL CHARACTERIZATION

Predictive modelling of the effect of processing variables on microstructure or the effect of microstructural properties only comes when the microstructure can be quantitatively defined. It is frequently found that the above effects can be quantified in terms of the size or size distribution of certain features in the microstructure. When the size range of important features is 1 to 1000 nanometers, SANS is a powerful and, often, unique tool in the quantitative characterization of microstructure. At present, the NBS SANS facility is capable of acquiring the raw data on computer, running it through various background correction and radial averaging schemes, and finally extracting an average Guinier radius of gyration. This last quantity can be related to the actual size of the scattering features if the shape is known, In the cases detailed below, it is evident that microstructural details in size ranges accessible to SANS are of considerable scientific and technological importance.

Characterizing Precipitate Distributions

Precipitate hardening is the most potent strengthening and toughening mechanism available to the alloy designer. Relatively small changes in precipitate distribution can cause large changes in yield strength or fracture toughness. Recently, an economic HSLA steel has been produced whose strength is developed by the precipitation of a copper-rich phase. While the properties can be optimized using purely empirical techniques, little insight is gained from such an approach. SANS has been used at NBS to investigate and characterize the precipitate concentration size distribution for various heat treatments[4].

Perhaps the first point to make is why choose SANS over other techniques like transmission electron microscopy (TEM). In the case of copper precipitates in iron, there is very little source of electron contrast because copper and iron have very similar electron scattering factors and because their atomic radii are nearly equal. The similarity in size means that there is little coherency strain which often provides a good outline of precipitates in TEM. When the precipitates of copper are coherent, there is also no structure factor difference to provide visibility in TEM. Hornbogen[5] has pointed out these obstacles to studying the Cu-Fe system with electrons. However, with neutrons, the scattering factor difference is greater and, in particular, there is a magnetic component to enhance the contrast. The iron matrix is magnetic and the copper precipitate is not. Neutrons are very sensitive to this source of

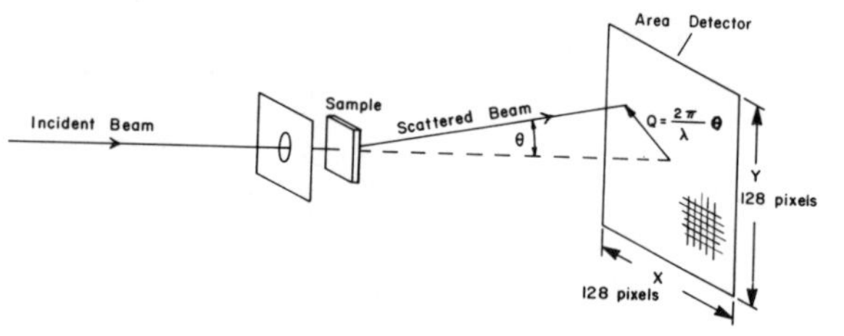

Figure 3. Scattering of beam by specimen is monitored on 2-D detector.

magnetic contrast because they have a magnetic moment and are scattered by variations of internal magnetic fields. These differences make SANS a much more effective tool for studying the Cu-Fe system than TEM, particularly when the copper precipitates are coherent.

A typical SANS output resulting from the transmission of a 0.5 nm wavelength beam of neutrons through a heat treated sample of this HSLA steel is shown in Figure 4. The output is a neutron intensity contour plot showing quantitatively how neutrons are scattered out of the main beam by precipitates. These data were taken with a horizontal magnetic field applied to the sample in order to align the magnetic domains in the iron matrix and thereby eliminate domain walls which can also be a source of scattering. The remaining scattering is due to the precipitates and appears anisotropic because the magnetic contrast between the precipitates and the matrix enhances the scattering only in the direction perpendicular to the applied field. When this scattering from the precipitates is plotted in a more conventional fashion (Figure 5), it becomes clear how the intensity falls off with scattering vector away from the main beam. The scattering vector, Q, is given by $4\pi\sin\theta/\lambda$ and is basically the normalized angle.

In Figure 6, similar curves are plotted for four different heat treatments. Two principles from scattering theory will assist in understanding these curves:

1. The greater the volume of precipitates, the greater will be the scattered intensity.
2. The smaller the size of the scattering precipitate, the greater will be the angle of the scattering.

Using these principles, we see that curve #21 (i.e., 90 minutes at 482°C (900°F)) lies noticeably below the other curves, yet crosses them at a high scattering angle. The lower intensity means that there is a lower volume fraction of precipitated material than at the

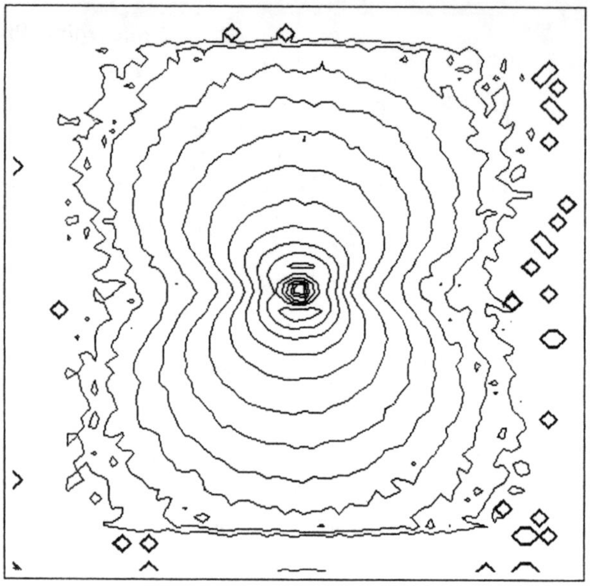

Figure 4. Typical scattering pattern on 2-D detector from an HSLA steel in an horizontal magnetic field.

high aging temperatures. The cross-over indicated that precipitates are smaller in this condition than in the other conditions. The other curves do not indicate a substantial difference in total scattered intensity. This suggests that precipitation is complete and we are dealing with the same volume fraction of precipitated material for aging temperatures 538°C (1000°F) to 649°C (1200°F). The main difference in these curves is where they cross each other. Note that the lower the temperature of aging, the farther out on the scattering curve is the cross-over. This indicates

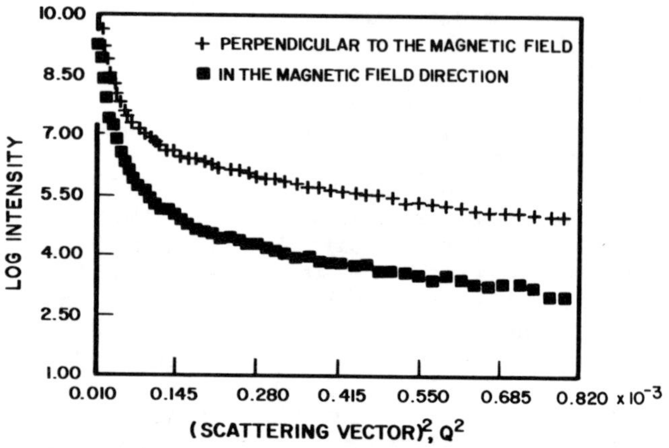

Figure 5. Quantitative scattering curves reduced from the pattern in Figure 4 in the directions indicated.

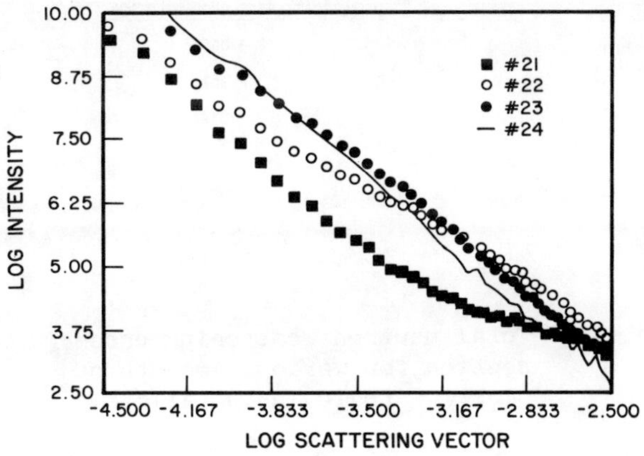

Figure 6. Scattering curves for the HSLA steel aged 90 minutes at 482°C (#21), 538°C (#22), 590°C (#23), and 649)C (#24).

that the lower the aging temperature, the smaller is the precipitate at 90 minutes aging time. The picture that emerges is entirely consistent with the Ostwald theory of the coarsening of precipitates.

In the course of the SANS analysis, the total neutron cross section (a quantity directly related to the transmission coefficient) was required. This quantity was determined by measuring the attenuation of a neutron beam by each specimen. It was found that the total neutron cross section correlated extremely well with hardness (Figure 7). This quantity was also sensitive to small variations in hardness to a degree that is rarely obtained by other microstructural characterization techniques.

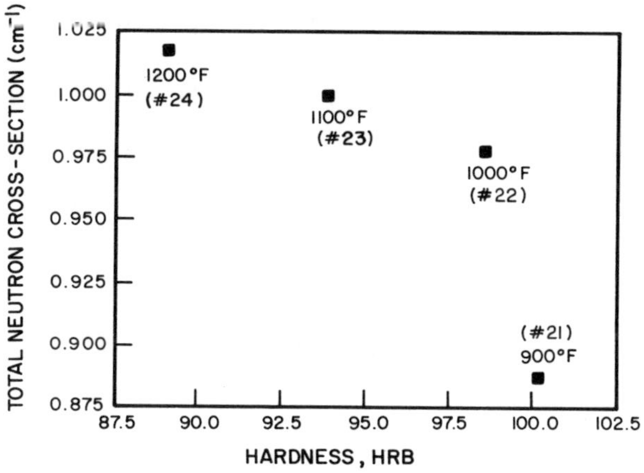

Figure 7. Correlation between total neutron cross section and hardness.

Such a study can be used to determine the kinetics of the precipitation reaction and to suggest alternative heat treatments. Moreover, a knowledge of the precipitate size distribution and mechanical properties can be used to evaluate and extend existing theories relating microstructure to mechanical properties. By knowing the precipitation kinetics and the relation between microstructure and properties, it is possible to reduce the amount of trial and error testing needed to determine the right heat treatment required for the desired properties.

Effect of Deformation Studied by SANS

The theory of plasticity routinely assumes that volume is conserved during plastic deformation. Recent density measurements on high strength, precipitation hardened metals (300 grade maraging steel and type 7039 aluminum alloys) have shown that there is a permanent volume expansion associated with plastic deformation in these alloys (Figure 8). This discovery has important consequences for engineers performing calculations of metal flow during forming operations, projectile penetratation, and other large scale plastic deformations. The question to be answered is: what microstructural features are changing to cause this volume expansion?

By subtracting SANS spectra before and after straining, the changes in microstructure caused by deformation can be deduced. Typical results are shown in Figure 9 and summarized in Table 1. In all cases, it was found that scattering was decreased by strain. These results are extremely significant in that they show that the volume change cannot be due to an increase in scattering centers such as microcracks or dislocations. Scattering can only be decreased in the systems studied here if some change in the precipitate concentration or size distribution take place. In

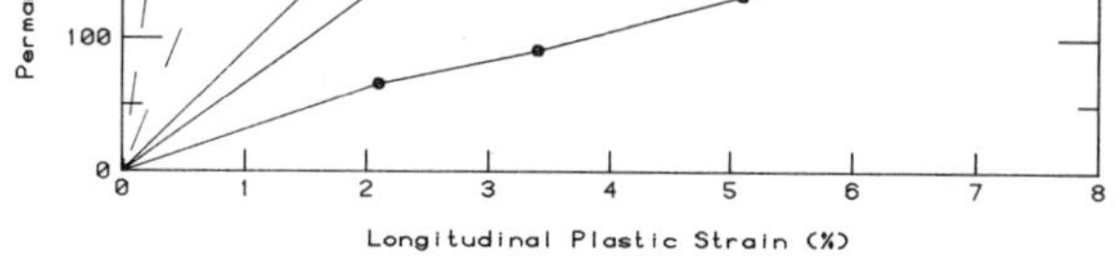

Figure 8. Permanent volume expansion associated with plastic strain in two precipitation hardened alloys.

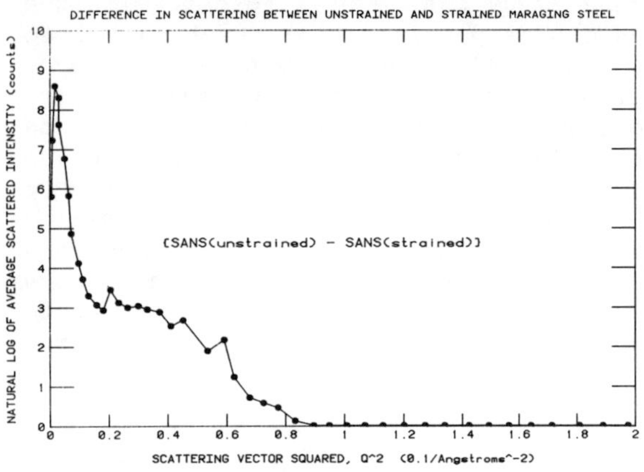

Figure 9. Scattering curve obtained by subtracting SANS spectra of strained material from the SANS spectra of unstrained material.

particular, deformation induced dissolution of precipitates would explain the scattering results and the volume change. Such a dissolution has been proposed as the cause of fatigue softening. Recently, Vogel et al[6]. have obtained TEM evidence demonstrating precipitate dissolution in persistent slip bands of Al-Mg-Zn, an alloy similar to the aluminum alloy studied here. Their interpretation is the same as ours: the precipitate size distribution is changed by plastic deformation. Specifically, small precipitates are dissolved by strain.

SANS Study of Creep Cavitation

At high temperatures, the failure of alloys and ceramics is frequently due to the slow nucleation, growth and linkage of grain boundary cavities. This time dependent process, called creep cavitation or creep rupture, is of increasing technological importance as ever higher temperatures are required for improved performance or high efficiency. In principle, cavities and precipitates are very similar from the point of view of SANS; both are density fluctuations. The approaches outlined above for the study of precipitates can be extended to the characterization of cavity sizes. Scientists studying cavitation have concluded that SANS is the only effective tool for studying the nucleation and early growth of cavities. We are presently using SANS to measure the rate of failure of Type 304 stainless steel and copper. The SANS facility at NBS is sensitive to cavities from their creation up to about 0.1 micrometer in diameter. This is found to be the earliest stage of failure and, in this size range, no other technique can give quantitative information about cavity size distributions.

MATERIAL	STRAIN (%)	TOTAL SCATTERING CROSS SECTION (cm^{-1})
Maraging Steel	0	3.0010
"	3.0	2.9611
"	7.0	2.9377
7039 Al	0	0.0659
"	2.5	0.0648
"	6.4	0.0635

Table 1. Total neutron scattering cross section for various amounts of plastic strain in two alloys.

SANS studies have been carried out by others on creep damaged stainless steel superalloys and copper. In every case, the scattering was observed to increase as the damage increased. Early interpretation of the results for the stainless steel and the superalloys was clouded by changes in the precipitate size distribution that occurred during the test[7,8]. Current SANS studies[9] of creep cavitation on Type 304 stainless steel have employed unstressed reference specimens and a carbide stabilizing heat treatment which have virtually eliminated the ambiguities arising from precipitate redistribution during the test. Also, microscopic studies have revealed intergranular cavities in specimens that have been tested long enough to produce cavities of optically resolvable size. Therefore, it is valid to attribute the neutron scattering solely to cavitation.

Type 304 stainless steel used in high temperature applications consists mainly of austenite grains and grain-boundary metal carbides. The carbide-austenite interface at

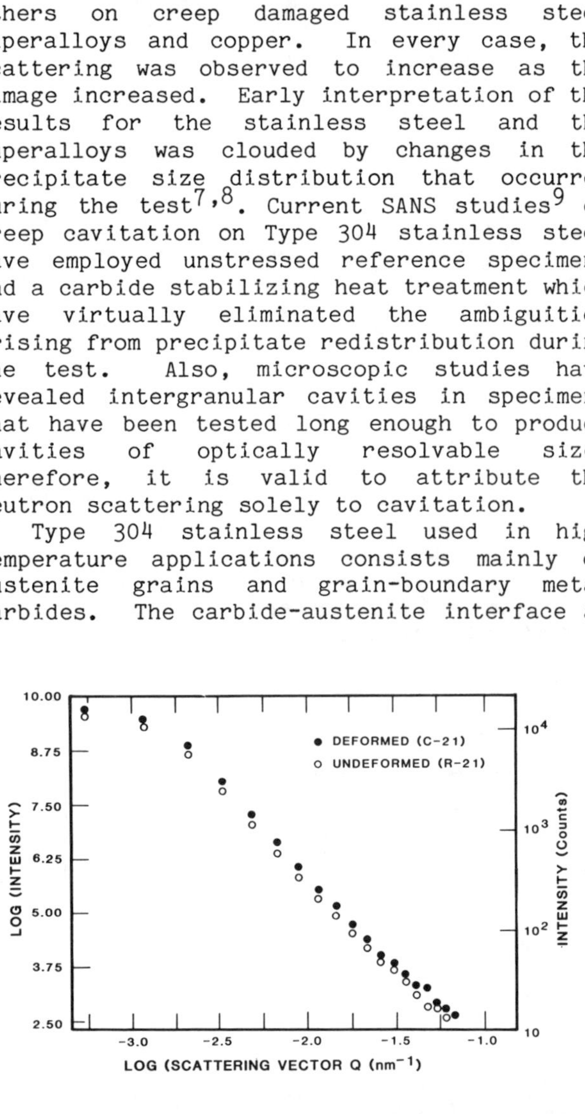

Figure 10. Scattering curves for creep damaged stainless steel and a reference (undamaged) sample of the same material.

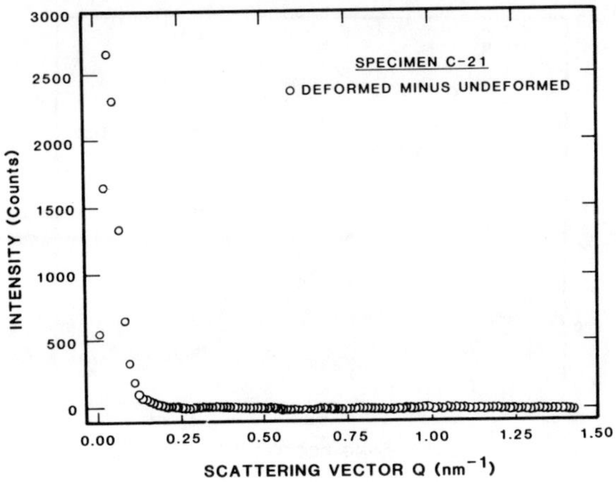

Figure 11. The difference between damaged and undamaged material as measured by SANS.

the grain boundary is a preferential site for cavity nucleation, which at high temperatures can be readily activated with low applied stresses by creep deformation and/or grain-boundary sliding. The resulting cavities then grow, link, and ultimately cause failure. The early stages of this process are extremely slow and the creep cavities, due to their small size, are not easily detectable or quantifiable. It is these early stages of nucleation and cavity growth that are examined here through the use of small angle neutron scattering. The structure after cavitation consists of grain-boundary cavities and two material phases (austenite and carbides). The unequivocal determination of cavity size, or size distribution, from a SANS spectra requires that the creep-induced changes in the carbide volume fraction, size distribution, and spacial distribution do not contribute significantly to the scattered intensity. By adopting special experimental procedures, the effects due to the carbides could be minimized. The following procedures were employed:

1. All samples were given a stabilizing heat treatment of 40 h at 775°C prior to creep testing. This resulted in a very coarse and stable carbide distribution.
2. A heat treated, but unstressed, reference sample was placed in the furnace with each creep specimen. In each case, the SANS spectrum of the unstressed sample was subtracted from that of the stressed specimen to obtain the SANS spectrum of the creep damage.

The SANS studies were carried out at the NBS reactor facility using a converging collimation and a neutron wavelength of 0.625 nm. With this choice of collimation and wavelength parameters, a scattering vector (Q) range of 0.05 to 1.5 nm^{-1} (0.005 to 0.15 A^{-1}) could be measured in these experiments. This Q range measures the scattering contribution of inhomogeneities in the size range of 4 to 120 nm. As noted previously, detailed theory and procedures for anlayzing SANS data have been described by Kortorz. A few relevant features will be discussed here. If the scattered intensity is denoted by I, the differential scattering cross-section is given by

$$d\sigma/d\Omega = cI/MTd$$

where d is the sample thickness, M is the total monitor count (or scattering time), T is the sample transmission, and c is a combination of experimental constants related to the SANS facility. Creep damage is related to the difference in differential scattering cross-sections between the crept sample (C) and the reference sample (R):

$$\frac{d\sigma_C}{d\Omega} - \frac{d\sigma_R}{d\Omega} = c[I_C/M_CT_Cd_C - I_R/M_RT_Rd_R]$$

Accordingly, all intensities in the following discussion have been normalized by sample thickness, transmission, and scattering time. We now assume that the entire difference in scattering is due to creep cavities alone. This is justified by the use of the special procedures described above. Figure 10 shows a scattering pattern for one set of creep and reference samples. This figure shows, as did all the samples, that the crept sample scatters more strongly than the reference sample. This result supports the assumption that cavities are the predominant scattering feature, because carbide coarsening would show up as decreased scattering (or scattering at unresolvable levels of Q). The data are replotted in figure 11 as the difference spectrum.

One parameter of major interest is the growth of average cavity size as a function of time under load. A Guinier analysis was performed on spectra like figure 11 to obtain an average cavity size. Guinier showed (see, for example, Ref. 3) that for a randomly distributed, monosized heterogeneity the intensity, I, and scattering vector, Q, relationship can be described by the equation

$$I = I_o\exp[-R_g{}^2Q^2/3]$$

Here R_g is the radius of gyration of the cavities and for spherical cavities it is related to the cavity redius (R) by the equation

$$R = Rg(5/3)^{1/2}$$

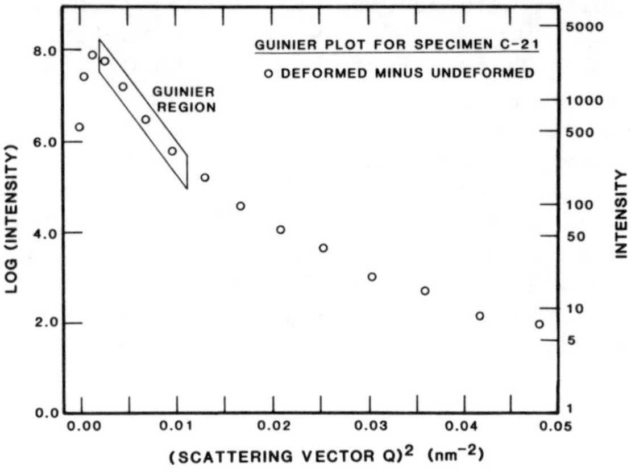

Figure 12. Guinier plot of preceding difference curve.

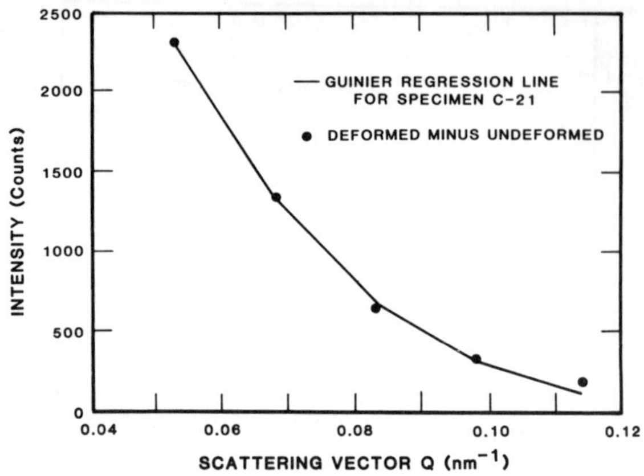

Figure 13. Regression line through data points that are in Guinier region (i.e., linear in Guinier plot).

Figure 12 is a Guinier plot (log I versus Q^2) for the difference spectrum of figure 11 giving information about the creep cavitation size. The cavity diameter obtained by this procedure normally overestimates the contribution of large cavities, yet it is a very useful parameter to study cavity growth as a function of creep time while degradation is in progress. Using a least square fit subroutine, the above equation for I was fit to the data points in the relevant Q range. A set of experimental data and the regression lines are shown in figure 13. (Note that this figure is plotted as I versus Q, so that the regression line is not straight.) A reasonably accurate fit was obtained in the selected Q range for this specimen as well as for the other creep specimens. The average cavity sizes derived for the various stresses and times under load at 600°C are shown in table 2.

Table 2 shows that doubling the creep time at the same stress apparently doubles the average cavity size, which is in agreement with the Chuang-Rice model, which predicts a constant cavity growth rate. Figure 14 shows cavity size as a function of creep time. If we include a point at the origin, the linear time dependence of the growth rate of cavities is again apparent but, in the absence of more data, only a limited significance can now be attached to the trend in figure 14. Specimens have been tested to intermediate times and their SANS spectra will be determined shortly.

From table 2, we see that specimen C-21, which was stressed at 145 MPa (21 ksi), has a higher average cavity size after 750 hours than specimen C-18 which was stressed at 124 MPa (18 ksi). After 1500 h, we find that the average cavity size at these two stress levels is not significantly different. This suggests that for longer creep times the size growth of cavities is less sensitive to small changes in applied stress. A more likely possibility, as noted by Weertman, is that when a sizable fraction of cavities grows large enough to scatter at Q values below the minimum resolvable Q in the experiments, the average size and volume fraction deduced from the standard SANS analysis are highly underestimated. Recent efforts[10] have been made to extend SANS measurements to larger size ranges for which other techniques have enough resolution to be effective. In this overlapping size range, SANS and these other techniques can be compared to provide some assessment of their validity in the nonoverlapping size ranges.

The most important development demonstrated here is that SANS is capable of measuring cavities at early times when they are only 20 to 30 nm in diameter. This must be close to the nucleation event. No other technique that we know of is capable of resolving this initiation phenomenon. Further work should provide direct experimental information that will test the assumptions of existing cavitation models. Since the models are used for

Sample	Creep Parameters: Applied Stress	Creep Parameters: Time (h)	Average Cavity Diameter - Guinier Approximation (nm)
C-21	145 MPa (21 ksi)	745	34.6
C-21	145 MPa (21 ksi)	1493	77.0
C-18	124 MPa (18 ksi)	789	26.6
C-18	124 MPa (18 ksi)	1537	76.3

Table 2. Growth of creep cavities as a function of time at constant stress in 304 stainless steel.

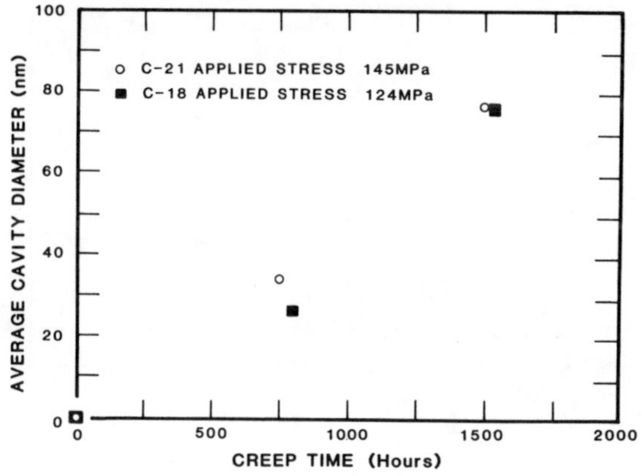

Figure 14. Cavity size as a function of creep time and stress.

predictive purposes, they must be verified. SANS currently appears to be the best technique to characterize creep damage in order to test these theories.

Besides being of use characterizing creep damage, SANS also has potential applications in NDE of creep damaged components. This is of interest to many electric generating facilities which, due to economic conditions, wish to run equipment beyond its intended lifetime. Since the creep damage of pressure vessels and steam pipe steel may be related to creep cavities, σ-phase precipitation, and carbide coarsening, these utilities would benefit greatly from a measurement technique which is sensitive to these forms of creep damage. SANS is such a technique. The information gained by SANS investigations of crept specimens can also be used to evaluate and extend existing theories of cavity nucleation and growth, thus improving the design engineer's predictive abilities.

Conclusion

The usefulness of SANS has been demonstrated by three examples of microstructural characterization: determining the precipitate distribution in a steel, measuring slight changes in microstructure of high strength alloys caused by plastic deformation, and following the growth of creep cavities in stainless steel. The success of these studies relied on certain strengths of the SANS technique. Among these strengths are the additional sources of neutron diffraction contrast, the penetrating power of neutrons which facilitates the examination of bulk specimens, and the ability to make SANS difference spectra. These three strengths made the above examples possible. There are many more examples of how SANS has been used at NBS and elsewhere for microstructural characterization. These would highlight other unique advantages of this non-destructive technique. Space does not permit a discussion of these. It is worthwhile to remember that, less than a century ago, x-radiography was in even a less developed state than SANS is now. X-rays are now an NDE tool of unequalled importance. Perhaps, in the future, SANS may join other NDE techniques to make reliable, high quality products in the most efficient way and to assure the safe operation of jets, reactors and other systems of a technologically oriented society.

Acknowledgements

The work reported here would not have been possible without the generous assistance of the Office of Non-Destructive Evaluation of the National Bureau of Standards, David Taylor Naval Ship Research and Development Center, and the U.S. Army's Ballistic Research Laboratory. The authors gratefully acknowledge this support. The author also thanks Dr. Richard Davis of NBS for performing the high precision density measurements.

References

1. A.C. Nunes, J. Appl. Cryst. Vol. 11, p 460 (1978).
2. C.J. Borkowski and M.K. Kopp, Rev. Sci. Instruments Vol. 46, p 951 (1975).
3. G. Kostorz, Neutron Scattering in Materials Science, Chapter 5 in Treatise in Materials Science and Technology, Vol. 15, Edited by H. Herman (Academic Press, New York, 1979).
4. G.E. Hicho, L.C. Smith, S. Singhal, and R.J. Fields, Journal of Heat Treating Vol. 3 #3, pp 205-212, (1984).
5. E. Hornbogen, Trans ASM Vol. 57, pp 120-132 (1964).
6. W. Vogel, M. Wilhem, and V. Gerold, Acta Met Vol. 30, pp 21-30 (1982).
7. M.H. Yoo, J.C. Ogle, J.H. Schneibel, and R.W. Swindeman, in Proc. 6th International Conf. on Strength in Metals and Alloys, Melbourne, Austrailia, Aug. 16-20, 1982 (Pergamon Press).
8. H. Walther and P. Pizzi, SANS for Nondestructive Testing in Research Techniques in Nondestructive Testing, Vol. IV, Edited by R.S. Sharpe (Academic Press, 1980).
9. E.R. Fuller, Jr., R.J. Fields, T.-J. Chuang, and S. Singhal, Journal of Research of NBS, Vol. 89, #1, pp 35-45, (1984).
10. K. Hardman-Rhyne, N.F. Berk, and E.R. Fuller, Jr., Journal of Research of NBS Vol. 89, #1, (1984).

ACOUSTIC EMISSION MONITORING OF LASER DRILLING

Roger B. Clough, Robert J. Schaefer, Haydn N. G. Wadley
Center for Materials Science
National Bureau of Standards
Gaithersburg, MD 20899

ABSTRACT

Laser drilling, because of its speed, controllability, and ability to penetrate tough or exceptionally hard materials, is a rapidly growing manufacturing method. However, beam power spikes and intervening disintegration products can disrupt or diminish energy transmission to the hole site, resulting in variable hole depth. To improve this situation, a combination of acoustic emission monitoring and laser settings has been investigated to improve hole depth prediction. The emission is suggested to be generated principally by liquid expulsion; intergranular cracking appears to be a secondary source. Acoustic emission shows promise as an experimental method for study of directed energy beam-material interactions.

LASERS ARE BEING increasingly used as sources of directed high energy beams for materials processing. Their primary applications are for surface modification (heat treatment[1] or rapid resolidification[2]), welding[3], cutting[3], and drilling[3]. Laser drilling, for certain types of applications, is a much preferred technique to traditional mechanical drilling because of its ability to: drill at high speed, make small precision apertures, penetrate tough or exceptionally hard materials, and to drill holes at shallow angles to surfaces. A major problem, however, in laser drilling is dimensional irreproducibility; hole diameter, shape, and particularly depth vary considerably from drill pulse to pulse[3]. A full understanding of this is complicated by the unavailability of in situ techniques for the study of laser/material transient interactions.

The irreproducibility problem has been recognized, in part, to be due to instability in the laser pulse and to the nonlinearity of laser/material interactions. While the pulse shape is approximately Gaussian in time, it contains numerous power spikes, resulting in variable energy transmission to the sample. In addition, disintegration products can interact with the incoming beam, causing scattering and a further variability of energy transmission. Both factors contribute to an unpredictable hole depth. Other contributions to variability arise if the bottom of a hole extends beyond the focal range of the laser or if liquid is expelled unevenly or runs back down the sides of the hole. Variation in hole depth can thus be a serious problem. Since the variations are random and cannot be controlled by adjustment of the settings on the laser, it is logical to attempt to develop a process sensor that can sense the depth drilled for each laser pulse.

Acoustic emission (AE), an in-situ, real time technique, is one of the few candidates for such a sensor. Acoustic emission in materials is usually caused by abrupt stress changes associated with the micromechanisms of dislocation motion, crack growth, and martensitic transformations. In this phenomenon (depicted in Fig. 1), rapidly varying stresses (i.e., force dipoles) at the emission source generate stress waves (elastic waves) which radiate throughout the material. These naturally generated waves can be remotely detected by a piezoelectric transducer attached to the part; by monitoring, quantitative information can be obtained about the source process.

The type of information that can be obtained about the source depends on the specific application. The surfaces of a typical piece being drilled cause a complex series of acoustic wave reflections and mode conversions (which eventually result in sample reverberation). This greatly complicates the detected signal and prohibits deconvolution methods of source reconstruction. Furthermore, due to these reverberations, the transient electrical signal from the transducer may persist for many milliseconds, and it is necessary to observe this signal for its entirety if accurate work is to be attempted. In situations such as these, the

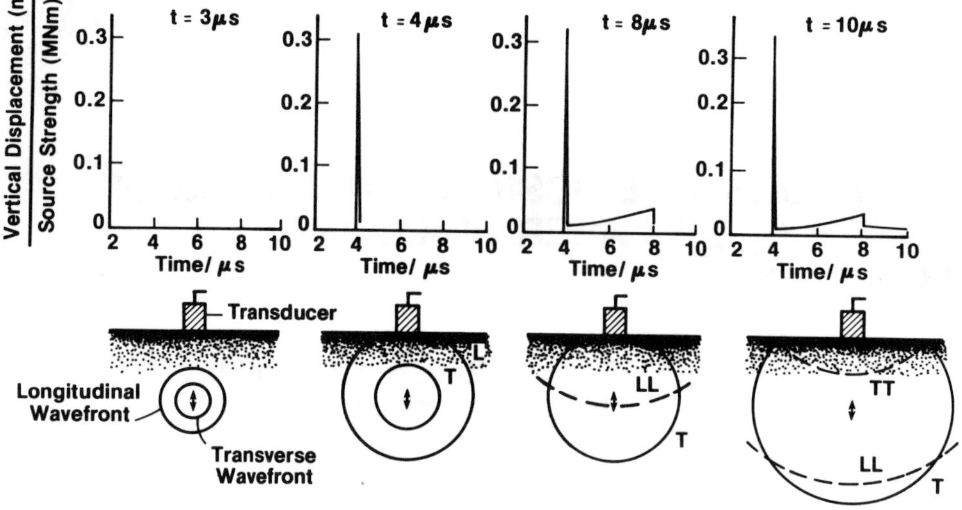

Fig. 1 Acoustic emission produced by a subsurface force dipole. Mode conversion at the surface is omitted for simplicity.

most promising approach is to measure a physically meaningful (preferably scalar) quantity of the signal and to calibrate the complete system (sample, transducer, and signal processing instrumentation) using a source of known monopolar or dipolar character[9].

The unloading of a vertical force associated with the fracture of a pencil lead was used to produce known monopolar sources. For sources of dipolar character, a small laboratory infrared (Nd:YAG) laser has been used to generate a predictable thermoelastic source at the location of interest on a sample. The energy of the ensuing acoustic emission is measured as it would be during laser drilling and a calibration curve is developed by varying the energy of the incident laser pulse (and hence its source strength). These simple calibration procedures obviate the need to fully understand the complexities of wave propagation in bounded samples, the mechanical/electrical conversion process of transduction, and the transfer characteristics of instrumentation (see Appendix) in order to deduce source properties. It enables determination of a scalar characteristic of the source that is useful for both determining the origin of the emission and developing a more detailed understanding of the micromechanisms of laser drilling.

EXPERIMENTAL PROCEDURE

LASER DRILLING--A schematic of the experimental apparatus used for acoustic emission monitoring of laser drilling is shown in Fig. 2. A Nd:Glass infrared laser (1.06 μm wavelength) capable of generating up to 200 J single pulses was used for the drilling. Incident pulse energies in the range of 20 to 50 J were used with controlled beam radii of between 0.35 to 1.60 mm. The energy of the laser pulse could not be measured directly. Instead the

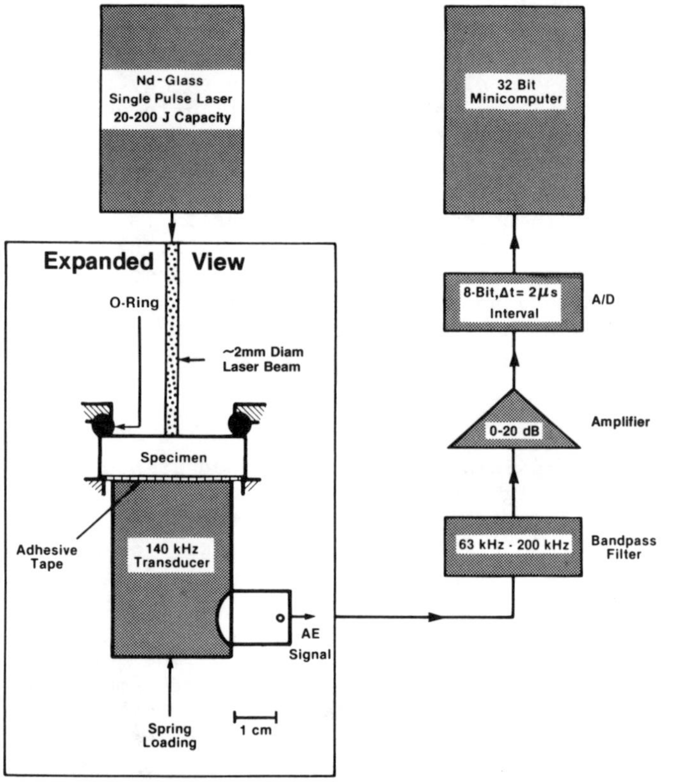

Fig. 2 Schematic of laser drilling and acoustic emission monitoring apparatus.

pulse energy density is reported in relative units (the square of the discharge voltage [5 to 8 kV] of the laser power supply divided by the square of the beam radius in mm); this is proportional to the incident energy density of the laser pulse itself). The average power density was determined from the Gaussian pulse half-width (1 ms). The power density for onset of drilling is a well-defined material constant[4]

(8.4×10^5 J/m^2 for Al-4.5%Cu) and can be used to calibrate the relative energy density measurements to obtain an absolute value of absorbed pulse energy (J/m^2).

Specimens of 2024-T4 aluminum were machined from plate material with dimensions of 2.5 x 2.5 x 0.63 cm. To produce a more standardized optical coupling with the laser, specimens were sandblasted, then cleaned with water and acetone.

ACOUSTIC EMISSION--For acoustic emission measurements, a low viscosity oil couplant was used between the specimen and a resonant type piezoelectric acoustic emission transducer (Fig. 2). Reverberation of the specimen can obscure closely spaced AE events and was reduced by plastic adhesive tape on the specimen underside. A typical drill pulse resulted in an electrical signal from the piezoelectric transducer prior to recording of the order of 1 volt and required no additional amplification. It was high-pass filtered at 63 kHz to eliminate low frequency signal components associated with the temporal character of the laser pulse itself. The main resonance of the transducer was at about 140 kHz; low-pass filtering at 200 kHz was used to reduce electronic noise and prevent signal aliasing. The signal was digitally sampled at 8-bits every 2 µs.

The acoustic emission signal energy was calculated from the digital record:

$$E_E = (\Delta t/R) \sum_{i=1}^{m} V_i^2$$

where V_i is the digital voltage at sampling times $i\Delta t$, Δt is the sampling time increment, R is a gain-compensated system constant, and m is the number of sampled points (2048).

CALIBRATION--In order to quantify the acoustic emission measurements, both monopole and dipole calibrations of the experimental system were made. Because of the limited measurement bandwidth, only a fraction of the source energy is in the measured bandwidth (63 to 200 kHz). In the Appendix we show that a number, α can be derived which accounts for differences in the spectra of natural and calibration sources.

The monopole source (i.e., a simple force applied externally to the body) calibration was achieved using the breaking of a mechanical pencil[8]. Its 0.5 mm lead, when broken against the sample, resulted in the release of an average force (determined from a beam balance) of 3.33 N over a period of 0.20 µs. This produced an average acoustic emission energy of $E_E = 2.02$ relative energy units. Thus, the magnitude of a monopole source can be deduced from its acoustic emission using the relation: $F = 2.34\alpha\sqrt{E_E}$ newtons (see Appendix for α). The smallest detectable event under drilling conditions (Fig. 7) had an acoustic emission energy, $E_E = 0.5$, corresponding to a monopole detection threshold of $F = 2.34(0.5)^{1/2} = 1.65$ newtons for a natural source whose time dependence was the same as the calibration ($\alpha = 1$).

Dipole sources are created by events located within the body. The dipole calibration source we have used was an infrared (1.06 µm wavelength) Nd:YAG laser pulse (25 ns halfwidth). It creates a rapid thermoelastic deformation near the sample surface. This, it can be shown[5], can be represented by three orthogonal dipoles whose strength is:

$$D = (3\lambda + 2\mu)\varepsilon\kappa E_L/K \quad (Nm)$$

where, for 2024 Aluminum alloy, $\lambda = 27.2 \times 10^9$ N/m^2 and $\mu = 52.8 \times 10^9$ N/m^2 are the Lamé elastic constants, $\varepsilon = 23 \times 10^{-6}$ K^{-1} is the coefficient of thermal expansion, $\kappa = 7.6 \times 10^{-5}$ m^2/s is the thermal diffusivity, K = 180.6 W/mK is the thermal conductivity, and E_L is the absorbed laser energy per pulse (taken to be 20 percent of incident energy) (J). By plotting measured acoustic emission energy versus absorbed laser energy, the calibration shown in Fig. 3 can be deduced[9]. The result is that the dipole strength of the source in Nm is $D = 0.0135\alpha\sqrt{E_E}$ with a dipole threshold (under drilling conditions, Fig. 7) of 9.54 mNm for a natural source whose time dependence was the same as the calibration source ($\alpha = 1$).

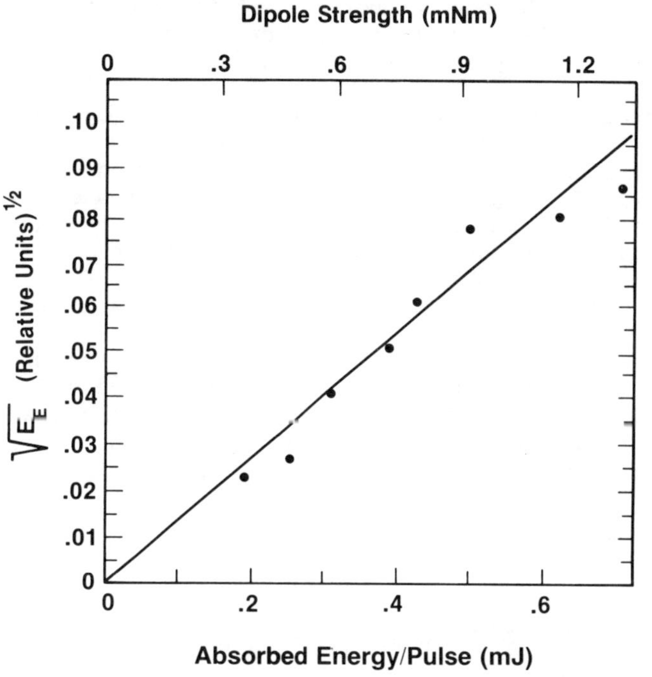

Fig. 3 Calibration of AE system using a pulsed Nd:YAG infrared laser.

RESULTS AND DISCUSSION

Figure 4a shows the cross-section of a typical laser drilled hole in 2024-T4 aluminum using a 1 ms duration pulse of energy density ~ 1.4×10^6 J/m^2. Visible on either side of the hole are lips of resolidified alloy. These

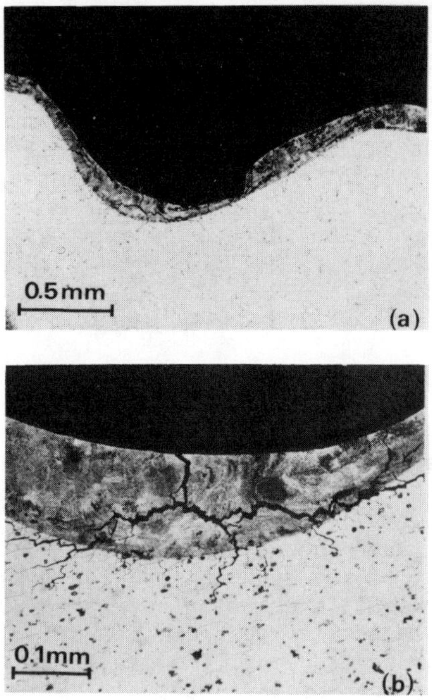

Fig. 4 Optical micrographs of 2024-T4 aluminum alloy, hole produced by ~ 1.4 x 10^6 J/m^2 pulse: (a) a section through a 1 mm deep hole (pulse length 1 ms) and (b) microstructural detail at bottom of hole shown above. Note liquation cracks at prior grain boundaries.

result from liquid metal being extruded up and out of the walls of the hole by the high pressure of vaporized metal (as discussed below). Figure 4b shows an expanded view of a region at the bottom of the hole. Grain boundary "liquation" cracks can be seen extending out from the substrate into hot tears (which form on cooling) in resolidified material at the hole bottom.

Figure 5 shows a typical acoustic emission signal produced by a laser drilling pulse of identical energy. The laser pulse itself had a duration of 1 ms. The remaining portion of the AE signal represents additional AE activity

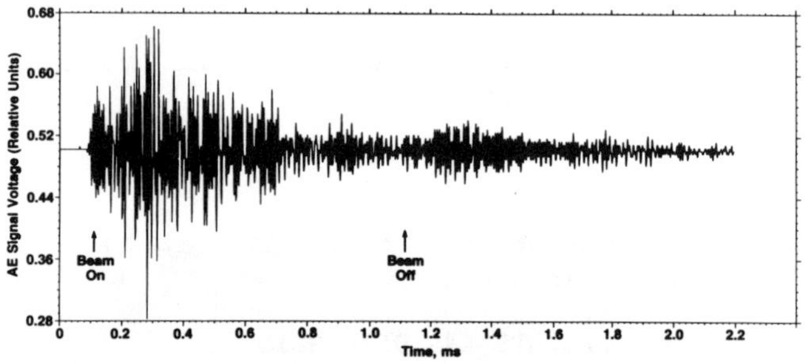

Fig. 5 Typical acoustic emission response to laser drilling of hole similar to that shown in Fig. 5.14. Pulse length = 1 ms (pulse energy density ~ 1.4 x 10^6 J/m^2).

during cooling. This is a crucial observation, since only dipole (internal sources) can be operative, i.e., cracking (hot tearing) and plastic flow during cooling. The liquation cracks observed in Fig. 4b occur predominantly during heating, and remain as significant possible sources of AE in that phase.

In Fig. 6 we plot hole depth versus the energy density of the drilling laser pulse. A great deal of scatter is observed in Fig. 6. It greatly exceeds the experimental uncertainty of determining the hole depth (± 5%) and the discharged laser energy density (± 10%). It is, we believe, a result of the variability of energy transmission to the sample, and the effect of this on hole depth, rather than a measurement error. The best fitting function relating hole depth and laser energy density (determined using least squares regression) was a quadratic function (superior to both linear and sigmoidal functions). The standard deviation of residuals for the best fit was 1.19 mm.

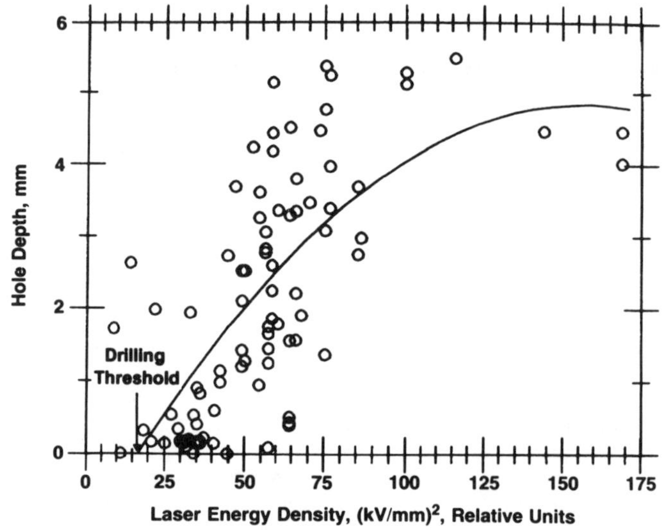

Fig. 6 Laser beam energy density versus depth of laser drilled holes. Quadratic fit. $L = -1.2318 + 0.077188X - 0.00024556X^2$, where L = hole depth (mm) and X = relative laser energy density $(kV/mm)^2$. The standard deviation of the residuals is 1.19 mm.

Figure 7 shows the experimental data for hole depth as a function of the acoustic emission signal energy for each laser pulse. A quadratic curve again was found to be the best fit. The standard deviation of the residuals for this curve (1.09 mm) is slightly better than that of Fig. 6, indicating that acoustic emission data alone is a slightly better method of estimating the hole depth than laser energy density. The variations are again not due to measurement errors alone, since accuracy is an order of magnitude better than the scatter. From the discussion presented below, it is most

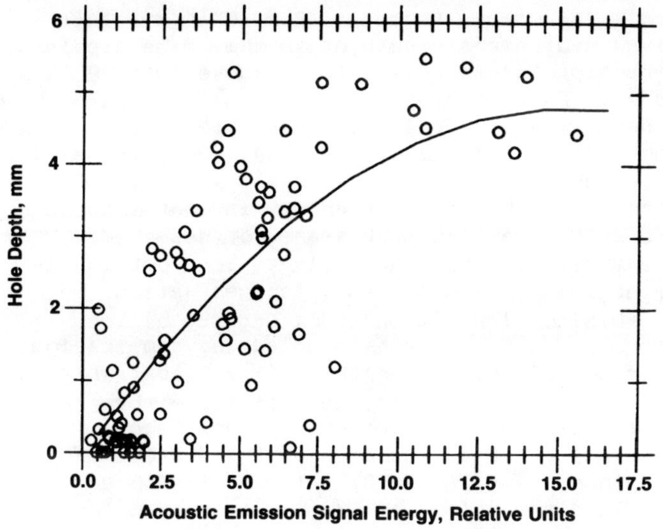

Fig. 7 Acoustic emission energy versus depth of laser drilled holes. Quadratic fit. $L = -0.055547 + 0.62606X - 0.020208X^2$, where L = hole depth (mm) and X = relative AE signal energy. The standard deviation of the residuals is 1.09 mm.

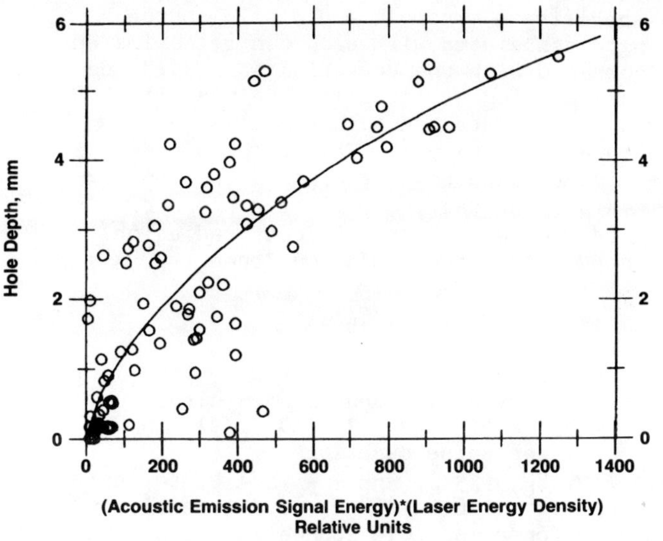

Fig. 8 Hole depth as a function of the parameter X = (acoustic emission energy)*(relative laser energy density). Polynomial fit. L = 0.059350 - 0.0064544X + 0.0444294X**(0.80240), where ** indicates exponentiation. The standard deviation of the residuals is 0.95 mm.

likely caused by irregularities (spikes) in energy generation, the nonlinear (chaotic) nature of plasma formation and liquid eruption, together with the irregular nature of crack growth.

Numerous combinations of acoustic emission and laser energy were tried for correlation with hole depth to produce a linear regression fit with lower residual errors. The best, shown in Fig. 8, is a plot of hole depth versus the parameter [acoustic emission energy]*[laser energy density]. The standard deviation of its residuals is 0.9 mm, an improvement of 25 percent in standard deviation from the plot against laser energy density alone.

The results presented above suggest two potential applications: one is the use of acoustic emission as a potential sensor for feedback control of laser drilling, the second is as a measurement technique to identify the complex sequence of processes that occur during the interaction of directed high energy beams with materials.

PROCESS CONTROL--The results above suggest the possibility of an acoustic emission process control sensor. Acoustic emission energy could be used to determine the depth of penetration (from a previously obtained calibration curve such as that of Fig. 3). Using microprocessor control, this can be achieved in considerably less than the cycle time of the laser. This can then be used to determine the laser pulse energy required to drill the next hole to a specified depth, or to adjust the laser energy if multiple pulses are used to drill at a given site. Perhaps the greatest problem with this simple feedback control approach is the uncertainty associated with determining the hole depth. As we have already seen this does not arise through errors of measurement. Extraordinary care was taken to reduce these to a minimum. Rather, it is believed, the uncertainty arises from the inherent variability of the laser pulse and laser material interactions, and imposes a fundamental limit on sensor accuracy.

DIRECTED HIGH ENERGY BEAM--MATERIAL INTERACTIONS--Directed high energy beams (e.g., electron and laser beams) are becoming increasingly important candidates for advanced materials processing. The high heating/cooling rates and steep temperature gradients make it possible to generate unique microstructures near the surface of net shape components in addition to more traditional welding and drilling applications. This localized short duration method of processing poses serious problems for conventional observation techniques. This has resulted in a rather weak understanding of the basic mechanisms involved. Since the speed of elastic wave propagation is high (6000 ms^{-1}), and acoustic emission of detectable magnitude accompanies microscopic events associated with the energy beam-material interaction, the acoustic emission technique offers interesting possibilities for in situ observation and assessment of some of the mechanisms involved.

Mechanisms that involve abrupt stress changes are the most likely to be detected. These include:

- o Radiation pressure of laser pulse (monopole)
- o Thermoelastic stresses (dipole)
- o Dislocation motion/crack growth (dipole)
- o Evaporation/ablation (monopole)
- o Liquid droplet expulsion (monopole)

The approximate monopole and dipole source strength associated with each can be estimated and compared with the detection thresholds determined earlier to establish the most likely origin of the detected acoustic emission.

Radiation Pressure. This is considered to give rise to undetectable signals in these experiments. The maximum radiation pressure is produced when all of the incident radiation is reflected. Then the radiation force is

$$F = 2E_L/c\tau$$

where

E_L = energy of incident laser pulse
c = speed of light (3×10^8 m/s)
τ = laser pulse duration

Typically, $E_L \sim 25$ J, and $\tau \sim 10^{-3}$ s, so that the force is $\sim 1.6 \times 10^{-4}$ N. Our detection threshold for a monopole source is 1.65α. Here $t_s \sim 0.5$ ms, $\omega^2 t_s^2 >> 1$ (see Appendix) $\alpha = 71.4$ so that detection threshold is 117 N. This process then makes a negligible contribution to the detected AE.

Thermoelastic Stresses. We believe this process is also undetectable in these experiments. The rise in subsurface temperature causes a thermal expansion which, because of surrounding elastic constraint, creates an internal stress. The laser pulse producing this temperature rise is approximately Gaussian, with a 1 ms duration. It also contains numerous power spikes, a characteristic feature of Nd:Glass lasers. Most of the energy can be considered deposited over a 1 ms duration so that most of the thermal stress spectrum produced will be at ≤ 1 kHz (i.e., 1/1 ms or lower). This is below the bandpass of the system (Fig. 2) and will not be detected. Only the small fraction of energy in the shorter duration (higher frequency) power spikes can create thermal stresses in the observation passband. We can only predict their contribution if we know how much energy they transmit to the sample.

One does not see pulse spikes of large size in the AE signal, which argues that they are not significant is a source of thermal stresses.

However, let us assume, as the worst case, that all of the energy is in the form of a single pulse with zero risetime. To calculate the associated dipole strength, we need to determine the depth of heating, and the rise in temperature. For a typical absorbed pulse energy of 10 J over 1 ms, heat flow theory[6] predicts a delay of ~ 13 μs for aluminum to reach its melting point. With a thermal diffusivity of $\sim 8 \times 10^{-5}$ m^2s^{-1}, in 13 μs heating extends ~ 33 μm below the surface. The thermoelastic source then can be approximated as a region 33 μm thick and 10^{-6} m^2 in area undergoing a mean temperature change of about half that required to cause melting (~ 300 K). This would give rise to a dipole strength $\sim (3\lambda + 2\mu)\varepsilon V\Delta T$, where λ, μ are Lamé constants (53 and 27 GPa, respectively, for 2024 aluminum), ε the thermal expansion coefficient (23×10^{-6} K^{-1}) and V the volume heated $\sim 33 \times 10^{-12}$ m^3. This gives a dipole strength of 48 mNm. The dipole detection threshold was found above to be $9.54 \times 10^{-3}\alpha$. In this case $t_s = 13$ μs so that (see Appendix) $\alpha = 1$ and the detection threshold is 9.5 mNm. Thus the event is barely detectable. And, since in reality, only a few percent of the incident energy is contained in the spikes, and AE spikes are not detected, thermoelasticity is unlikely to be a detectable mechanism of acoustic emission generation.

Dislocation Motion/Crack Growth. Dislocations and cracks can be modeled[5] as combinations of dipoles whose strength is $(\lambda + 2\mu)bA$, where b is defect plane displacement (i.e., either the dislocation Burgers vector or crack face opening) and A is the area of the defect. For aluminum, $(\lambda + 2\mu) = 107 \times 10^9$ N/m^2, for a single dislocation, $b \sim 3 \times 10^{-10}$ m, and $A \sim 10^{-10}$ m^2 (10 μm loop). Thus, the dipole strength for a single dislocation is $2\ 10^{-8}$ Nm. Since the detection threshold is 9.54×10^{-3} Nm (if $\alpha \sim 1$) about 5×10^5 dislocations must move cooperatively to give a detectable signal and this is unrealistically large.

Cracking, (both of the liquation and hot tearing types) however, because of the large crack face displacements, produces more energetic emission. During drilling, a compressive thermoelastic stress field is produced at the hole bottom, which is bounded by tensile stresses, the cause of cracking. Instead of a single crack opening at the drilling velocity, craking, especially of an intergranular type, is discontinuous and made up of many rapid small events not unlike a zipper. Examination of Fig. 4 indicates that a typical increment has a grain-sized face area of ~ (25 μm) x (25 μm) and an opening of ~ 1 μm. Thus, the dipole strength of a single crack increment is on the order of 6.68×10^{-5} Nm.

If these open at ~ 50% shear velocity, then $t_s \cong 20$ ns and $\alpha = 1$. Since the dipole detection threshold is then 9.53 mNm, it is clear that individual events would not be detected. Metallography indicated that several hundred of these cracks form during drilling. Their combined signals could approach the detection threshold and may account for the weak signals following pulsing.

Evaporation/Ablation--Evaporation or ablation of material could possibly make a significant contribution to the AE, even though only a small quantity of material is evaporated, due to the high anticipated plasma pressures. If we estimate a maximum pressure of 10 atmospheres, there would be a resultant (monopole) force of 3.1 N exerted on the base of a 2 mm diameter hole. The risetime of the source would be about that of the laser pulse (0.5 ms). Then $\alpha = 71.4$ so that the monopole detection threshold is 107 N. This mechanism therefore makes a negligible contribution to the detected AE.

Liquid Expulsion--Figure 9 schematically illustrates liquid expulsion during drilling.

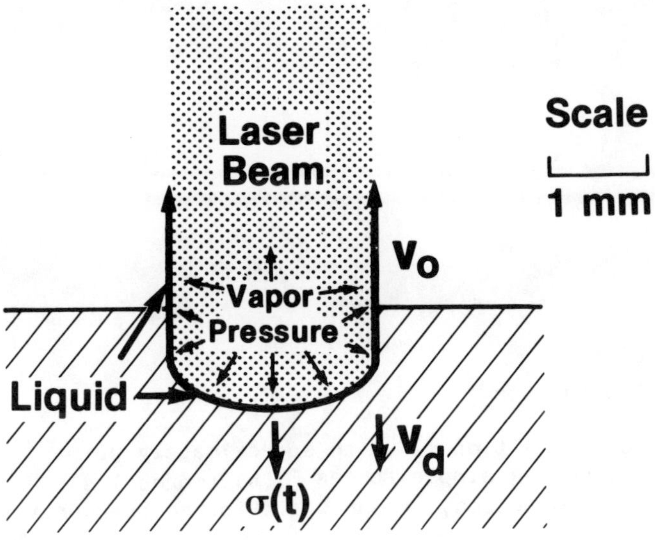

Fig. 9 Model of laser hole drilling mechanism.

Liquid metal at the bottom of the hole is anticipated to expell around the rim by the high vapor pressure of ablated material. This produces a monopole reactive force on the sample due to the momentum imparted to the liquid droplet.

If v_d is the velocity of drilling and A is the circular area of the hole, then the amount of liquid produced in a time dt will be $\rho A v_d dt$, where ρ is the material density. Because of the thinning of liquid on the walls due to vapor pressure, the velocity v_o of the expelled liquid can be expected to be higher than v_d, perhaps of the order of the speed of sound in air. On the average, the force produced with the expulsion of liquid is

$$F(t) = d(mv_o)/dt = (\rho A v_d dt) v_o/dt = \rho A v_d v_o.$$

For a 1 mm hole radius, with $\rho = 2700\ kg/m^3$, $v_d = 3$ m/s, and $v_o \geq 330$ m/s (the acoustic velocity in air) we obtain a force of ≥ 8.2 N. It is difficult to accurately predict the risetime of this source. Taking a typical hole depth of 3 mm, we obtain $t_s \leq 10\ \mu s$ if we assume the droplets have a speed of $\geq 330\ ms^{-1}$. This gives $\alpha = 1$ so that the monopole detection threshold $F = 1.65\ \alpha = 1.65$ N. Thus, liquid expulsion could account for a significant fraction of the detached acoustic emission. Furthermore, since droplet expulsion can be irregular and is likely to be tied to beam power spikes, it could account for the irregularities in AE, and for the observation that the emission increased with the hole depth.

CONCLUSIONS

Acoustic emission, measured passively during laser drilling, has been found to provide a basis for determining the depth of a hole produced by a laser drilling pulse. The technique is sufficiently fast that it could serve the role of a sensor in a closed-loop feedback control system for laser drilling. Its accuracy, while better than that from pulse energy predictions, is not ideal. It is believed to be limited by the inherent stochastic nature of the processes involved. The physical processes involved in laser drilling are complex and not at all well understood.

Acoustic emission can be used to shed some light on this. The emission seems to originate mainly from liquid droplet expulsion. The contribution of crack growth is secondary and will vary with the metallurgical condition of the material being drilled (for example, its heat treatment and prior deformation[7]). High purity aluminum is not expected to crack[7], but would produce substantially the same liquid expulsion. Liquid expulsion will not be dependent on substrate conditions. Thus, acoustic emission, coupled with systematic microstructural variation and metallography, may provide an enlightening means for separating these different phenomena and advancing the study of the micromechanics of laser drilling in particular and of directed energy beam-material interactions in general.

ACKNOWLEDGMENTS

This work was supported by the Defense Advanced Research Projects Agency under DARPA Order No. 4275. The project monitors for this work were Lt. Col. J. Jacobsen and Capt. S. Wax. We wish to thank Dr. C. Spiegelman for assistance in linear regression calculations and C. Brady for metallographic assistance.

REFERENCES

1. "Applications of Lasers in Materials Processing," Second Conference, E. Metzbower, Ed., ASM, Metals Park, OH (1984).

2. Mehrabian, R., Int. Metals Rev. 27, 185 (1982).

3. Rykalin, N., A. Uglov, and A. Kokord, "Laser Machining and Welding," O. Glebov, Translator, Pergamon Press, NY (1984).

4. Sekhar, J. A., S. Kou, and R. Mehrabian, Met. Trans. A, 14A, 1169 (1983).

5. Wadley, H., J. A. Simmons, and C. Turner, "Predictive Modeling of Quantitative Acoustic Emission Waveforms," **Rev. of Prog. in Quant. Nondestructive Evaluation**, Vol. 3B, D. Thompson and D. Chimenti, Eds., Plenum Press, NY 683-697 (1984).

6. Hsu, S. C., S. Kou, and R. Mehrabian, Met. Trans. B, 11B, 29 (1980).

7. Clough, R. B., F. S. Biancaniello, and H. N. G. Wadley, "Acoustic Emission Studies of Electron Beam Melting and Rapid Solidification," in **Applications and Development of NDE for Use in Materials Processing,** O. Buck and S. Wolf, Eds., ASM, Metals Park, OH (1984).

8. Hsu, N. N., J. A. Simmons, and S. C. Hardy, Materials Evaluation, 35(10), 100 (1977).

9. Wadley, H. N. G., and R. B. Clough, "Scalar Acoustic Emission Calibration," submitted for publication.

APPENDIX
Source--Signal Relationships

In experiments, the voltage, V(t), has been measured at position r due to approximately point sources at location r'. These sources can be modelled either as simple forces $F_i(t)$ (monopoles) or force couples $D_{ij}(t)$ (Dipoles). Using the transfer function approach, we can write in the frequency domain:

$$V(r,r',\omega) = \begin{cases} T_m(\omega)\ A(\omega)\ G_{mi}(r,r',\omega)\ F_i(\omega) & \text{for monopole sources} \\ T_m(\omega)\ A(\omega)\ G_{mi,j'}(r,r',\omega)\ D_{ij}(\omega) & \text{for dipole sources} \end{cases}$$

where $G_{mi}(r,r',\omega)$ is the Heaviside Green's tensor representing the displacement at r in direction χ_m due to a simple force step at r' applied in direction χ_i. The notation $_{,j'}$ is used for spatial differentiation with respect to the χ_j'-th component. $T_m(\omega)$ is the transducer (and associated instrumentation) transfer function relating displacement in direction χ_m to scalar voltage. $A(\omega)$ accounts for the absorption of ultrasonic amplitude by internal friction and other loss mechanisms.

In our experiments, we have calculated an acoustic emission energy by integrating the square of the voltage measured in the frequency band ω_o-ω_1. Thus the quantity of interest is $V(\omega)V^*(\omega)$ where $V^*(\omega)$ is the complex conjugate of $V(\omega)$.

Consider first the term $A(\omega)A^*(\omega)$ common to both expressions. From experiments, it is known that the signal decays exponentially with the form:

$$A(t) = (1 - H(t))A_o\ e^{-t/ta}$$

where t_a is a relaxation time and H(t) the unit Heaviside function. Taking Fourier transforms we find:

$$A(\omega)A^*(\omega) = \frac{A_o^2}{\omega^2(1 + t_a^2\omega^2)} \quad \text{for} \quad \omega \neq 0$$

The constant A_o is obtainable by calibration.

Now consider the forms of the monopole and dipole Green's tensor. It can be shown that above dc their spectra are dominated by $1/\omega$ terms for the monopole, while dipole spectra frequency is independent in the far field. Since the directional character of the source is lost after multiple reflections in a bounded sample, we can write:

$$G_{mi}(\omega)G^*_{mi}(\omega) \cong \frac{G^2_{mono}}{\omega^2} \qquad \text{for monopole sources}$$

and

$$G_{mi,j'}(\omega)\ G^*_{mi,j'}(\omega) \cong G^2_{dip} \qquad \text{for dipole sources}$$

where G_{mono} and G_{dip} are constants experimentally obtainable.

Finally, consider the likely temporal form of the monopole (magnitude F_o) and dipole (D_o) sources. If we assume these have a variable risetime t_s:

$$F(t) = H(t)\ F_o(1 - e^{-t/t_s}) \qquad \text{for monopole sources}$$

and

$$D(t) = H(t)\, D_o(1 - e^{-t/t_s}) \qquad \text{for dipole sources}$$

Thus, for $\omega \neq 0$:

$$F_i(\omega)F_i^*(\omega) = \frac{F_o^2}{(t_s^2\omega^2 + 1)\omega^2}$$

and

$$D_{ij}(\omega)D_{ij}^*(\omega) = \frac{D_o^2}{(t_s^2\omega^2 + 1)\omega^2}$$

Assuming $T_m(\omega)$ to be a constant T_o between ω_o and ω_1 and zero otherwise we obtain, for the measured energy: for $\omega \neq 0$,

$$E_E = \int_{\omega_o}^{\omega_1} V(\omega)V^*(\omega)d\omega = \begin{cases} T_o^2 A_o^2 \; G_{mono}^2 \; F_o^2 \; \phi_{mono}(t_s) \\ T_o^2 A_o^2 \; G_{dip}^2 \; D_o^2 \; \phi_{dip}(t_s) \end{cases}$$

where

$$\phi_{mono}(t_s) = \int_{\omega_o}^{\omega_1} \frac{d\omega}{(1 + t_a^2\omega^2)(1 + t_s^2\omega^2)\omega^6}$$

and

$$\phi_{dip}(t_s) = \int_{\omega_o}^{\omega_1} \frac{d\omega}{(1 + t_a^2\omega^2)(1 + t_s^2\omega^2)\omega^4}$$

Typically, $t_a > 1$ ms and $t_s < 1$ μs and 63 kHz $\leq \omega/2\pi \leq$ 200 kHz so $t_a^2\omega^2 >> 1$ and $t_s^2\omega^2 << 1$. Thus, for abrupt sources:

$$\phi_{mono}(t_s) = \int_{\omega_o}^{\omega_1} \frac{d\omega}{t_a^2\omega^8} = \frac{1}{7t_a^2}\left(\frac{1}{\omega_o^7} - \frac{1}{\omega_1^7}\right) \sim \frac{1}{7t_a^2\omega_o^7}$$

and

$$\phi_{dip}(t_s) = \int_{\omega_o}^{\omega_1} \frac{d\omega}{t_a^2\omega^6} = \frac{1}{5t_a^2}\left(\frac{1}{\omega_o^5} - \frac{1}{\omega_1^5}\right) \sim \frac{1}{5t_a^2\omega_o^5}$$

For slower sources such that $\omega^2 t_s^2 >> 1$ and $\omega^2 t_a^2 >> 1$, we have:

$$\phi_{mono}(t_s) = \int_{\omega_o}^{\omega_1} \frac{d\omega}{t_a^2 t_s^2\omega^{10}} = \frac{1}{9t_a^2t_s^2}\left(\frac{1}{\omega_o^9} - \frac{1}{\omega_1^9}\right) \sim \frac{1}{9t_a^2t_s^2\omega_o^9}$$

and

$$\phi_{dip}(t_s) = \int_{\omega_o}^{\omega_1} \frac{d\omega}{t_a^2 t_s^2\omega^8} = \frac{1}{7t_a^2t_s^2}\left(\frac{1}{\omega_o^7} - \frac{1}{\omega_1^7}\right) \sim \frac{1}{7t_a^2t_s^2\omega_o^7}$$

For slower sources the risetime of the source becomes significant.

If sources of known monopole or dipole strengths (F_o or D_o) and risetimes (t_s) are caused to occur in a sample, then all the parameters that relate the magnitude to the measured energy can be determined. This calibration would then be valid for unknown source strengths provided it was known that they were either monopole or dipole in character and their time dependence was the same as the calibrating source. Should the unknown sources have a different risetime t_s to that of the calibration source the final expressions above provide an approximate scaling factor $\alpha = \sqrt{\phi_{cal}(t_{cal})/\phi_s(t_s)}$ so that we may write:

$$F_s \equiv \alpha F_{cal} \quad \text{monopoles}$$

$$D_s \equiv \alpha D_{cal} \quad \text{dipoles}$$

for using system calibration. Interestingly, we note that by taking the ratio of ϕs, the system decay time constant t_a cancels out, so that the results are independent of this.

REAL TIME IDENTIFICATION OF MATERIAL CONSTITUTIVE RELATIONSHIPS DURING METAL FORMING

Kim A. Stelson
University of Minnesota
Minneapolis, MN 55455

ABSTRACT

Automated machines can be equipped with sensors to detect variations in workpiece material properties that can cause unacceptable variations in the final product. The machine can act as a material testing machine during the early stages of deformation and "learn" the properties of the workpiece. This information can subsequently be used to control the process or for quality assurance. The technique has been applied to pressbrake bending and upset forging.

INTRODUCTION

Considerable progress has been made toward the goal of a flexibly automated unmanned factory. Such a factory could make small lots of parts on general purpose fabrication machinery. The increased flexibility caused by software automation would enable such a factory to directly manufacture custom parts designed on a computer-aided design (CAD) system. By reducing costly design and development costs as well as direct labor and specialized tooling and equipment requirements, a computer integrated manufacturing (CIM) system can bring the economies of scale of mass production to smaller lots of parts. Such a computer integrated manufacturing system could respond to the ever increasing demand for new products that must be produced quickly to compete effectively in markets characterized by rapid technological change.

In spite of the complexities and some remaining fundamental difficulties, unmanned CIM systems for metal cutting are being rapidly developed. Comparable systems are not being realized for metal forming. As an example, consider the highly automated Yamazaki machine tool plant in Florence, Kentucky. The metal cutting operations in this plant are completely unmanned, but the sheet metal forming operations are done in the traditional "one man/one machine" manner. The reason for the discrepancy between the development of CIM systems for machining and forming is not that unmanned flexible automation is undesirable for metal forming; it is simply not possible at the present time.

The almost exclusive reliance on machining as the fabrication technology for computer integrated manufacturing is unfortunate. Forming has a number of advantages of cutting. Little or no scrap is produced in metal forming processes. In some cases, desirable mechanical properties are imparted to the workpiece by strain hardening. Metal cutting can be seen to be a serial (one point at a time) interface for shape information transfer from the tool to the workpiece, and metal forming can be seen to be a parallel (many points at one time) interface for shape information transfer from the tooling to the workpiece. As anyone with computer interfacing experience knows, all other things being equal, parallel data transmission is faster than serial data transmission, and by the same reasoning, metal forming requires less machine time than metal cutting.

This brings us to the fundamental question: If metal forming is such an attractive fabrication process, why isn't it being more vigorously pursued as a candidate for flexibly automated manufacturing? The most common answer to this question is that metal forming processes often require specialized tooling, making them too expensive for small lot fabrication. But this answer is only valid if one insists on being able to make any conceivable part on a flexibly automated system. Many metal forming processes require no specialized tooling. Moreover, manufacturing cells composed of a number of these processes could make many (but, unfortunately, not all) useful parts. For example, a sheet metal fabrication cell consisting of automated shearing, plasma or laser cutting, pressbrake forming, punching, and spot welding could make many useful sheet metal parts.

If such a combination of processes is attractive, why has it not been more vigorously pursued? The answer is that the metal forming process for this cell (pressbrake forming), as currently practiced, is not sufficiently well controlled to permit unmanned operation. (Incidently, some of the other processes have control problems as well.) The source of the problem is that variations in the material properties and frictional characteristics of the incoming material stock are so large that unacceptable variations in the final shape of the product result. Maintaining very precise control of the properties of the stock is likely to be prohibitively expensive, if possible at all. However, an alternative approach is to equip automated forming machines with sensors to detect the variations in workpiece material properties that cause unacceptable variations in the final product. The machine can act as a material testing machine during the early stages of deformation and "learn" the properties of the workpiece. This information can subsequently be used to control the process or for quality assurance.

The purpose of this paper is to review recent efforts to use in-process measurement from metal forming processes to control the process so that uniform results are obtained in the presence of substantial variations in the material, geometric, and frictional properties of the workpiece. The closed-loop control of pressbrake bending and upset forging will be examined as examples.

MODELLING AND IDENTIFICATION

Most efforts in metal forming analysis have concerned themselves with the problem of modelling rather than the problem of identification. Although modelling and identification may concern the same process under the same assumptions, because different variables are assumed to be known or unknown, a different approach may be needed. Because the known and unknown variables are interchanged, the identification problem is also known as the "inverse" modelling problem. In modelling, one assumes that the initial configuration, constitutive relationship, and boundary conditions of a particular body that will be deformed are all known. The desired solution is the response of the body. The desired response is usually expressed in terms of variables that can be measured such as forces, displacements, or velocities. In identification, the response of the body is known, but the constitutive relationship or the boundary conditions are assumed to be unknown. The desired solution is the unknown constitutive relationship or boundary condition. Thus, the technique of using in-process measurements (the response) to find unknown material properties (constitutive relationships) or frictional conditions (boundary conditions) can be seen to be an identification problem.

Identification problems are, in general, much more difficult to solve than modelling problems. Established techniques exist for the treatment of identification problems for linear dynamic systems [1]. However, these techniques cannot be used with the geometric and material nonlinearities associated with the plastic deformation of metals. Modelling problems can be formulated in terms of ordinary or partial differential equations, and well established theorems on existence and uniqueness can indicate when a modelling problem is well posed, even when the problem is nonlinear. The notion of a well posed modelling problem can also be understood on the basis of physical intuition. Unfortunately, the notion of a well posed identification problem is not clearly understood, either formally or intuitively, and there are no clear guidelines to indicate, for example, how many measurements would be needed to determine an unknown boundary condition or constitutive relationship.

Even if an identification procedure is well posed, experimental and computational difficulties are more likely to arise with identification than with modelling. Modelling involves the integration of differential equations, and integration tends to smooth any high frequency noise present in experimental data. Conversely, because it is the "inverse" of modelling, identification involves differentiation, and differentiation tends to amplify any noise present. Extreme care must be taken when smoothing experimental measurements used for the purposes of identification so that unwanted noise is rejected without losing significant information.

As another example of some of the difficulties encountered in identification, as opposed to modelling, consider the finite element method, a widely used numerical method for modelling deformation problems (see, for example, [2]). Although the finite element method is most widely applied to small strain, linear elastic problems, it can be generalized to account for the geometric and material nonlinearities encountered in metal forming. In spite of its wide use to model metal deformation, the finite element method is rarely used for identification. An exception is the work of Manahan [3]. One reason for the lack of use of the finite element method for the "inverse" problem is that the computational costs become prohibitive. The "inverse" problem can only be solved by iteratively solving a large number of "forward" problems. At each iteration, the values of the unknown parameters are altered until, it is hoped, convergence occurs. Because of the computational costs of solving nonlinear problems using the finite element method, investigators often are forced to use less powerful but more economical methods. The computational requirements are even greater for identification problems.

This is not to say that identification techniques have not been used successfully to identify unknown constitutive relationships in complex loading situations. Nadai has shown how to identify an unknown stress-strain law by using torque-twist data from a torsion test on a cylinder [4], or by using moment-curvature data from a bending test on a beam [5]. Bridgman [6] has shown how to estimate post necking stress-strain behavior by using shape and force data from a tensile test, and Kobayashi and Thomsen [7] have measured the shear stress on the shear plane from force measurements taken during machining.

All of these well known examples were developed before computers were widely used for the real time closed-loop control of manufacturing processes, and were intended primarily as tools to compare the constitutive relationship found from a uniaxial tensile test to the constitutive relationship found in more complicated loading situations. With the current pervasive use of computers for the control of manufacturing processes, however, these classical approaches take on a greater significance, because they provide a basis for real time identification of workpiece constitutive relationships. Therefore, a solid physical foundation is provided to use in-process measurement for closed-loop adaptive control of deformation processes in the presence of frictional and material property variations. Recent applications of these concepts for the closed-loop control of pressbrake bending and upset forging are described in the following sections.

PRESSBRAKE BENDING

Pressbrake bending is one of the most common sheet metal forming operations. The process is an ideal candidate for flexible automation, because a reasonable range of bend angles, thicknesses, and materials can be handled with one set of tooling. However, as currently practiced, pressbrake bending cannot be satisfactorily automated. The lot to lot variations in thickness and material properties of commercial sheet metal products cause an unacceptable variation in the final angle of the bends. Therefore, manual operation is necessary to set up the machine for each lot.

Recent developments in the adaptive control of pressbrake bending could lead to a commercial pressbrake that can automatically control the final bend angle by identifying the thickness and material properties of each part. Such a pressbrake is capable of automated set up, even when significant material and thickness variations are present. Although such a machine has a number of advantages for stand alone operation, its main contribution would be that it would make possible an unmanned, flexibly automated, sheet metal fabrication cell.

By measuring the punch force during forming, it is possible to use the pressbrake as a materials testing machine. Thus, the thickness and material property variations can be detected from the early part of the bend cycle. This data can be used to control the punch reversal position of the press so that the desired final bend angle results. Stelson and Gossard [8] have shown how to implement such a real time identification and control scheme for an elastic-plastic material with no strain-hardening. Stelson [9] has further extended the concept for an arbitrariy stress-strain curve. Based on this identification method, an elastic, power-law, strain-hardening, adaptive pressbrake controller has been demonstrated [10]. The identification and control algorithms are simple and fast enough that they can be implemented in a high level language such as FORTRAN or PASCAL on a microprocessor without modifying the forming rate of the press.

A comparison between the identification equations of Nadai [5] and Stelson [9] reveals that the methods are analogous. In other words, the equations that Nadai derived to find the stress-strain relationship from the moment-curvature relationship have an identical structure to the equations Stelson derived to find the moment-curvature relationship from the force-displacement relationship. The analogy is in the form of a dual since the generalized forces and velocities are switched. The reason for the analogy can be seen physically: As bending progresses, Nadai calculates the change in bending moment as stresses propagate inward from the edge of the sheet. Similarly, Stelson calculates the change in displacement as curvature propagates outward from beneath the punch.

In order to control the press, the identification procedure, which calculates the material properties of each sheet, must be combined with a bending model, which calculates the required punch reversal position to give a desired unloaded bend angle. The model must be accurate enough to account for all relevant effects, but simple enough to allow real time calculation without interfering with the production rate of the pressbrake. Relevant effects include permanent deformation and springback in both the punch wraparound and the free section of the sheet as well as rolling and sliding phenomena at the die sheet interface. As was mentioned in the previous section, the finite element is a powerful but computationally intensive approach to modelling metal deformation. With currently available hardware and software, it is not possible to use the finite element models such as [11] for the real time control of pressbrake bending. Thus, the necessity of developing simple yet accurate models for the real time control of deformation processes is apparent.

UPSET FORGING

Upset forging is another process where the final result is influenced by variations in material and frictional properties. In-process sensing of these properties can be used to avoid fracture. A cylindrical billet is often upset or squeezed between two plattens as a preliminary step toward making a finished part. Upsetting can be done either hot or old, and a number of upsetting steps with annealing between each is usually necessary to achieve sufficient shape change while avoiding fracture. As currently practiced, many unnecessary steps may be performed in the process, but if the capability to predict fracture before it occurs could be demonstrated, the number of upsetting steps can be minimized. It must be emphasized that the point of fracture is dependent on the material frictional properties of the workpiece, and billet to billet variations in these properties can significantly alter the point where fracture occurs.

The need for simple but accurate models of axisymmetric upset forging is apparent, but a similar dilemma as in the case of pressbrake bending exists: If a model is simple enough to be implemented in real time, it often neglects important effects. On the other hand, accurate but computationally intensive methods cannot be utilized in real time. In the case of upsetting of cylinders, the slab method of Sachs [12] and Siebel [13] provides a simple method to estimate the forming force including friction effects. The method neglects inhomogeneous deformation that exhibits itself as bulging when friction is present. This assumption makes the model inaccurate for height to diameter ratios greater than about 0.4. Unfortunately, this is the regime of greatest technological interest. Computationally intensive schemes such as the finite element method [14,15] or the matrix method [16] are accurate but too slow for real time control.

In an effort to develop a simple model that could take inhomogeneous deformation into account, both Saluja, et al., [17] and Ettouney and Hardt [18] developed the same model independently. This approximate analysis is similar to the analysis of the stresses in the neck of a tensile specimen due to Bridgman [6] except that in upset forging the axial stress is compressive rather than tensile. The bulging is outward in upset forging whereas the bulging is inward in the neck of a tensile specimen. With appropriate changes in sign due to the differing signs of the stresses and radii, the two analyses can be seen to be identical. Ettouney and Hardt demonstrated that the method accurately predicts the material stress-strain relationship from the measured punch force-displacement curve by comparing their results to the modified Bridgman test [19], a standard test to determine the compressive stress-strain curve where the specimen is frequently remachined during the test to eliminate the effects of inhomogeneous deformation.

Although the Bridgman type correction factor for the upsetting of cylinders [17,18] corrects for the inhomogeneous deformation (bulging) that is caused by friction, the coefficient of friction does not appear in the equations. The effect of friction manifests itself indirectly through the part geometry. Therefore, if one measures the part geometry and the force-displacement curve during upsetting, the stress-strain curve can be found, but the coefficient of friction remains unknown. In an attempt to estimate the coefficient of friction, as well as the stress-strain law during upsetting, Ettouney and Stelson [20] have developed a modified slab method. The method replaces the normally straight slabs with curved slabs to account for the inhomogeneous deformation. Because the stresses at the die-workpiece interface appear in the equilibrium equations, the effect of friction is explicitly shown. In contrast, because the equilibrium equations for the Bridgman type correction factor involve only stresses at the equator, the friction does not appear explicitly.

Using these analytical tools, one can visualize an upset forging press equipped with force, displacement, and shape (solid-state camera) measurement. The stress-strain curve can be found using a Bridgman type correction factor. The coefficient of friction can be found by equating the forming force found using the Bridgman type correction factor with that found from the modified slab method. This information can be combined with a fracture theory to predict failure. As was shown in [18], Thomason's theory of plastic instability is an effective predictor of fracture before it occurs [21,22]. The theory is an extension of the well known fact that a tensile test of a ductile material exhibits plastic tensile instability (necking) before fracture occurs. In the case of upset forging, the tensile hoop stress at the equator of the bulged specimen causes the instability that proceeds fracture.

SUMMARY AND CONCLUSIONS

The previous sections have concerned themselves with two examples of the automatic control of metal forming: pressbrake bending and upset forging. Similar techniques have been applied to other processes, for example, roller bending [23] and straightening [24]. The examples exhibit characteristics that one can expect to find as one attempts to automate other metal forming processes, namely, that variations in workpiece material, geometry, and frictional characteristics prohibit the use of open-loop control for flexible automation. Since, for many applications, metal forming has a number of advantages over metal cutting, one would wish to overcome these barriers to acceptable flexible automation of metal forming.

In an attempt to overcome these difficulties, a number of processes have been instrumented for closed-loop control. Measurements taken from the process can be used to identify the unknown constitutive, frictional of geometric properties. Applicable techniques for using in-process measurements to find these unknown properties are not straightforward, since the problem involves identification as opposed to modelling. Further, although the finite element method is a powerful tool for modelling the deformation of metals, it cannot currently be used for real time control, nor is it easy to invert the method to perform the reverse modelling or identification procedure. However, classical techniques have been developed for this identification of workpiece constitutive laws from in-process measurement 4-6 . These techniques provide the basis for identifying workpiece material characteristics from brakeforming 9 and axisymmetric upset forging 17,18 . Endeavors along similar lines are currently being pursued by the author and others in an attempt to realize reliable, flexible automation of metal forming processes.

REFERENCES

1. Graupe, D., "Identification of Systems," Van Nostrand Reinhold, New York (1972).
2. Armen, H. and R. F. Jones, Jr., eds., "Applications of Numerical Methods to Forming Processes," A.S.M.E., New York (1978).
3. Manahan, M. P., Nuclear Tech., 63, pp. 295-315 (1983).
4. Nadai, A., "Theory of Flow and Fracture of Solids," Vol. 1, pp. 347-352, McGraw-Hill, New York (1950).
5. Ibid., pp. 356-359.
6. Bridgman, P. W., "Studies in Large Plastic Flow and Fracture," pp. 9-86, McGraw-Hill, New York (1952).
7. Kobayashi, S. and E. G. Thomsen, Trans. of A.S.M.E., J. of Engineering for Industry, 81, pp. 251-262 (1959).
8. Stelson, K. A. and D. C. Gossard, Trans. of A.S.M.E., J. of Engineering for Industry, 104, pp. 389-393 (1982).
9. Stelson, K. A., Trans. of A.S.M.E., J. of Engineering for Industry, 105, pp. 45-53 (1983).
10. Stelson, K. A., "An Adaptive Pressbrake Control for Strain-Hardening Materials," in D. E. Hardt and W. J. Book, eds., "Control of Manufacturing Processes and Robotics Systems," pp. 159-170, A.S.M.E., New York (1983).
11. Oh, S. I. and S. Kobayashi, Int. J. of Mech. Sciences, 22, pp. 583-594 (1980).
12. Sachs, G., "Spanlose Formung der Metalle," Springer Verlag, Berlin (1931).
13. Siebel, E., "Die Formgebung in Bildsamen Zustande," Verlag Stahleisen, Duesseldorf (1932).
14. Lee, C. H. and S. Kobayashi, "Analysis of Simple Upsetting of Solid Cylinders by FEM," in S. Kobayashi, ed., "Analytical Prediction of Defect Occurrence in Simple and Complex Forgings," AFMl-TR-69-114B, Air Force Materials Laboratory, WPAFB, Ohio (1959).
15. Nagamatsu, A., T. Murota, and T. Jimma, Bull. of the J.S.M.E., 14, pp. 331 (1971).
16. Lee, C. H. and S. Kobayashi, Trans. of A.S.M.E., J. of Engineering for Industry, 95, pp. 865-873 (1973).
17. Saluja, S. S., P. C. Pandey, and S. Dalela, Ninth North American Manufacturing Research Conference Proceedings, p. 153 (1981).
18. Ettouney, O. and D. E. Hardt, Trans. of A.S.M.E., J. of Engineering for Industry, 105, pp. 161-167 (1983).
19. Brigman, P. W., Trans. for A.S.M., 32, p. 553 (1944).
20. Ettouney, O. and K. A. Stelson, Twelfth North American Manufacturing Research Conference Proceedings, pp. 133-140 (1984).
21. Thomason, P. F., Int. J. of Mech. Sciences, 10, p. 501 (1968).
22. Thomason, P. F., Int. J. mech. Sci., 11, p. 187 (1969).
23. Hardt, D. E., M. A. Roberts, and K. A. Stelson, Trans. of A.S.M.E., J. of Dyn. Sys., Meas., and Control, 104, pp. 317-322 (1982).
24. Lee, M. Y. and K. A. Stelson, "Adaptive Control for a Straightening Process," in K. A. Stelson and L. M. Sweet, eds., "Sensors and Controls for Automated Manufacturing and Robotics," A.S.M.E., New York, 1984, pp. 153-166.

A MICROCOMPUTER BASED SENSOR SYSTEM FOR RESISTANCE WELDING STUDIES

D. G. Waters, K. C. Lee, R. J. Mayhan
The Ohio State University
Columbus, Ohio

D. W. Dickinson
Edison Welding Institute
Columbus, Ohio

Abstract

The hardware and programs for an inexpensive add-on type microcomputer based sensor system designed for use in the study of the short-lived phemonena which occur in resistance welding are described. Three sensors which monitor the electrode voltage, current, and displacement were connected to a 75 kVA, single-phase spot welder. The outputs of the sensors are input to an Apple computer through a multichannel A/D converter and the digitized data are stored in the computer. After a weld, the data can be plotted on the display screen, printer, or strip chart recorder. The data can also be processed according to any desired algorithm to study various characteristics of the weld phenomena.

Typical outputs of the sensors are presented and discussed. Examples of processed data shown include the workpiece dynamic resistance, energy input to the weld, and pre-weld resistance. Finally, some preliminary results of an on-going study of spot welding galvanized steel are given. Here the discussion centers on irregularities in the dynamic resistance curve which occur in the first few cycles of the weld.

FEEDBACK CONTROL has been advocated as one method of maintaining the quality of resistance welding over a wide range of conditions. While a great deal of effort has been expended in this area of research, as yet no one single method or control algorithm has emerged as the definitive candidate for this purpose. It seems evident that before a viable control method can be devised a much better understanding of the metallurgical, mechanical, and electrical phenomena occurring during a spot weld is needed. To aid in achieving such an understanding, more detailed and accurate measurements of these relatively short-lived phenomena are required. Heretofore most measuring apparatus utilized conventional analog techniques and suffered from a major compromise. If the systems were easy and convenient to use they only produced a gross measure of what took place during the weld. On the other hand, if the systems did yield more detail and accuracy, the convenience element was lost, making numerous measurements of the weld process over a wide range of conditions impractical. Because of this it was decided that a computerized, digital technique was needed to allow for convenient and simple measurements with a high level of detail and accuracy.

This paper describes a microcomputer based sensor system which can be used in the study of resistance welding. A 75 kVA, 60 Hz., 240 Volt, single-phase spot welder was equipped with three analog sensors to monitor electrode voltage, current, and displacement. The outputs of the sensors are preconditioned and then input to a multichannel A/D converter and the digitized data are synchronously stored in the microcomputer. After the weld, the raw data can be output on a video display terminal, printer, or conventional strip chart recorder, or stored on a floppy disk for later use. Additionally, the data can be processed according to any desired algorithm to study various characteristics related to the weld process.

The first part of the paper is a description of the system, its hardware and software, and the basis of its operation. Some examples of typical data and the types of processing that can be performed are illustrated. In the second part, data from an on-going series of tests of spot welding of galvanized steel common to the automotive industry are presented. This particular example was chosen because it illustrates a finding that might easily have been overlooked using conventional measurement methods.

SYSTEM DESCRIPTION

The sensor system has three major parts - the sensors and associated interface circuits, the computer system, and the software written to perform various tasks. Since many of the needs of the resistance welding study are not unique, much of the system could be used as a data collection tool in other research areas.

The major component of the system is the Apple II+ personal computer. The Apple has 64 kilobytes of memory on board. As purchased the computer has the programming language BASIC installed, but it is not used at present for the data acquisition software. Instead, the languate Apple PASCAL was acquired and the programs were written in PASCAL. The Apple is also equipped with a video monitor, two disk drives, and a printer.

In addition to the Apple II+, the data acquisition system has the sensors, some analog interface circuits, an analog-to-digital converter system, a pair of digital-toanalog converters, and a strip chart recorder. A more detailed description of most of these components is given later. Fig. 1 shows a block diagram of the complete system. A description of the computer programs is given in another part of this paper.

The system is designed to perform four major activities:

1) data collection,
2) processing data,
3) data storage and retrieval,
4) printing and plotting of data.

Data collection from up to six sources at sampling rates of up to 5000 samples per second per channel can be performed. The use selects both the number of channels and the sampling rate desired. Each data collection includes at least two signals - the weld current and weld voltage - while up to four other channels may be added. A single data collection may contain up to a total of 10240 samples from all channels. Table 1 shows the data storage capacity for different numbers of collected channels and sampling rates. (The storage capacity is given as the number of cycles of the 60 Hz. welding current that can be collected). Since a spot weld is usually completed in less than 30 cycles, a sufficient amount of data can be collected even at the fastest sampling rates.

Processing of the data after collection can include averaging, filtering, doing various calculations, etc. At present three forms of such signal processing are performed. The present processing is "rectification" of the voltage and current data. The welding current used is 60 Hz. AC, and so the resulting weld voltage is also an AC signal. It was determined that for display and analysis only the absolute value of the weld current and voltage is required. Displaying only a rectified form of those signals makes quick visual analysis of the displayed data simpler and allows improved resolution in the displayed data. Therefore, the voltage and current data are converted to their absolute values after collection.

The second processing step is the calculation of the dynamic resistance of the welded workpiece. The resistance is calculated as the ratio of the weld voltage and current at each of the weld current peaks (twice per 60 Hz. cycle). The reason for restricting the calculation to the current peaks only is the presence of an undesirable induced voltage component in the sampled voltage signal. That component is caused by the strong, varying magnetic field generated by the voltage sense wires along with the actual weld voltage. However, at the times of the peaks in the weld current, when the rate of change of the current is zero, the field and therefore, the induced voltage are also zero. by using only the voltage values at the times of the current peaks to calculate the dynamic resistance, error due to the induced voltage component is negligible.

Numerical integration of the current waveform to find its RMS value is the third data processing step. The technique used is trapezoidal integration and is performed on the squares of the weld current values between peaks in the waveform. The total of all the halfcycle integrals from the first peak to the last welding current peak is used to calcualte the true RMS value of the current over the weld time. Additionally, each half-cycle integral is multiplied by the dynamic resistance value for that interval to calculate the energy input during that half-cycle. A total of all the energy values up to that time is also computed to give the cumulative energy input.

Collected data can be stored on floppy diskettes for future use. Any previously stored data can be read back into the Apple memory for display and analysis. The collected data is stored in a disk data file along with some identifying information, including the data collection parameters such as sampling rate, number of channels, and number of cycles collected. The length of a data file depends on the amount of data collected; on the average about 10 data files can be stored on a diskette.

The data in the Apple memory can be displayed in several ways. The data from any two channels can be played back at a slow enough rate for display on the strip chart recorder. By varying the chart paper speed the level of detail of the displayed data can be selected. The data is output to the recorder at a known rate so that the time scale of the displayed data is known and so that recordings of different data can be accurately compared. The result of the dynamic resistance calculations can also be displayed in the same manner.

The data can be plotted on the system's video monitor or printer. At present only the dynamic resistance and energy values can be plotted, but programs are planned to enable plotting of any of the input data.

A third display method is a printed table of the values of the current, resistance, and energy at the times of the peaks in the current waveform. Such a table provides more exact knowledge of the data without the need for interpolation from plots.

Examples of the display results are shown in Figs. 2-4. In Fig. 2, examples of the various chart recorder plots available to the user are shown. The first two plots are of the weld voltage and current for a typical weld. Note the rectification of the signals, as previously explained. Below those plots are six more on a compressed time scale, showing the weld voltage and current, relative tip displacement, dynamic resistance, and energy, all for the same weld. Fig. 3 is a printed table of some of the weld data. Finally, Fig. 4 shows the plots that can be created using the printer.

HARDWARE DESCRIPTION - The analog-to-digital converter (A/D) system used is designed for use with Apple II computers and is manufactured by Interactive Structures,Inc. It provides 16 input channels, of which only one can be sampled at a given instant. Eight input voltage ranges are selectable. The weld data collection system uses only the first six channels and a fixed input range of +/- 5.0 volts. The data conversion is performed with 12-bit resolution, dividing the 10 V total input range into 4096 intervals of 2.4 mV. A conversion is performed by sending the A/D a command containing the desired channel and range and then reading the result after a delay of about 20 microseconds. During data collection the desired input channels are converted sequentially at each sample time.

The digital-to-analog converters (D/A's) provide signals in the range of +/- 5 volts for the chart recorder. The D/A's have 8-bit resolution, dividing the 10 V range into 256 intervals of 39 mV, which is adequate for data display purposes.

At present, three sensors have been installed on the spot welder to detect weld current, voltage, and displacement. Fig. 5 is an illustration of the sensor installation. The voltage sensor consists of two shielded wire leads attached to the welding electrodes. The actual electrical contact is made through a wide wire braid that is clamped to the welding electrode tip. The braid provides a large contact area and the flexibility required by the vertical movement of the electrodes. The area of the loop formed by the two sense leads and the welding electrodes is kept to a minimum to reduce the amount of 60 Hz. voltage induced in the loop by the large magnetic field generated by the welding current in the electrodes. Also, the electrical contact is made as close to the workpiece as possible to reduce the resistive voltage drop in the electrode between the sense lead and the workpiece.

The welding current is sensed by a toroidal coil that is installed on the lower welding electrode. The output of the coil is proportional to the rate of change of the current through the center of the coil. The manufacturer of the coil specifies the output as 0.2 volts per 1000 amps at 60 Hz. Although similar to those used for various AC ammeters, the coil was constructed to specifications for this project.

The displacement sensor is a linear variable displacement transformer (LVDT). An LVDT is basically a transformer with a core that is moved in and out of the coils by an actuator. If supplied with a constant-voltage AC input, the output of the LVDT is an AC signal with amplitude corresponding to the position of the actuator. By installing the LVDT between the welding electrode conductors it is possible to detect the changes in the distance between the tips by monitoring the changes in the LVDT output voltage.

At present, the analog interface circuitry consists of three sections, one for each sensor. The input stage of each section is a differential amplifier to reduce the amount of electrical noise in the signals. The output stage of each section is a differential amplifier to reduce the amount of electrical noise in the signals. The output of the voltage-sense-lead differential amplifier is input to an amplifier with gain adjustable from 0.1 to 10. Since the weld voltage is normally in the range of a few tenths of a volt to a few volts, the amplifier gain range allows amplifying the voltage signal to near the full input range of the A/D.

The current sensor circuit is identical to that of the voltage channel except for the inclusion of an integrator circuit. The integrator is needed since the toroid coil output is proportional to the time derivative of the welding current rather than to the current. A variable gain amplifier is again provided to boost the signal output to the A/D.

The circuit designed for use with the LVDT consists of two parts, the driver and the detector. The driver is an oscillator that provides a 5 kHz. sinewave with constant amplitude for the LVDT input. The detector receives the output from the LVDT and generates a voltage proportional to the magnitude of that output. That signal is then connected to one of the A/D inputs for conversion to a digital value.

SOFTWARE DESCRIPTION. The major program used in data acquisition is called WELD. With WELD the user can collect new welding data, store data, retrieve old data, or display data using the chart recorder. Two other programs are also available to display collected data in different formats. Those programs are called DPRINT and PLOT. The program WELD will be described first, followed by a discussion of the other two. The descriptions cover the program organization, in order to show the type of programming that is needed for a sensor system of this type.

As previously mentioned, the bulk of the programming is in APPLE PASCAL. However, two subroutines written in assembly language for the 6502 microprocessor are used within the PASCAL programs to perform time-critical tasks. The two tasks are 1) the control of the A/D converter to allow sampling of the input signals at the desired sample rate, and 2) control the D/A converters when plotting data on the chart recorder. Writing those two routines in assembly language provided the control of their execution speeds required for accurate timing of the data collection and plotting functions. The remaining tasks are programmed in PASCAL.

A major goal for the software (beyond having it perform the requisite tasks properly) was to make it easy to use by a relatively untrained user, because the system will be used by people other than the authors for welding research. As a result, the program includes quite a bit of work to protect collected data and to guide the user through the process of recovering from several common types of errors. Also, the software makes extensive use of messages to lead the user through the desired process and onscreen "menus" list the options available to the operator at various times.

The program WELD consists of a main program, five subprograms, and a few utility subroutines used by all parts of the program. The utilities perform such frequently used activities as reading input from the user and clearing the CRT screen and they also define a few data formats used by the other parts of the program. The main program first initializes certain variables used in the program and then displays the main menu which gives the operator the choice of collecting, retrieving, storing, or displaying data, or of quitting the WELD program entirely. In response to the user's selection of any of the first four options, the main program jumps to the subprogram NEWDATA, OLDDATA, STORDATA, or VIEWDATA, respectively. After running any of the subprograms, the main program again displays the main option menu. Another subprogram called CALCULATE, described later, is called by the main program whenever new data has been entered into the computer.

The subprogram NEWDATA carries out several tasks sequentially. Before new welding data can be collected, the parameters for that data collection must be entered. Therefore, NEWDATA first asks the user for the desired sampling rate, number of channels to be collected, number of welding cycles, amplifier gain settings, and other needed information. The user is given a chance to change any of the values entered before the actual data collection takes place. Next, some set-up work is done to prepare for collection and the user is informed when the system is ready. The user then presses a key to begin the collection process.

No data is actually collected immediately, however. Instead the weld voltage signal is continuously monitored through the A/D. When the magnitude of the voltage first exceeds about 0.3 V, the data collection begins and then continues for the desired length of time. This allows a delay between when the operator presses the key on the computer and when the weld is started. The delay is needed to allow the user to start the welder, and because a certain amount of "squeeze" time before the weld current is turned on is usually programmed into the welder controller. By waiting for the appearance of a sufficient weld voltage the program avoids storing a large amount of data before the weld actually starts. The 0.3 V threshold is low enough to not cause the loss of any significant data at the beginning of the weld.

Once the data collection is completed the voltage and current signal data are "rectified" as described earlier and the locations of the peaks in the current waveform are determined. The dynamic resistance calculations are then carried out. Next, a copy of the data is stored in a temporary data file on the program diskette. This file is meant as a safeguard in case the data is erased from the computer memory before the user stores it on a data diskette. Normally the data remains in the computer memory until new data is collected or the user exits the program. The user is then given a chance to store the data in a file on a data diskette. Finally, the subprogram CALCULATE is executed to perform integration of the current waveform and to calculate the input energy and the RMS weld current. The calculated values are then displayed on the APPLE screen.

If the operator opts to retrieve an old data file rather than collect new data, the subprogram OLDDATA is used. OLDDATA offers the user a choice of fetching the temporary "safety" file just mentioned or retrieving any other data file from another diskette. In either case the data retrieved replaces whatever data was previously in the computer memory. The new data can then be displayed or even stored again in a different file. The program checks to make sure that the data was read back correctly and provides error messages and recovery instructions if some error occurs. Before again displaying the main menu, the main program executes the subroutine CALCULATE to do the integration of the current data just retrieved from disk.

The subprogram STORDATA is called if the user selectes the data storage option from the main menu. In STORDATA the operator is asked for the name of the diskette and data file to use. The data in the computer memory is then transferred to the specified file. Once again appropriate error testing is done. STORDATA is also called by the subprogram NEWDATA if the user wishes to store the data right after it is collected.

If it is desired to display the data that is in the computer memory, the data display option is selected from the main menu, which causes teh subprogram VIEWDATA to be activated. VIEWDATA presents a menu offering the choice of plotting either the calculated dynamic resistance data or the data from any two sensors on the chart recorder. If the resistance data is desired, the user has a choice of plotting the half-cycle or total energy values on the recorder's second channel. The data is scaled appropriately for plotting and then the user is prompted to start the recorder and then press a key on the computer. The scaled data is then sent to the recorder input via the D/A converters and the scale used is shown on the screen. When complete, the user is asked if another plot of the same data is desired (possibly at a different chart recorder speed).

In order to plot the collected data rather than the dynamic resistance, the operator is prompted for the channel numbers of the desired two signals. The operation is then the same as for the resistance plot.

The other two data display programs, DPRINT and PLOT, will now be described. Both are separate programs, rather than subprograms called by program WELD, mainly due to the memory space required by WELD. DPRINT is used to print a table of the current, resistance, and energy data. The user of the program specifies the diskette data file to be used and that data is read into the computer memory. The operator is given a chance to see some of the data collection parameters used when the data was collected, to see if it is the correct data file. If a printed table is then desired, the appropriate data is transmitted to the computer's printer.

The program PLOT utilizes the basic computer graphics procedures that are part of APPLE PASCAL. Using those procedures the program creates a scaled plot of the dynamic resistance or energy data from a chosen data file. The scales for the graph can either be determined by the computer or manually entered by the user. The plot is first shown on the computer monitor and then the operator is asked if a printed copy is desired. The user also has several options available for printing the graph with different orientations, sizes, etc.

CHARACTERIZATION OF THE DYNAMIC RESISTANCE CURVE

To illustrate the capabilities of the sensor system we have chosen to present some preliminary data from a study of spot welding of galvanized steel. The study is focussed on the dynamic resistance curves obtained during the weld process.

The characterization of the dynamic resistance curves while welding plain carbon steel and High Strength Low Alloy (HSLA) steels which do not have a surface coating has been previously accomplished (1). This characterization has shown a direct relationship between the shape and values on the dynamic resistance curve and all the appropriate metallurgical changes occurring during the formation of an acceptable nugget. These changes include:

a) surface breakdown
b) asperity softening
c) base metal temperature rise
d) first melting
e) nugget growth to acceptable size
f) expulsion upon exceeding acceptable parameters.

Unfortunately, welding coated steel is more complex. The surface and interdiffusion region between the coating and the steel substrate now become important factors. In addition, irregularities occur so rapidly that most present weld-monitoring equipment is unable to determine the true shape of the resistance curve in the early stages of weld formation. The computerized sensor system we described, however, computes the workpiece resistance every half-cycle of weld current. This is sufficient to verify the typical dynamic resistance curve for hot-dipped, galvanized steel.

Consider the typical resistance curve as shown in Fig. 6. The area of interest, which differs from that for uncoated steel, is the small peak occurring between times T1 and T3 early in the process. From the observation of various test samples it has been determined that this peak is related to zinc surface melting and alloying. At time T1, surface breakdown and asperity softening have occurred so that the resistance dramatically drops from the initial contact resistance. At this point, however, zinc surface melting at random locations on the faying surface is beginning to occur. Accompanying the initial melting is a rapid diffusion and alloying of the molten zinc into the iron substrate to produce a solid iron-zinc layer. By time T2, the alloy layer at the faying interface is relatively complete. Because of the alloying, the layer exhibits the increase in the total resistance as seen at time T2. As energy input continues to time T3, heating continues, producing a temperature above which a sufficient quantity of the alloy layer melts or partially melts and is squeezed out of the weld zone by the action of the electrode force. (The alloy layer has a lower melting point than steel but higher than zinc). At this point an iron-to-iron substrate contact is made and the resistance decreases to a level near that expected for uncoated steel with faying surface temperature equivalent to that at time T3. From time T3 until weld termination the resistance curve is similar to that for uncoated steel. The region between T1 and T3 is dependent upon the coating type and condition. The extent of this region can affect the nugget growth characteristics. More complete results of the studies of the dynamic resistance curve for various galvanized steels are the subject of another paper. However, three typical curves obtained for different weld times are shown in Fig. 7. Similar resistance peaks occurring in the early stages of the weld have been observed in almost all of the tests performed to date with galvanized steel.

From the curves in Fig. 7, other features of the sensor system can be noted. First observe from the typical voltage and current traces of Fig. 2 that extra, small pulses (one cycle each) of current are present before and after the main pulse of welding current. The pulses are used to determine the pre-weld resistance (contact resistance) and the post-weld resistance variation. In Fig. 7, the pre-weld resistance is printed above the dynamic resistance plots. Note the dramatic drop in resistance that occurs within the first half-cycle as indicated by the difference in the pre-weld resistance value and the first plotted value of resistance.

The post-weld resistances are shown on the graphs with no connecting lines drawn between them, to distinguish them from the resistance values during the actual weld time. These values should be related to the cooling rate of the nugget after the weld current is extinguished. The magnitude of the current pulses used before and after the weld for resistance measurement was chosen to minimize any heating of the workpiece during those pulses.

Finally, note that in the last curve ("U14", 14 cycles of current) the expected drop in the resistance at expulsion clearly shows in the resistance plot. In addition, the expulsion point is indicated in the electrode displacement curve for the same weld, shown in Fig. 8.

SUMMARY

A microcomputer based sensor system for monitoring the resistance welding process has been described. Examples of the type of data that can be easily collected or calculated by the system were given. They include electrode voltage, current, and displacement as well as processed data such as workpiece dynamic resistance, pre-weld and post-weld resistance, and energy delivered to the workpiece. Some preliminary results concerning irregularities in the dynamic resistance curves for spot welding of galvanized steel were presented and discussed.

REFERENCES

Dickinson, D.W., Franklin, J.E., Stunya, A., Characterization of Spot Welding Behavior by Dynamic Electrical Parameter Monitoring", Welding Journal, (June 1980).

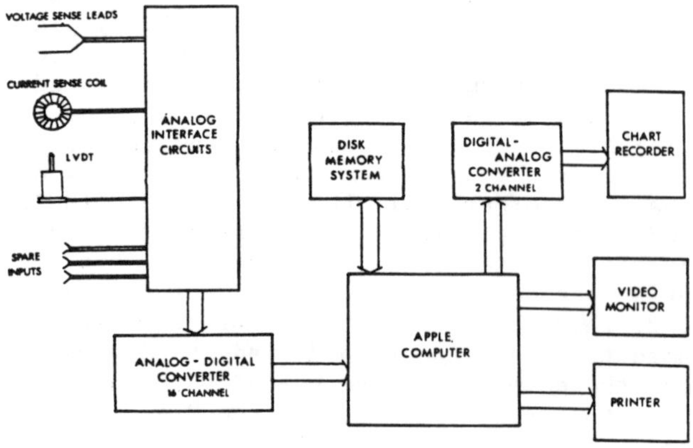

Fig. 1. System Block Diagram.

Table 1

DATA STORAGE CAPACITY

SAMPLE FREQ. (kHz)	SAMPLES/ CYCLE	MAX. NO. OF CYCLES IF NUMBER OF CHANNELS IS: 2	3	4	5	6
1	16.7	300	199	149	119	99
1.5	25	203	135	102	80	67
2	33.3	154	102	76	61	50
3	50	101	67	50	---	---
4	66.7	75	---	---	---	---
5	83.3	60	---	---	---	---

Based on a total of 10240 samples.

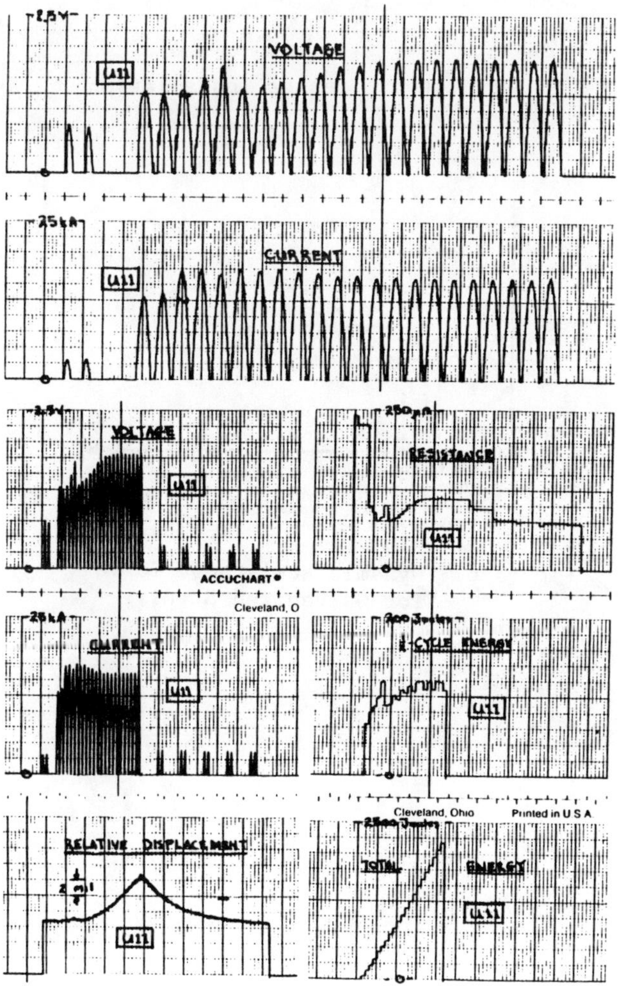

Figure 2. Strip Chart Outputs.

CURRENT, RESISTANCE, and ENERGY DATA

From File: U11

PEAK #	CURRENT (kA)	RESIST. (u-Ohms)	ENERGY (kJ)	SAMPLE #
1	3.17	246.5	0.00	4
2	3.21	232.3	0.00	28
3	13.33	101.1	0.07	100
4	13.98	91.2	0.15	125
5	17.50	77.5	0.23	149
6	17.49	83.8	0.33	173
7	16.99	102.9	0.45	198
8	17.03	77.0	0.54	223
9	17.50	80.4	0.63	247
10	17.44	83.4	0.74	271
11	17.30	89.1	0.83	296
12	16.98	95.0	0.94	320
13	16.74	103.7	1.05	345
14	16.59	104.4	1.16	369
15	16.47	110.7	1.27	394
16	16.24	112.6	1.39	418
17	16.25	114.0	1.51	443
18	16.48	113.9	1.62	468
19	16.19	115.2	1.74	492
20	16.44	114.2	1.85	517
21	16.09	115.5	1.97	541
22	16.41	113.6	2.09	566
23	16.38	114.4	2.19	591
24	16.36	113.8	2.19	615
25	3.71	103.3	2.19	741
26	3.85	97.1	2.19	766
27	3.97	81.8	2.19	889
28	3.97	79.7	2.19	914
29	3.98	75.8	2.19	1037
30	3.86	77.5	2.19	1061
31	4.03	71.8	2.19	1184
32	3.67	75.4	2.19	1208
33	3.76	75.3	2.19	1331
34	3.94	74.9	2.19	1356

Figure 3. Data Table.

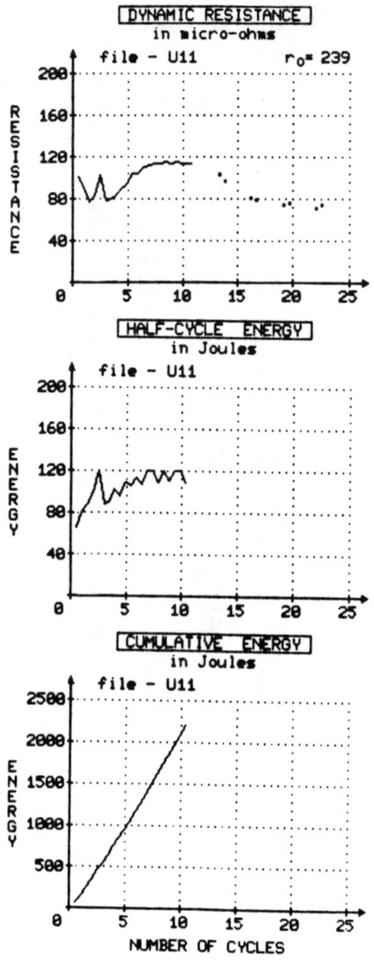

Figure 4. Printer Plots.

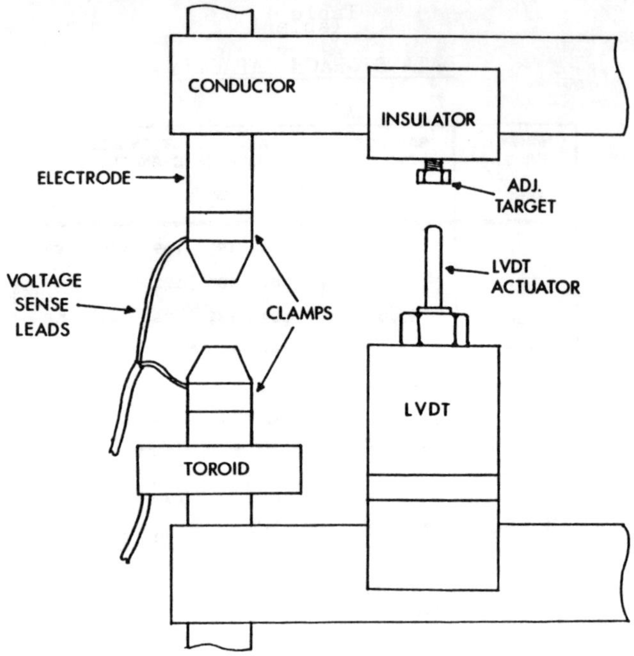

Figure 5. Sensor Installation.

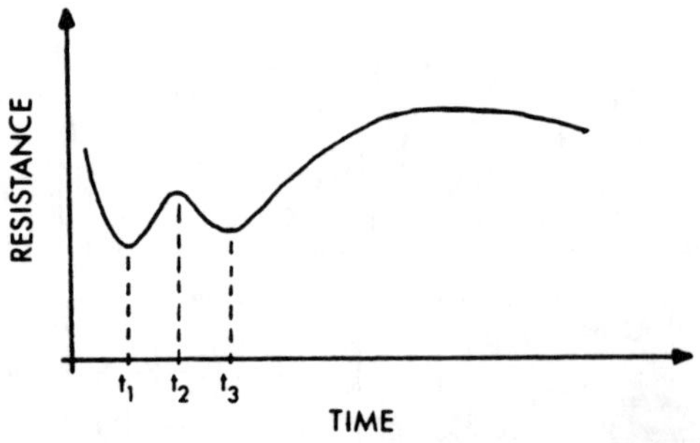

Figure 6. Dynamic Resistance Curve for Galvanized Steel.

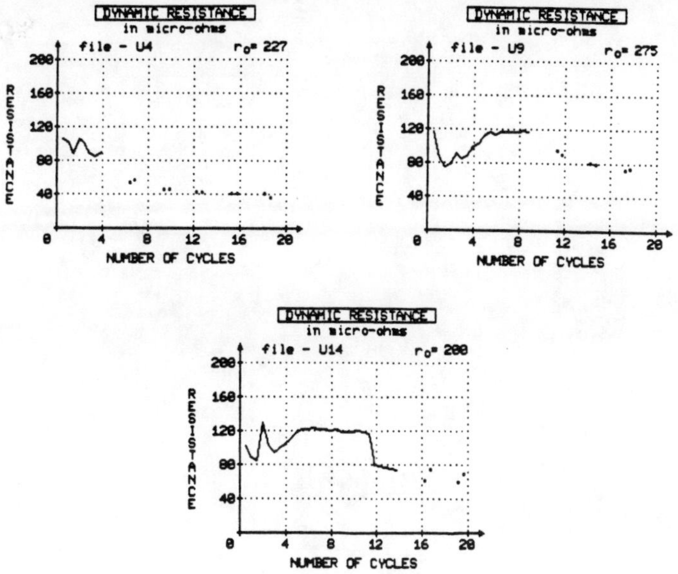

Figure 7. Experimentally Obtained Dynamic Resistance Curves for Galvanized Steel.

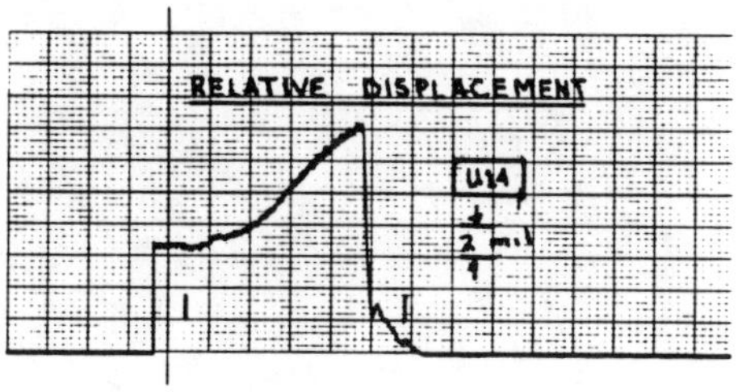

Figure 8. Electrode Displacement.

PROCESSES FOR ALTERING WELD RESIDUAL STRESSES IN AUSTENITIC STAINLESS STEEL PIPING

Joseph C. Danko
American Welding Institute
Knoxville, TN

ABSTRACT

Incidents of intergranular stress corrosion cracking of welded austenitic stainless steel piping in boiling water reactors have had a major impact upon plant availability and economics. One of the key contributors to this problem is the tensile weld residual stresses. Several engineering solutions, based upon processes that produce compressive residual stresses in the weldment, have been developed. These include heat sink welding, last pass heat sink welding, weld overlay, and induction heating stress improvement. Some of these have been implemented in plants under construction and in operation.

DURING THE PAST DECADE, the number of incidents of intergranular stress corrosion cracking (IGSCC) in welded austenitic stainless steel piping in boiling water reactors (BWRs) has increased significantly[1]. This has seriously affected plant reliability, availability, and economics. Currently, the problem is well understood, and engineering solutions for the mitigation are available. The purpose of this paper is to review the problem of pipe cracking and to describe those solutions that were based upon altering the weld residual stresses in the piping systems.

NATURE OF PIPE CRACKING

Stress corrosion cracking in types AISI 304 and 316 stainless steel piping in BWRs has been confined to the weld heat-affected zones. The cracking morphology is intergranular, with cracking patterns being predominantly of a circumferential orientation that are related to axial stresses and some axial cracks that are a result of hoop stresses acting upon the weldment.

Intergranular stress corrosion cracking has appeared in a number of piping systems including: (1) recirculation, (2) core spray, (3) reactor water cleanup, (4) control rod drive return, and (5) isolation condenser system. A schematic of a boiling water reactor with one of two loops of the recirculation piping system and the core spray line is illustrated in Figure 1.

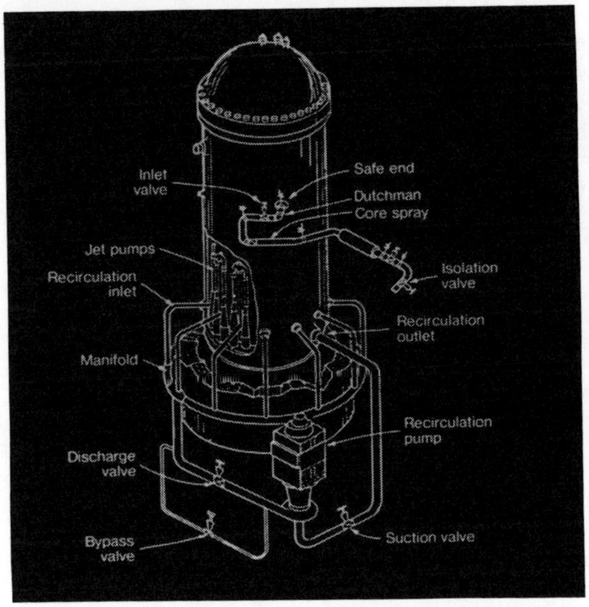

Figure 1 - Schematic of a boiling water reactor recirculation piping system

Cracking has been reported in 304 and 316 stainless steel pipes ranging from 77 mm (3-inch) to 710 mm (28-inch) in diameter. The IGSCC has been found by leaks in the pipes, ultrasonic inspection (UT), and dye penetrant (DP) tests. Over 800 incidents of cracking have been reported worldwide. In spite of this large number of incidents, there has never been a reported pipe severance. A leak-before-break concept has evolved over the years and has been accepted by the industry[2] and the Nuclear Regulatory Commission.[3,4] Pipe cracking in boiling water reactors does not constitute a safety issue or hazard to the safety of the public.

MODEL FOR IGSCC

A model[5] for the intergranular stress corrosion cracking in welded 304 and 316 stainless steel was developed under a research and development program jointly sponsored by the Electric Power Research Institute and the Boiling Water Reactor Owners Group. This model requires the coincidence of three major factors: (1) weld sensitization, (2) the boiling water reactor coolant, and (3) tensile stresses above yield stress of the base material. The boundary conditions for the three factors and the region of IGSCC are illustrated in the ternary-type diagram of Figure 2 for a temperature at 288°C (550°F), the operating temperature of a BWR. The boundaries show that IGSCC will not occur below 20 parts per billion (ppb) of dissolved oxygen in the coolant, at stresses below the yield stress and levels of sensitization of two coulombs/cm^2 based upon the electrochemical potentiokinetic reactivation technique[6] or an A262 practice A stepped structure.[7]

Using this model, engineering solutions commonly known as pipe remedies were identified, developed, and qualified for field use. The stress-related remedies that involve altering the weld residual stresses are reviewed in this paper.

STRESS-RELATED PIPE REMEDIES THAT ALTER WELD RESIDUAL STRESSES

A number of processes have been developed that alter the weld residual stresses in austenitic stainless steel pipes. How these stresses are altered is best understood by examining the as-welded stresses. The as-welded residual stresses in 100 mm (4-inch), 250 mm (10-inch), and 660 mm (26-inch) diameter 304 stainless steel have been measured.[8] Axial residual stresess as a function of the distance from the weld centerline as determined by the combination of X-ray and strain gauge measurements for the three pipe diameters are shown in Figure 3.

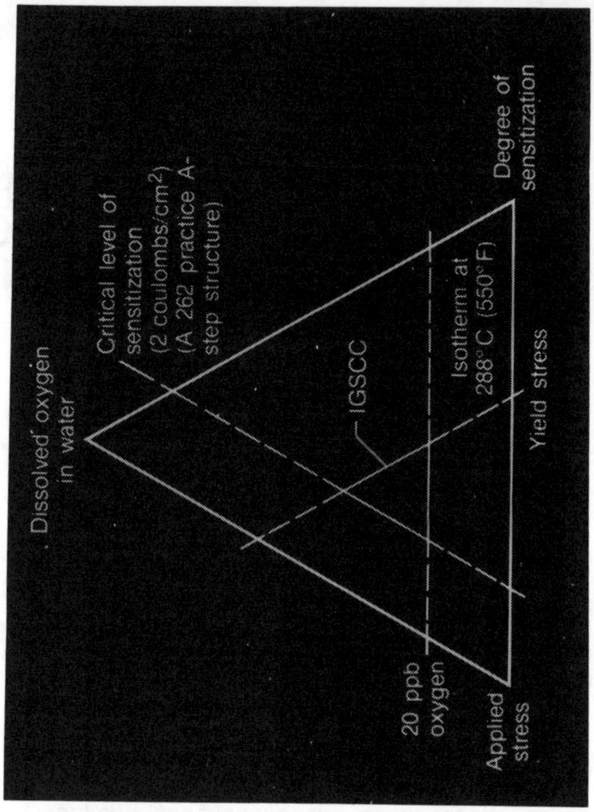

Figure 2 - Factors that contribute to the intergranular stress corrosion cracking in welded types 304 and 316 stainless steels

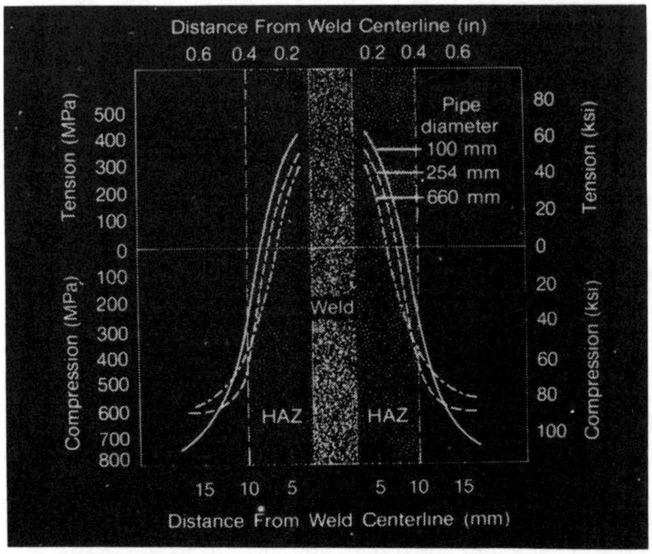

Figure 3 - Peak axial residual stresses on the internal surface of welded 304 stainless steel pipes

The peak tensile surface residual stresses are pipe size dependent with the highest stresses in the 100 mm, intermediate in the 250 mm, and the lowest in the 660 mm diameter pipe. These peak tensile stresses occurred in the heat-affected zone. The variation of the weld residual stresses as a function of pipe diameter correlate very well with the service performance of the 304 and 316 stainless steel pipes. IGSCC propensity decreases with increasing pipe diameter. IGSCC has occurred in large diameter pipes 600 mm, but the time required for cracking is significantly increased relative to the smaller pipes.

To alter the weld residual stresses, the processes that have demonstrated this capability include heat sink welding, last pass heat sink welding, induction heating stress improvement, and weld overlay for repair. Each of these processes introduce compressive residual stresses on the pipe inner surface and partial through-wall in the heat-affected zone. The weld overlay process also provides a new pressure boundary over the IGSCC defect. The presence of the compressive residual stresses prevents the initiation of IGSCC while for welded pipes with cracks, the propagation may be arrested, depending upon the crack depth.

The various processes for altering the weld residual stresses are described in the following sections.

HEAT SINK WELDING - This process involves the welding of the root pass and the first several layers to provide structural support of the pipe weld. The subsequent weld layers are deposited while flowing or spray water is applied on the inside of the pipe. With this process, favorable compressive residual stresses result on the pipe inside surface and partially through the wall. The peak axial surface residual stresses of an as-welded and heat sink welded pipe of 250 mm (10-inch) diameter 304 stainless steel are shown in Figure 4. This illustrates the considerable alteration of weld residual stresses produced by heat sink welding. Another benefit of heat sink welding is the reduction in the level of sensitization.[9] However, the sensitization level is not below the critical value to avoid IGSCC.

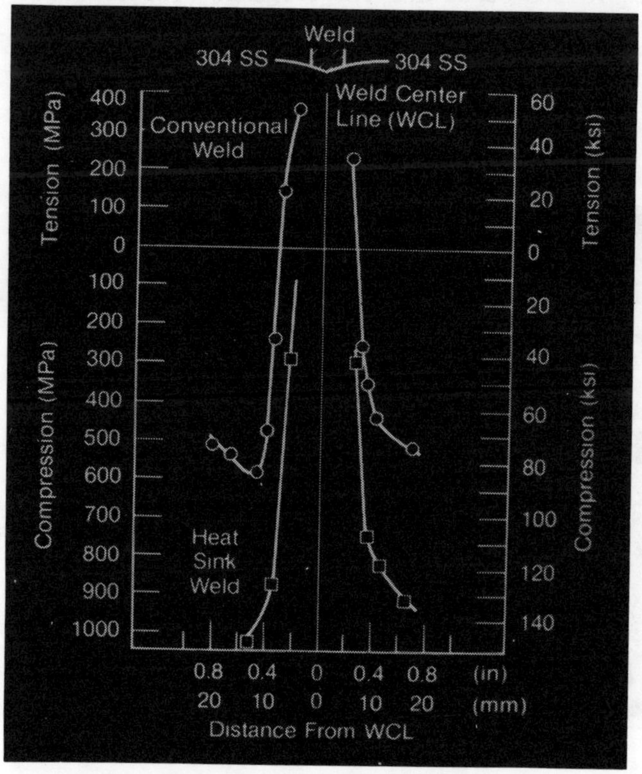

Figure 4 - Peak axial surface residual stresses in an as-welded and heat sink welded 250 mm (10-inch) 304 stainless steel pipe

LAST PASS HEAT SINK WELDING - The only difference between this process and heat sink welding is the introduction of water cooling when the last pass or crown weld is applied. A high weld heat input is used to establish a proper temperature gradient across the pipe wall. The net result on the weld residual stresses is the same as that of heat sink welding. This is illustrated in Figure 5, a plot of the weld residual stresses of a 610 mm (24-inch) diameter 304 stainless steel pipe weld. The inside surface of the pipe as well as substantial through-wall stresses are highly compressive.

INDUCTION HEATING STRESS IMPROVEMENT - This process, conceived, developed, and implemented in Japan,[10,11] involves induction heating the outside surface of the pipe weldment to approximately 500 to 550°C (930 to 1020°F) while maintaining the inside surface temperature at 100°C (212°F) with flowing water. The temperature gradient across the wall thickness results in a small amount of plastic strain in the stainless steel and the original weld tensile residual stresses are removed and replaced with compressive residual stresses. Typical compressive residual stresses imparted by induction heating stress improvement of a 406 mm (16-inch) diameter 304 stainless steel pipe weldment are illustrated in Figure 6.

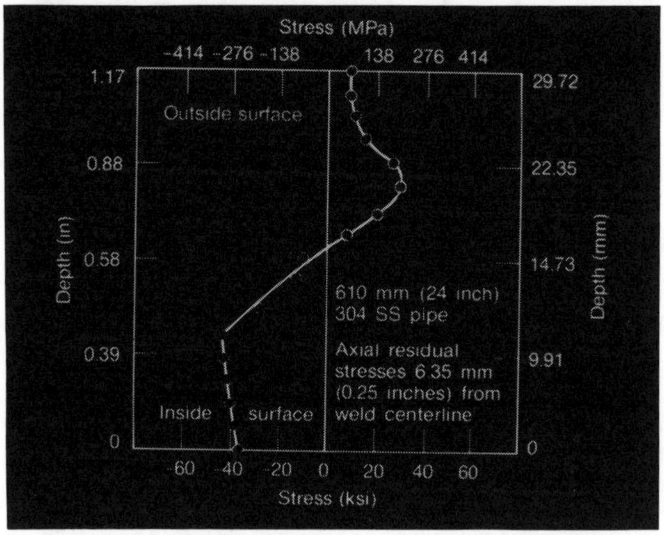

Figure 5 - Weld residual stresses in a last pass heat sink welded 635 mm (24-inch) 304 stainless steel pipe

The benefits of induction heating stress improvement for the mitigation of IGSCC have been demonstrated in full size pipe tests.[12] In fact the process also provides benefits for welded 304 stainless steel pipes with shallow existing intergranular stress corrosion cracks.

WELD OVERLAY FOR REPAIR - Welded pipes of 304 stainless steel with existing intergranular stress corrosion cracks have been repaired with the weld overlay process.[13] A 308L weld metal with controlled delta ferrite content of 8 percent minimum is deposited over the original weldment. The thickness of the overlay deposit may vary, but most overlays are of a thickness that provides structural reinforcement that satisfies the original pipe stress design of the ASME Code. Application of the overlay is performed while the reactor coolant remains in the pipe. Under these conditions, a heat sink welding operation is performed. As a result of the radial and axial shrinkage, compressive residual stresses are produced. Schematics of various weld overlays are shown in Figure 7.

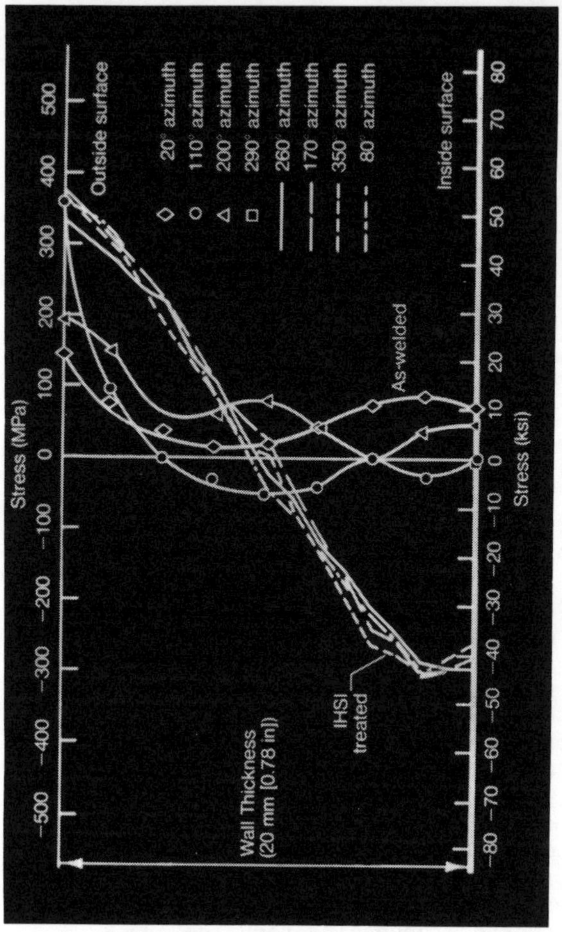

Figure 6 - Residual stresses in an induction heated stress improvement processed 406 mm (16-inch) 304 stainless steel pipe

Weld overlay repairs have saved the utilities significant cost savings since the repairs are performed with reactor coolant in the lines and pipe removal is not required. In addition, the use of automatic machine welding reduces the radiation exposure of the skilled personnel involved in the repairs. Currently, the Nuclear Regulatory Commission has accepted the weld overlay repairs for one fuel cycle operation. Pipe tests are in progress to determine the capability for multicycle operation.[14]

IMPLEMENTATION

Of the various stress-related pipe remedies that alter the residual stresses in welded 304 and 316 stainless steel pipes, only the induction heating stress improvement process has been

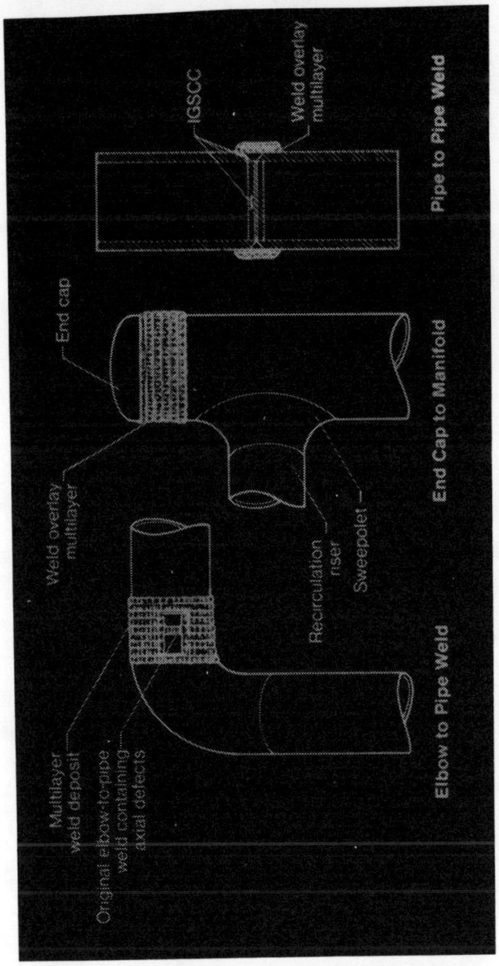

Figure 7 - Schematic of weld overlay pipe repairs

implemented in both operating and new boiling water reactors. Heat sink welding has found limited application in the repair and replacement of piping in operating plants. Last pass heating sink welding has been recently qualified as a pipe remedy and has not been implemented at the time of this publication. Weld overlay for repair has found extensive use in operating plants. The implementation of these processes is presented in Table 1.

Table 1 - Implementation of Stress-Related Pipe Remedies

	Operating Plants	New Plants
Heat Sink Welding	•	
Last Pass Heat Sink Welding*		
Induction Heating Stress Improvement	•	•
Weld Overlay	•	

*Recently qualified for field use.

SUMMARY

Incidents of intergranular stress corrosion cracking in welded pipes of 304 and 316 stainless steel in boiling water reactors continue and have adversely affected plant reliability and economics. Pipe cracking does not constitute a safety problem but is one of economics. The mechanism of cracking is well understood and has been used to identify engineering solutions for the problem. A number of these solutions involves the alteration of the weld residual stresses by processes such as heat sink welding, last past heat sink welding, induction heating stress improvement, and weld overlay for repair. These processes impart compressive residual stresses in the pipe weldment and for unflawed pipes initiation of cracking is prevented. With the exception of last pass heat sink welding, the processes have been implemented in operating and new plant construction.

REFERENCES

1. J. C. Danko, Boiling Water Reactor Owners Group Research Program on Pipe Cracking: 4 years later, Corrosion 84 Paper 162, NACE April 1984

2. H. H. Klepfer et al., Investigation of Cause of Cracking in Austenitic Stainless Steel Piping, Report NEDO 21006-1, General Electric Company July 1975

3. Nuclear Regulatory Commission, Investigation and Evaluation of Cracking in Austenitic Stainless Steel Piping of Boiling Water Reactor Plants, NRC Report NUREG-75/067 NTIS, October 1975

4. Nuclear Regulatory Commission, Investigation and Evaluation of Stress Corrosion Cracking in Piping of Light Water Reactor Plants, NRC Report NUREG-0531 NTIS, February 1979

5. R. E. Hanneman, Models for the Intergranular Stress Corrosion Cracking of Welded Type 304 Stainless Steel Piping in BWR, EPRI-WS-79-174 Volume 1, May 1980

6. W. L. Clarke, V. M. Romero, and J. C. Danko, Detection of Sensitization in Stainless Steel Using Electrochemical Techniques, Corrosion 77, Paper No. 180, NACE 1977

7. A. P. Majidi and M. A. Streicher, Potentiodynamic Reactivation Method for Detecting Sensitization in AISI 304 and 304L Stainless Steels, Corrosion Volume 40 No. 8, August 1984

8. Studies on AISI Type 304 Stainless Steel Piping Weldments for Use in BWR Aplications, NP-944 EPRI Report RP 449-2, December 1979

9. N. Hughes, Qualification of Solution Heat Treatment, Corrosion Resistant Cladding and Heat Sink Welding of 304 Stainless Steel Pipe Welds, EPRI WS-174, Volume 1, May 1980

10. T. Shimizu, K. Enomoto, S. Araki, and T. Ikegami, Induction Heating Stress Improvement for Welded Pipe and Its Effectiveness, EPRI WS-79-174, Volume 1, May 1980

11. N. Ihki, S. Araki, K. Wakasa, S. Nakayama, and A. Tanaka, Actual Proof Test of Induction Stress Improvement of Large Diameter Pipe, EPRI WS-79-174 Volume 1, May 1980

12. D. C. Bertossa, Induction Heating Stress Improvement Qualifications, EPRI WS-79-174 Volume 1, May 1980

13. D. R. Pitcairn, Advances in Weld Overlay Repair Technology, EPRI NP-3684-SR Volume 3, September 1984

14. J. C. Danko and R. L. Jones, Overview of Boiling Water Reactor Owners Group Research Program to be published in Corrosion 85 Paper NO. 100

CONTROLLING WALL THICKNESS OF SEAMLESS TUBES BY THE ELECTROMAGNETIC ACOUSTIC TRANSDUCER (EMAT) GAGING METHOD

L. O. Forstrom
U.S. Steel Corporation
Lorain, Ohio

M. L. Mester
U.S. Steel Corporation
Monroeville, Pennsylvania

L. T. Spragins
MagnaSonics
Albuquerque, New Mexico

ABSTRACT

An electromagnetic acoustic transducer (EMAT) produces ultrasonic waves in steel without the use of water or oil couplants. In the production of seamless tubular products, such a device would be useful for in-process control of wall thickness if installed immediately after the hot mill cooling table. Tests were therefore made to evaluate a prototype EMAT wall thickness gage. The EMAT was mounted in-line with a gamma ray back-scatter gage and 5-1/2- and 4inch OD tubes were gaged. The EMAT demonstrated the ability to measure normal cyclic wall variations and to detect one area missed by the gamma gage. The system reproduced measurements well during multiple scans of a tube and maintained calibration for 24 hours.

IN ORDER TO CONTROL WALL THICKNESS of seamless tubes during the manufacturing process, early feedback of wall thickness measurements is necessary, and the earlier this feedback occurs the better. The electromagnetic acoustic transducer or EMAT offers certain advantages for gaging cold tubes, plus it promises possibilities of gaging hot tubes.

Maxfield and Fortunko[1)] describe "a simple EMAT element composed of a wire carrying a dynamic current, $I\omega$, and a source of strong magnetic bias field, H_o." The current, $I\omega$, induces dynamic eddy currents, $J\omega$, in the surface of the metal conductor. The presence of H_o causes deflection of the moving electrons in a direction defined by the vector crossproduct of $J\omega$, and H_o. The resultant Lorentz forces, F, generate ultrasonic energy that propagates into the bulk of the metal away from the wire. Whittington[2)] says the Lorentz forces cause the metal surface to vibrate in sympathy with the applied dynamic current, and this is "analogous to a moving-coil loudspeaker where the eddy currents flowing in the metal surface represent the currents flowing in the loudspeaker coil." This simple EMAT is shown in Figure 1. Since an EMAT

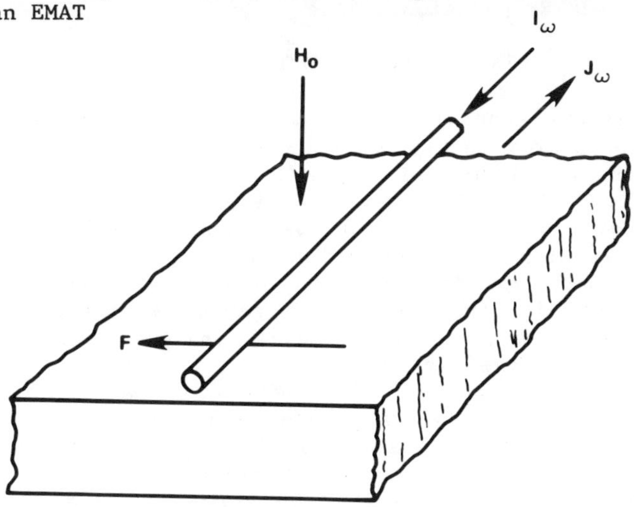

Figure 1. A Simple EMAT element.

produces ultrasonic waves in steel without the use of water or oil couplants, it eliminates problems associated with couplants such as air bubbles, water treatment, and freezing. An EMAT thickness gage inherently has faster response time and faster operation than the radiation gages frequently used to gage tubes. Although EMAT's have been under development for the last 10 years or more, only recently have domestic companies offered commercial units for sale. In the production of seamless tubular products, such a device would be useful for in-process quality control if installed immediately after the hot mill cooling table.

Tests were made to evaluate a prototype EMAT wall thickness gage. The specification sheet for this gage is shown below:

SPECIFICATION - Tube System (Non-Rotating)

Thickness Gauge Model TG-1 and TG-2

Wall Thickness........0.2 in. to 1.0 in. (TG-1)
1.0 in. to 3.0 in. (TG-2)

Measurement Range.....Nominal Thickness Setting
± 0.10 in. (TG-1)
± 0.30 in. (TG-2)

Accuracy..............± 1%

Tube Speed...........Maximum: 500 ft/min for 100%
Longitudinal Coverage
Minimum: None

Tube Temperature......0°-150^oF (Standard Wear Surface)
0°-500^oF (High Temperature Wear Surface)

Magnetic Field........Encircling Coils

Surface Preparation...None

EQUIPMENT AND PROCEDURES

A single channel system was used to scan a line approximately 2.54 cm (one inch) wide along the length of 14 cm (5-1/2-inch) OD by 6.1 mm (0.244-inch) wall and 10.2 cm (4-inch) OD by 6.6 mm (0.262 inch) wall seamless tubes in an off-line inspection area. The air-operated meander-coil transducer was installed in-line ahead of a gamma ray backscatter gage, and the normally rotating gamma gage was fixed in position for a direct comparison of results. The gamma unit has been operating for several months and has proven to be useful in determining light wall. Two copper meander coils (transmitter and receiver) are deposited on a thin film of high molecular weight polyethylene plastic backed by sponge rubber to form the flexible, contour-following transducer shown in Figure 2A. The transducer was mounted within a large (40.6 cm [16-inch] ID) dc solenoid used to produce a longitudinal field of 300 gauss at the tube surface. (Figure 2B).

Power input to the solenoid was 570 watts (7.6 amperes dc at 75 volts dc) while the EMAT required only 0.300 watts ave. at 60 volts dc. The EMAT power requirements for cold steel are much less than the megawatt pulse-power used in a high temperature gaging application reported by the Japanese. The combination of a longitudinal dc field and a transverse ac current through the meander transmitter coil results in a force acting radially through the tube wall to produce ultrasonic shear waves at a 45-degree angle with a frequency of 1.79 MHz. When the system is calibrated at the nominal wall thickness, it reads deviations from this value by measuring phase differences (± 180°) of the received pulses.

SYSTEM CALIBRATION

A 14 cm (5-1/2 inch) OD by 6.1 mm (0.244 inch) wall calibration tube normally used to calibrate the gamma gage was also used to calibrate the EMAT system. A hand ground area designated as 5.5 mm (0.220 inch) wall was used as one calibration point, but we also selected a second as-rolled area measured by a portable ultrasonic gage to read an average of 5.7 mm (0.226 inch) with a high reading of 5.9 mm (0.237) and a low reading of 5.5 mm (0.221) [(range = 0.4 mm (0.016 inch)].

Calibration for the 10.2 cm (4-inch) OD x 6.6 mm (0.262 inch) wall tubes was performed using a tube standard also containing a hand ground thin area. Figure 3 shows a calibration chart of the EMAT response during two rotations of the tube standard. The light wall 5.9 mm (0.234 inch) and heavy wall 6.5 mm (0.259 inch) areas were used to adjust the range of the instrument. The calibration chart shows remarkable similarity for both rotations, considering that the tube was rotated by hand with a very bumpy motion. The pneumatic transducer mount followed the tube motion very well.

RESULTS

While scanning a number of 14 cm (5-1/2inch) OD tubes, a naturally thin area predicted by the EMAT to be less than 5.5 mm (0.220 inch) thick was marked and measured by the portable ultrasonic gage in the longitudinal direction to average 5.5 mm (0.219 inch) thick (11 readings) with a high reading of 5.9 mm (0.235) and a low reading of 5.4 mm (0.216) [range = 0.5 mm (0.019 inch)].

Figure 4 shows the generally good reproducibility of the system during multiple scans of the same 10.2 cm (4-inch) OD tube even though the conveyor caused some tube rotation. Maximum spread between readings is about 0.225 mm (0.009 inch). Also, the cyclic wall variations that normally occur along the tube length are evident in these charts. Correlation with the gamma gage charts is illustrated in Figure 5 which shows overlays of EMAT charts and gamma gage charts for two tubes scanned simultaneously by both gages. Very good agreement is evident, especially at the cyclic variations. There are two obvious disagreements at both ends of the chart for tube No. 6, but experience has shown that end measurements by the gamma gage are made inaccurate by tube ends not being held firmly in both pinch rolls. During scans by the EMAT, a "step" response was generated whenever a thick or thin area fell outside the range of the instrument. This step is shown for tube No. 6 in Figure 5.

Figure 6 shows a computer generated comparison of multiple light-wall scans of the EMAT with thickness measurements made by the portable ultrasonic gage. A total of 41 wall thickness measurements were taken at increments of one foot along the tube. Agreement for the five sets of EMAT data is good (as shown also in Figure 4), but the ultrasonic readings diverge from the EMAT readings as the tube progresses through the unit along the conveyor. This divergence could be caused by tube rotation imparted by the conveyor, the effects of which would be cumulative as the tube moves linearly. The difference between the EMAT and ultrasonic readings are also attributable to some extent to the difference in areas being measured by each gage. The EMAT gages an area approximately 1.3 cm (1/2 inch) long x 2.54 cm (1 inch) around the circumference, whereas the ultrasonic gage measures only a narrow line about 6.24 mm (1/4 inch) long. This could account for an EMAT tendency to read higher in the vicinity of a narrow, thin-wall area. A third reason for this variation could be in the method used to make each measurement. The ultrasonic readings were obtained along a straight line, but the EMAT readings were taken along the path provided by the conveyor which sometimes rotated the tube slightly by varying amounts.

EMAT readings for the thick-wall scans shown in Figure 7 essentially overlap readings of the ultrasonic gage, except for the tube ends. This variation could be caused by the solenoid field being distorted by the tube ends. This variation is also apparent in Figure 6 at the beginning of the tube. There is no apparent effect of tube rotation during the thick-wall scans as there was during the light-wall scans.

A comparison of EMAT responses at pipe speeds of 0.269 M/sec (50 ft/min) and 0.645 M/sec (120 ft/min) was also made. The same piece of 4-1/2 inch x 0.262 inch wall tube was scanned along the same line twice at 120 FPM and once at 50 FPM. Recorder charts of the two scans at 120 FPM when superimposed show very good reproducibility of the unit as previously mentioned. From the response to the cyclic variations, it is apparent that tube speed did not show any appreciable change within this limited speed range. Test speed was limited by the conveyor where the prototype unit was installed.

STATISTICAL ANALYSIS OF DATA

An analysis of variance was done for the three sets of 41 thick-wall measurements (T1, T2, T3) and the five sets of 41 light-wall measurements (L1, L2, L3, L4, L5). Table I shows that the repeatability among the three groups of 41 thick-wall EMAT measurements and among the five groups of 41 light-wall EMAT measurements is good.

The standard deviations from their respective means are 0.0020 for the thick-wall group and 0.0027 for the light-wall group; the composite measure of repeatibility was 0.0025. The overlays of the EMAT measurements have already demonstrated this repeatibility, however, these calculations statistically verify the overlays.

Comparing each of the 41 ultrasonic readings to the three sets of thick-wall EMAT readings (T1, T2, T3) and testing for a statistically significant difference in the two methods of measurement reveals that there is none (Table II).

However, when the same comparison is done for the ultrasonic measurements vs. light-wall EMAT readings (L1, L2, L3, L4, L5), there does exist a significant difference between the two methods of measurement. This has been illustrated in Figure 6 and an explanation offered in the third paragraph of the results section.

SUMMARY AND CONCLUSIONS

The feasibility of utilizing the EMAT method to gage tube walls has definitely been shown, at least for short runs. Durability of the transducer, the tendency for scale buildup to cause liftoff, the ease of changing transducers, and the time required to calibrate new transducers would have to be established over longer time periods. However, the ability of the system to measure normal cyclic wall variations, to detect not only light-wall areas found by the gamma gage but also one area missed by the gamma gage, is impressive. Also, the ability of the system to repeat measurements during multiple scans of the same tube and to maintain its calibration for more than 24 hours were clearly demonstrated. Table III lists the advantages and disadvantages of this unit.

Table I

ANALYSIS OF VARIANCE

THICK-WALL AREAS (T1,T2,T3)

NO. OF READINGS = 123

	SUM OF SQUARES	DEGREES OF FREEDOM	MEAN SQUARE	STD DEV.
SSQ TOTAL	.00391919	122	.000032229	.005677
SSQ BLOCKS	.003583333	40	.000089583	.009465
SSQ REPEATA-BILITY	.000348586	82	.000004251	.002062*

* - Repeatability among 3 identical groups of 41 EMAT measurements.

LIGHT-WALL AREAS (L1,L2,L3,L4,L5)

NO. OF READINGS = 205

	SUM OF SQUARES	DEGREES OF FREEDOM	MEAN SQUARE	STD. DEV.
SSQ TOTAL	.016266	204	.0000797	.0089295
SSQ BLOCKS	.015028	40	.0003757	.0193830
SSQ REPEATA-BILITY	.002239	164	.0000076	.0027484**

** = Repeatability among 5 identical groups of 41 EMAT measurements.

TABLE II

ANALYSIS OF VARIANCE - TEST OF SIGNIFICANCE

THICK-WALL MEASUREMENTS (T1,T2,T3)

OVERALL AVERAGES (INCLUDING LONGITUDINAL DIFFERENCES)

UT = .27234
EMAT = .27286

SOURCE	DEGREES FREEDOM	SUM OF SQUARES	MEAN SQUARE	F-VALUE
ROWS	40	.23638E-02	.59096E-04	4.334
COLUMNS (UT, EMAT)	1	.55849E-05	.55849E-05	.410*
ERROR	40	.54541E-03	.13635E-04	
TOTAL	81	.29148E-02		

* = This indicates that there is <u>no</u> significant difference between the UT and EMAT readings.

LIGHT-WALL MEASUREMENTS (L1,L2,L3,L4,L5)

OVERALL AVERAGES (Including Longitudinal Differences)

UT = .25232
EMAT = .25982

SOURCE	DEGREES FREEDOM	SUM OF SQUARES	MEAN SQUARE	F-VALUE
ROWS	40	.54083E-02	.13521E-03	4.265
COLUMNS (UT, EMAT)	1	.11539E-02	.11539E-02	36.399**
ERROR	40	.12680E-02	.31701E-04	
TOTAL	81	.78303E-02		

* = There <u>does exist</u> a significant difference between the two methods. It is estimated at .00750.

TABLE III

EMAT ADVANTAGES AND DISADVANTAGES

SHOWN BY TESTING AT LORAIN

ADVANTAGES	DISADVANTAGES
UNIT IS COMPACT AND REQUIRES MINIMUM SPACE.	EMAT'S APPEAR TO BE FRAGILE AND MAY WEAR OUT.
NO COUPLANT REQUIRED.	ELECTRICAL NOISE MAY CAUSE ERRONEOUS IMPULSES.
CALIBRATION IS RELATIVELY QUICK AND SIMPLE.	
CALIBRATION HOLDS FOR AT LEAST 24 HOURS.	
NO WARM-UP TIME NEEDED.	
NO PERFORMANCE CHANGE FOR TUBE SPEEDS OF 50 FPM AND 120 FPM.	
CAN EVALUATE FULL LENGTH OF TUBE IF TUBE IS STOPPED AT FRONT END AND SOLENOID HAS END-COMPENSATION CIRCUIT.	
ACTUAL WALL READINGS CAN BE DETERMINED FROM THE CHARTS.	
EMAT RESPONSES ARE NOT SENSITIVE TO PIPE STRAIGHT-NESS.	
RELATIVELY SAFE (NO RADIATION HAZARD) AND EASY TO OPERATE.	

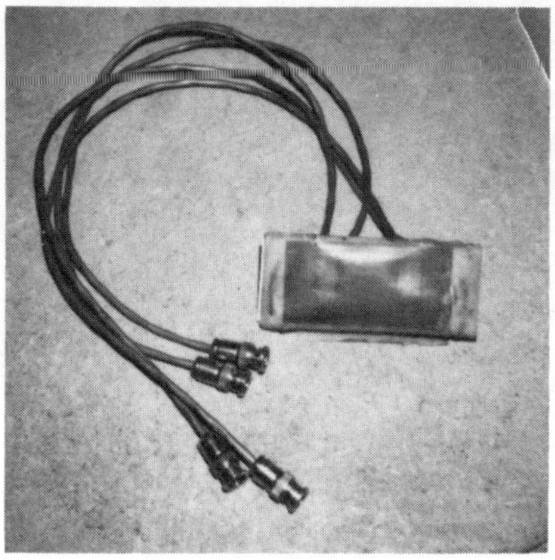

Figure 2A. A Flexible EMAT.

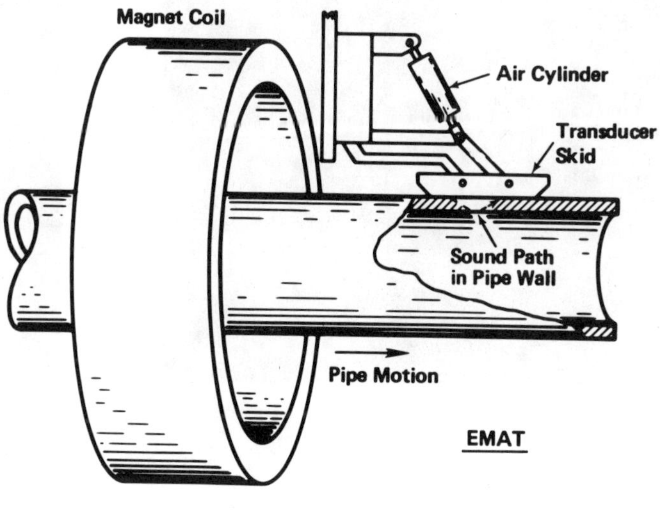

Figure 2B. An EMAT Mounted Within a DC Solenoid.

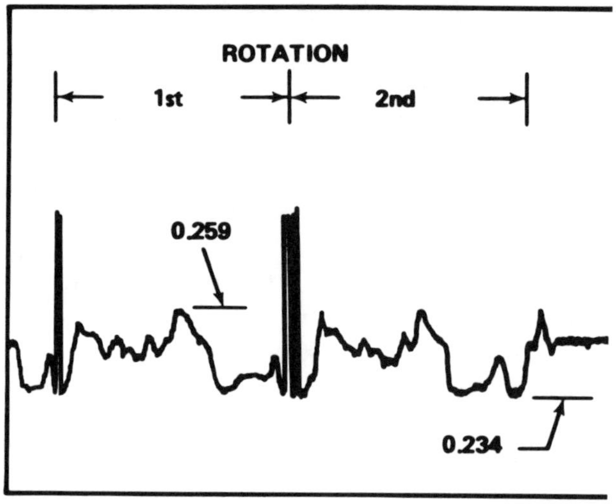

Figure 3. A Calibration chart of the EMAT Response.

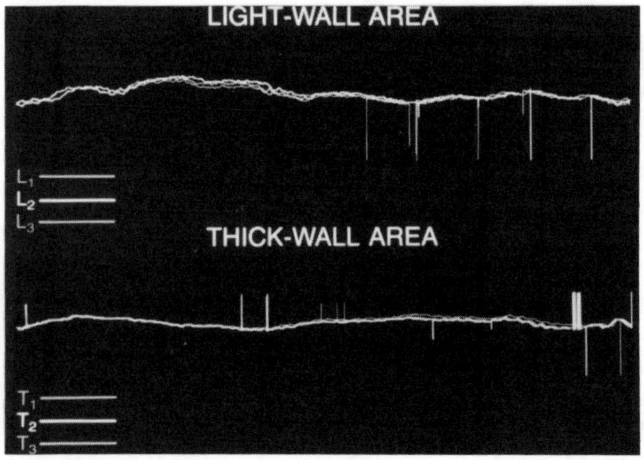

Figure 4. EMAT Response During Multiple Scans of the Same Tube.

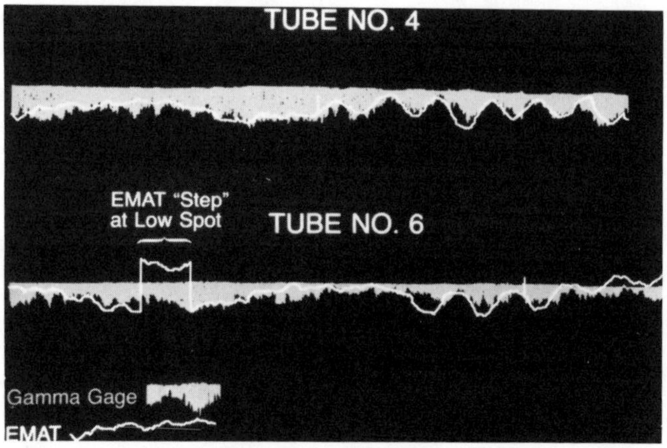

Figure 5. Overlays of EMAT and Gama Gage Charts.

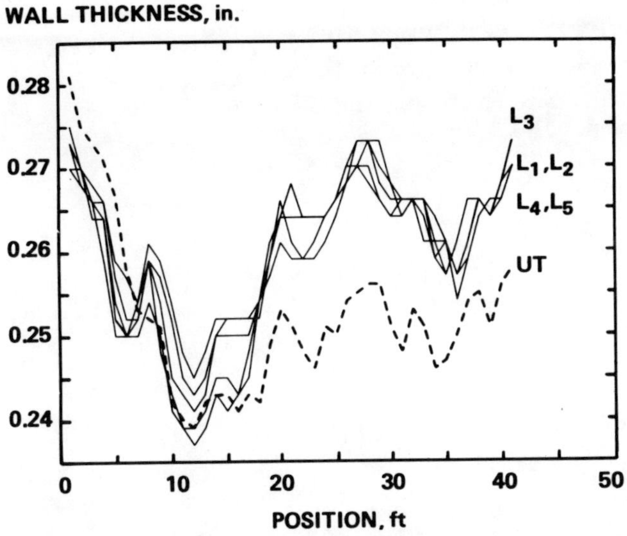

Figure 6. Comparison of EMAT and Ultrasonic Thickness Measurements for Light-Wall Scans.

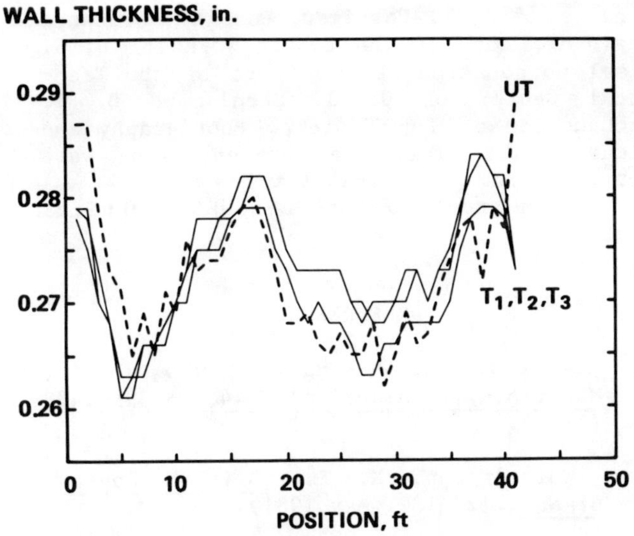

Figure 7. Comparison of EMAT and Ultrasonic Thickness Measurements for Thick-Wall Scans.

ACKNOWLEDGEMENTS

The authors were able to plan, organize, and perform these tests, statistically analyze the data, and present the results in graphical form through the capable efforts of W. L. Posey, Manager- Pipe Mills Metallurgical, and J. T. Corrigan, Quality Assurance Engineer, both of the Lorain Works of U. S. Steel Corporation, D. M. Scott of the Technical Center of U. S. Steel and D. T. MacLauchlan of MagnaSonics. Photography and visual aids for the presentation were obtained from E. M. Vrabel and A. L. Armstrong, Jr. of the USS Technical Center.

REFERENCES

1. Maxfield, B. W. and Fortunko, C. M., Materials Evaluation, 44, 1399-1408 (1983).

2. Whittington, K. R., British Journal of NDT, 127-132 (May 1981).

OPTICAL GENERATION AND DETECTION OF ULTRASOUND FOR FLAW DETECTION AND ITS POTENTIAL FOR IN-LINE PROCESS/QUALITY CONTROL

James W. Wagner
Dept. of Materials Science and Engineering
The Johns Hopkins University
Baltimore, MD 21218

In principle, optical techniques for ultrasonic testing and inspection may be used to perform the same tasks for which more conventional, contact transducer techniques are already used. Further, optical techniques offer distinct advantages over contact techniques in that they are potentially non-contacting, require no couplant, may be easily scanned over the surface of an object, are point sensing, and may provide large bandwidth for accurate time-of-flight measurements. Unfortunately, the practical application of optical systems for generation and detection of ultrasonic signals has been limited by their relatively high cost and poor sensitivity when compared with commercially available, contact transducer systems. In reviewing some of the basic principles involved, one can identify areas where further development in the optical techniques may likely result in more sensitive systems at a more reasonable cost.

Consider the case of a hot steel slab being processed from an ingot in a primary rolling mill. To improve productivity it is desirable to shear off from either end of the slab only that amount of steel which contains an internal defect such as those resulting from primary pipe or fishtail. To aid in this process, it is proposed that an ultrasonic "pulse-echo" system be employed to determine the extent of the defect in either end of the slab. As shown in Figure 1, if a transducer is positioned over a defect-free portion of the slab (position A), the acoustic signal observed when a contact piezoelectric transducer is used shows only an echo from the far side of the slab. At position B, the defect causes a portion of the acoustic signal to be reflected back to the transducer at a point in time earlier than the arrival of the echo from the back wall. Thus the extent of the defect may be determined by finding the position along the slab at which the early echo disappears. For practical reasons, simple piezoelectric contact transducers cannot be used to examine hot steel during production without elaborate provisions for cooling and coupling.

A far more desirable approach would be to use optical techniques for generating and detecting ultrasonic signals to inspect for these defects. As shown in Figure 2, an optical system differs from a contact transducer type pulse-echo system in that it must use separate instrumentation for generating and detecting the ultrasonic signals. Typically, a high energy, pulsed laser is used to generate the ultrasonic waves, and an interferometer incorporating a lower power CW laser may be used to detect surface vibrations at the point of interest.

LASER GENERATION OF ULTRASOUND

In general, one considers three modes by which elastic waves are generated using a pulsed laser[1,2,3]. These three modes are illustrated schematically in Figure 3. In the case of a free surface exposed to a short pulse of laser light, elastic waves are generated as a result of rapid, localized heating and expansion of a small volume of material at the metal surface. Since the surface is not constrained, the volume is free to expand normal to the surface resulting in a

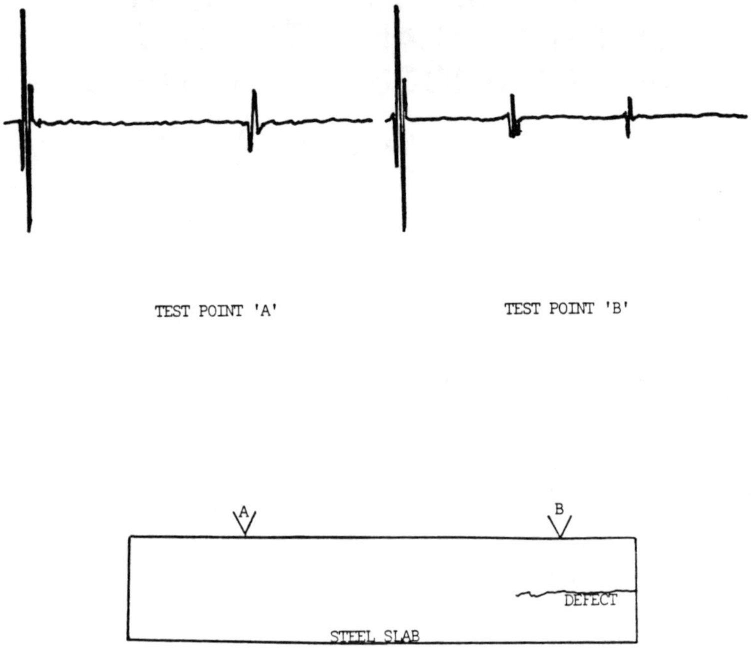

Figure 1: Defect location using pulse-echo testing.

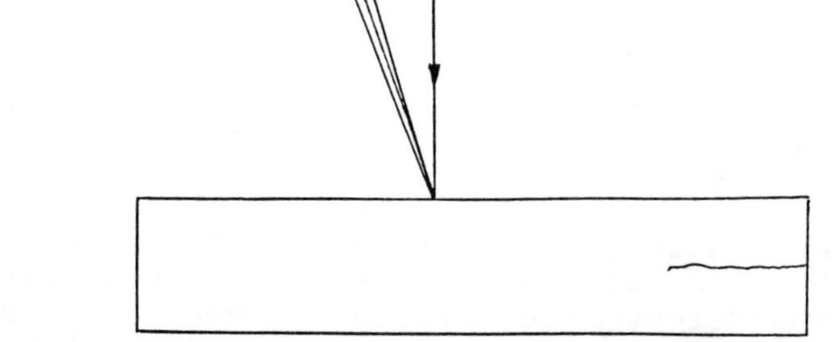

Figure 2: Ultrasonic pulse-echo arrangement
using optical generation and detection.

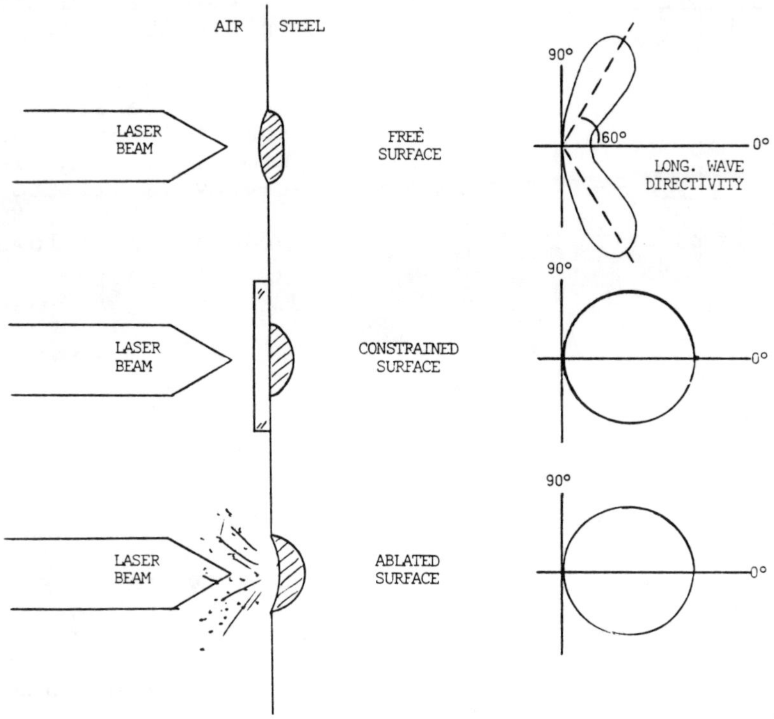

Figure 3: Optical generation of acoustic waves.

very small p-wave component propagating in a direction normal to the surface. By constraining the surface with some transparent material, like silicone grease, for example, a perpendicular p-wave component some 40dB larger than that generated in the free surface case is produced[3]. In lieu of constraining the surface, one may deliver a laser pulse sufficient energy to ablate a portion of the surface material. The resulting momentum transfer again contributes to a large perpendicular p-wave component. It is interesting to note that of the three cases discussed, only the free surface case is truly noncontacting and nondestructive. Unfortunately, this free surface case also provides the weakest perpendicular p-wave component and is, therefore, least suited for use in the pulse-echo detection scheme described earlier. Ablative generation of acoustic waves, however, causes insufficient damage to be of concern in the primary processing of steel slabs.

OPTICAL DETECTION OF ULTRASOUND

The most common approaches to optical detection of ultrasound incorporate optical interferometers of one form or another. One of the simplest interferometers is the Michelson interferometer shown schematically in Figure 4. The laser provides a beam of polarized, coherent light which is split into two beams. One of the beams is reflected from the object surface, the other from a reference mirror. They recombine at the beamsplitter and are directed to a photodetector. The photodetector converts the intensity into a current signal which is amplified and displayed. Since the light is coherent, the intensity at the photodetector varies as a result of the interference between the recombined beams. This interference intensity varies periodically as a function of the optical path length difference between the two legs of the interferometer. Therefore, vibrations on the specimen surface will cause a change in path length difference which in turn causes a change in detected intensity resulting in a fluctuation of the detector output current.

Michelson interferometers with some modifications have been used to measure acoustic signals in laboratory conditions. Since the output intensity varies periodically for a linear change in path length, it stands to reason that there must be some optimum path length difference at which the interferometer will be most sensitive to the very small path changes caused by surface vibration. In order to maintain this optimum path difference, most

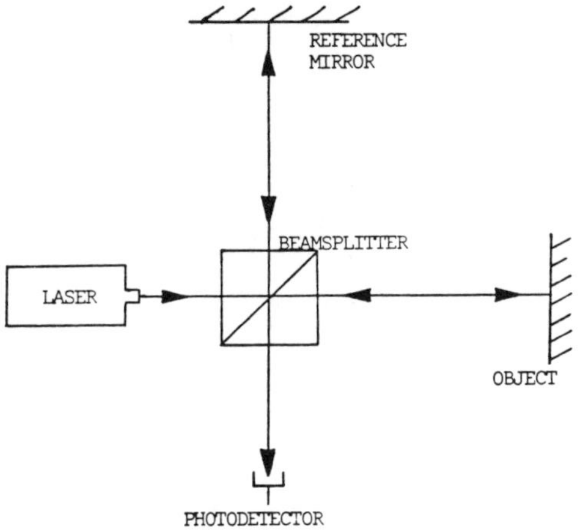

Figure 4: Michelson interferometer.

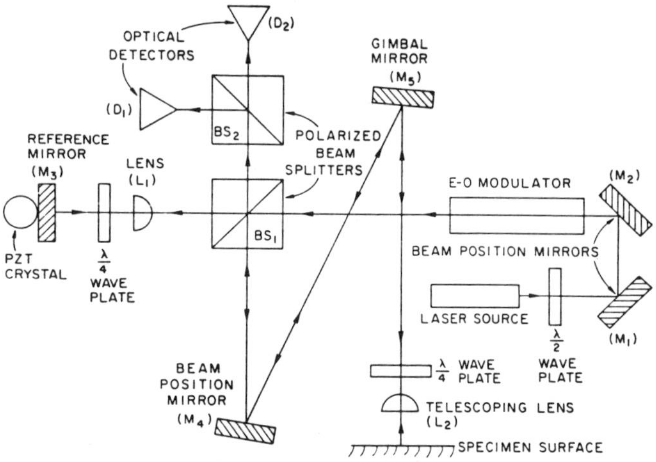

Figure 5: Practial path-stabilized Michelson interferometer.

practical Michelson systems employ some form of optical pathlength stabilization. The interferometer system shown schematically in Figure 5 has been used to study acoustic emissions phenomena in steel specimens. Its design incorporates a PZT crystal for controlling the position of the reference mirror and an electro-optic crystal to provide high speed corrections to the optical path length difference. Both of these devices are driven by signals fed back from the photodetectors.

SENSITIVITY OF OPTICAL ULTRASONIC SYSTEMS

As mentioned earlier, the relatively poor sensitivity of optical systems for detecting ultrasonic signals has been a major issue restricting their use. Fortunately, as will be shown, there appear to be several avenues of development available which may contribute to significant improvements in sensitivity.

First, consider the sensitivity of a path stabilized Michelson interferometer for detecting small surface motions, x. The interferometer's sensitivity is limited by its signal to noise ratio (SNR) which may be expressed as follows[4]:

$$SNR = \frac{\eta P_0}{2h\nu B}\left(\frac{4\pi}{\lambda}x\right)^2 \frac{M^2 \sin^2 \frac{4\pi}{\lambda}X}{1 + M\cos\frac{4\pi}{\lambda}X}.$$

where

η = quantum efficiency in carriers/photon

q = electron charge

P = incident light intensity

h = Plank's constant

ν = light frequency

λ = wavelength of light

X = stabilized path difference

x = path change due to surface vibration.

By solving this expression for x, the following is obtained:

$$x = [\textit{Path Sensitivity Term}] \times \sqrt{\frac{Bandwidth}{Power_D}}$$

It should be noted that this is a general expression applicable to many interferometric detectors. The pathlength sensitivity term varies for quadrature dual detection or for various forms of heterodyne interferometers[5], but the basic dependence on power and bandwidth remains the same. From this equation it is clear that two direct means for minimizing x are to decrease the signal bandwidth and to increase the light intensity falling on the detector. Indeed it is this later approach which has been used by one German manufacturer of an optical system for ultrasonic testing. In order to increase the amount of light on the detector, the manufacturer employs a long pulse width, high power Nd:YAG laser in the interferometer portion of the system. The use of this extremely high intensity light source helps to ensure that sensitive measurements can be made even on rough or highly absorptive surfaces. Unfortunately, this represents a fairly expensive solution to the sensitivity problem.

A second improvement which can be made in the interferometer system is to reduce the signal bandwidth of the processing electronics. This solution is not as straightforward as it appears, however. For although the minimum detectable x is indeed reduced as the bandwidth is narrowed, the system is sensitive only to those frequency components of the surface vibration which are within the bandwidth of the processing electronics. This is not a problem for contact transducer pulse-echo systems since the band-limiting generally occurs within the transducer. If the same transducer is used to both transmit and receive, all of the source energy returning to the specimen surface is within the bandwidth of the receiving system and is detected. In the case of pulsed laser generated sound waves, however, the frequency spectrum of the generated acoustic pulse ranges from DC to frequencies up to 30 to 100 MHz. Consequently, by narrowing the bandwidth of the receiver (interferometer) system, only a small portion of the laser-generated acoustic energy will be detected.

The above observation leads to the conclusion that perhaps some effort should be given to the development of more appropriate laser sources for acoustic testing. Using the proper lasing medium and cavity design, relaxation oscillations may be induced in the output pulses of some lasers[6]. These damped oscillations often occur in the low megahertz regime indicating a strong possibility that they may be used as sources for narrow band thermoelastic wave generation. Other alternatives such as pulse modulation techniques might also prove useful. Once narrow band acoustic signals are generated, bandwidth limiting in the detection system becomes a viable means by which sensitivity to acoustic signals may be improved.

CONCLUSIONS

Several approaches have been discussed for improving the performance of optical systems for noncontact generation and detection of ultrasound. At present the only commercially available systems, with sensitivity comparable to that of contact transducer systems, achieves its sensitivity at considerable cost. By pursuing some of the alternative means for improving sensitivity, such as bandlimiting both the opto-acoustic source and the interferometer system, it appears possible to develop practical systems capable of performing process control and quality assurance functions using optical generation and detection of ultrasound.

REFERENCES

1. Aindow, A.M., J. Acoustic Soc. Am. 69(2), 449-455 (1981)

2. Dewhurst R.J. et al., J. Appl. Phys. 53(6), 4064-4071 (1982)

3. Hutchins D.A. et al., Ultrasonics 19, 103-108 (1981)

4. Kwaaitall, T.H. et al., J. Phys. D: Appl. Phys. 13, 1005-1015 (1980)

5. Whitman R.L. and A. Korpel, Appl. Opt. 8(8), 1567-1575 (1969)

6. A. Yariv, "Introduction to Optical Electronics", pp147-153, Holt, Rinehart, and Winston, New York (1976)

INTEGRATED CONTROL OF THE STEEL SLAB REHEATING PROCESS USING ON-LINE ESTIMATION OF FURNACE AND SLAB TEMPERATURE PROFILES

Brian G. Frock, James R. Cook and Dave F. Ellerbrock
Armco Inc.
Middletown, OH 45043

ABSTRACT

Optimal control of the steel slab reheating process involves a tradeoff among the conflicting objectives of production, quality and energy usage. The problem is made more difficult when slabs of different chemistries, geometries and initial thermal profiles are reheated. This paper describes work-in-progress toward development of an on-line system for control of the reheating process. The system combines in-process measurements with a mathematical model of the heat transfer process to predict the through-thickness temperature distribution of the slab at any time during reheating. Information from the rolling process will be used to provide feedback corrections to the control system.

MANY CURRENT SYSTEMS rely upon an experienced human operator to control the reheating process in order to deliver a slab with an average temperature and a temperature profile which are within prescribed ranges. These properly heated slabs must be delivered at a rate which is often dictated by the downstream processes, and with a scale jacket which is easily removable in the early stages of rolling. The reheating process is extremely complex and is controlled by maintaining zone temperature setpoints or fuel flow rates which have been established by experience and which may vary from operator to operator. Because there are a number of independently controlled zones within a furnace and because the thermal energy transfer process is very complex, many different combinations of zone temperature setpoints may produce a slab with acceptable metallurgical and thermal properties. However, few of these combinations are energy efficient. The control problem is further complicated by unscheduled delays in subsequent processing steps which cause unanticipated and sudden changes in delivery rates (pacing), and by the unreliable nature of uncorrected feedback information which is currently being provided to the furnace operator.

The use of an on-line, real-time computer control system with an imbedded thermal energy transfer model and in-process nondestructive measurements of slab surface temperatures will allow the reheat process to be controlled in an improved manner subject to system constraints. In addition, the computerized control system will be able to use downstream information (surface temperatures and rollability measures) to correct for inaccuracies in the thermal energy transfer model.

REHEAT PROCESS

The function of the reheat process is to heat slabs to a specified temperature and to deliver them to the hot strip mill at an acceptable rate. In addition to the temperature and delivery rate (pacing) requirements, a surface oxide layer must be developed which is thick enough at extraction to include all surface defects and which is loose enough to be entirely removed so that it is not rolled into the finished product.

The conditions under which the reheat process must operate are dynamic and highly variable. At input the average slab temperature may vary from 20°C (68°F) if cold charging is being employed to over 550°C (1022°F) if hot charging is being used. The internal temperature distributions of the slabs may also be quite different from slab to slab at the time of entry (charging). The pacing of slabs through the furnace may change suddenly due to both scheduled and unscheduled delays in subsequent processing steps. Slab thicknesses range from 165 mm (6.5 in) to approximately 267 mm (10.5 in) while slab widths may range from 610 mm (24 in) to 2032 mm (80 in). Thus, the reheat process must function subject to very adverse operating conditions.

A typical slab reheat furnace is shown in Figure 1. The furnace is characteristically divided into a top and bottom charge zone, a top and bottom heat zone, and a top and bottom soak zone. A waste gas flue is located in the charge zone very near the slab entry door. Once the slabs are inside the furnace thermal energy is transferred by: (1) radiation from the hot face elements in the roof, side walls and floor of the zones, (2) by radiation from the burner flames, and (3) by convection from the surrounding air. While heating does take place in the charge zone, the furnaces are designed so that most of the thermal energy transfer takes place in the heat zone. This permits more efficient use to be made of the thermal energy in the air since this energy can be used to heat the slabs in the charge zone. The soak zone functions mainly to redistribute the thermal energy within the slab. The furnace illustrated in figure 1 has an unfired solid hearth. In this type of furnace all thermal energy input in the soak zone comes from burners in the roof, while the solid hearth generally functions as a "semi-adiabatic" lower boundary for the slab.

A typical average temperature for a slab as a function of its position in the furnace is illustrated in figure 2. The temperature of the slab rises rapidly in the charge and heat zones then levels off toward the end of the heat zone and remains relatively constant through the soak zone. Through-thickness temperature distributions for the slabs at specific locations are illustrated in figure 3. The effect of the soak zone and its solid hearth on the temperature distribution within the slab are evident in this figure.

The entire process is energy intensive, requiring the addition of approximately 200 KCal/Kg (0.75 MMBTU/ton) to heat a slab from an average temperature of 20°C (68°F) to an average temperature of 1274°C (2325°F). At a pacing rate of 181600 Kg (200 tons) per hour, the furnace must be capable of inputing at least 3.6×10^7 KCal (150 MMBTU) per hour just to heat the steel. This does not include the energy losses through the walls, floor and ceiling, nor the energy losses due to the exit of hot air through the flue. Taking these losses into account, the efficiency of a reheat furnace is about 30%. Thus, approximately 1.2×10^8 KCal (500 MMBTU) per hour would be used during reheating at the specified pacing rate.

EXISTING CONTROL SYSTEM

The slab reheating process at Armco is currently controlled by human operators who must rely on past experience to heat the slabs to the required rolling temperatures. Control is exercised by specifying temperatures (called zone temperature setpoints) for each of the five zones. Once the operator has set these temperatures, microcomputers compare them to actual zone thermocouple readings and independently adjust the fuel flow rates in the zones to maintain the setpoints. Fuel/air ratios and furnace pressures are also controlled by microcomputers.

The independent temperature setting and controlling of so many zones presents the human operator with an overwhelming variety of choices. There are many different combinations of zone temperature setpoints which can lead to nearly identical average temperatures of the slabs at the time of the slab's extraction from the furnace. The problem of choosing and changing setpoints in response to changing downstream processing requirements could be overcome if some nondestructive measures of the slab's average temperature could be provided during the reheating process. However, no such measurements are currently available and the operator has almost no information with regard to the actual temperature of any of the slabs in the furnace. Controlling the process is made even more difficult by the lack of immediate feedback information at the time of a slab's extraction from the furnace.

The feedback information which the operator receives from subsequent processing steps is quite often unreliable because of the variability of these processing steps. An examination of the hot strip mill (see figure 4) will help to illustrate this point. After extraction from the furnace the slab passes through a vertical scale breaker which loosens the scale formed during reheating. The slab then passes through a series of roughing stands where it is reduced in thickness. High-pressure water sprays at the entry side of the roughing stands cool the surface of the slab while they remove the scale jacket. Thickness reductions during roughing add heat to the slab, whereas radiation losses cool it between stands. Some conductive cooling of the slab also occurs during contact with the rolls in the roughing stands and on the roll table. In addition to heat losses during normal processing, the slab may be delayed between stands due to problems which are occurring downstream. This leads to further heat losses from the slab. Time delays also occur in the finishing stands where the bar, as the slab is called after exiting the roughing stands, is further reduced to a strip. At the present time no measurements are being made of the average temperature of the slab during the roughing process. The only techniques being used are noncontacting measurements of surface temperatures by optical pyrometers at the third and final roughing stands. Surface temperature measurements are also being made at the exit of the final finishing stand.

Since stringent temperature constraints are imposed on the bar at its exit from the last roughing stand and on the strip at its exit from the last finishing stand, it is these temperatures (measured by noncontacting optical pyrometers) that the operator relies on for his feedback control information. It should be obvious from the discussion in the preceding paragraphs that these temperatures are only loosely related to the average temperature of the slab at the time of its extraction from the reheat furnace and that their unrestricted use by the operators could lead to serious control problems. It should also be obvious from the previous discussion that the existing state of control of the reheat process results in the very inefficient heating of the slabs. Some of the slabs will be overheated, some will be underheated, and those that are delivered to the hot strip mill with an acceptable temperature distribution may not have been heated efficiently. Overheating the slabs wastes energy, decreases furnace life and can produce a scale jacket which cannot be entirely removed during subsequent processing steps. Underheating the slabs results in the downgrading of the material for metallurgical reasons.

PROPOSED CONTROL SYSTEM

A schematic of the planned control system is illustrated in figure 5. The system will include: (1) an off-line mathematical model of the reheat process, (2) an on-line, real-time mathematical model of the reheat process which will predict the through-thickness distribution of temperatures of the slab at any location in the furnace as well as at extraction, (3) noncontacting slab surface temperature sensors within the furnace to correct the on-line model's predictions of temperatures during the reheat process, and (4) nondestructive sensors at subsequent processing steps to provide feedback information for further corrections to the on-line model.

MATHEMATICAL MODELS - The mathematical models are of primary importance to the reheat process control system. An off-line version which has been completed will be used to: (1) develop and evaluate reheat furnace firing strategies, and (2) develop a real-time, on-line reheat process control model. The off-line version is one-dimensional and allows the slab's upper and lower surfaces to exchange thermal energy with the external environment by radiation (roof and floor elements only), by natural convection[1,2], and, in the case of the solid hearth, by conduction[3]. Other sources of radiation (flames, sidewalls, and hot gases) are ignored. Thermal energy exchange within the slab takes place by conduction, where explicit finite difference equations are used to calculate temperatures at twelve uniformly spaced locations through the thickness of the slab. The model requires specification of the slab's through-thickness temperature distribution at the time of charging. This allows it to accommodate both hot and cold charging so long as a reasonable approximation to the slab's through-thickness temperature distribution at the time of charging can be obtained.

The off-line model requires more information with regard to roof and floor element temperatures than can be supplied by the zone thermocouples. To overcome this deficiency an experimental program is being carried out to correlate roof and floor element temperatures (as measured by an optical pyrometer) with zone thermocouple readings and fuel flow rates. There are also several important model parameters which are not well known, e.g., furnace roof and floor element emissivities, slab surface emissivities, flame emissivities, and gas temperatures. These values can be estimated by comparing the model's predictions of the slab's through-thickness temperature distributions with experimentally measured through-thickness temperature distributions of the slab. The unknown parameters can then be adjusted so as to minimize the error between the predicted and actual temperature distribution.

Actual through-thickness temperature distributions are being measured with the use of a six-channel recording device called a Thermophil STOR. This device rides through the furnace in the end of a modified slab (see figure 6). Six thermocouples are imbedded in the slab at predetermined locations, and temperatures from each of the thermocouples are sampled and stored in the Thermophil STOR's memory. Data accumulated in this manner for several runs through the furnace will be used to tune the model by adjusting the appropriate model parameters.

After the off-line model has been tuned it can be used to generate a number of efficient slab heatup (time-temperature) curves[4] for the range of furnace operating conditions which are likely to be encountered. The simplified, real-time, on-line model, which would also be developed from the off-line model, will be used to predict each slab's internal temperature distribution and its average temperature during actual furnace operation. The calculated average temperature will then be compared to the appropriate heatup curve and corrections will automatically be made to fuel flow rates, zone temperature setpoints and pacing so that the calculated average temperatures of the slabs match those of the appropriate heatup curve.[4]

INITIAL TEMPERATURE OF SLABS - Cold-charged slabs present no special problems for the model since their internal temperatures can be assumed to be that of the surrounding air. Hot-charged slabs, however, do present a problem. A knowledge of their through-thickness temperature distributions is important in order for the model to make accurate predictions of the slab's temperature and in order to operate the furnace in an efficient manner. These temperatures might be obtained by: (1) using a mathematical model to calculate them or (2) using a nondestructive temperature sensor to measure the required temperatures. No mathematical model is currently available, nor does an adequate nondestructive sensor exist. However, an AISI sponsored effort to develop such a sensor is underway.[5] This sensor will infer the through-thickness temperature distributions by measuring the transit times of ultrasonic waves through the thickness of the slabs. Until the sensor is developed or an adequate mathematical model becomes available, optical pyrometer measurements of surface temperatures must be relied upon to estimate the temperatures of hot-charged slabs.

IN-PROCESS SENSORS - Because of simplifications which will be required for the on-line model and because of the dynamic conditions under which it must operate, there may be significant errors in the model's temperature predictions under certain operating conditions. These errors can be reduced by providing the model with actual nondestructive measurements of the slab's surface temperature at specific locations within the furnace. These surface temperature measurements will be made by optical pyrometers which will "view" the slab's surfaces through water-cooled, air-purged sight tubes. The model can then use these measured temperatures to correct its own predictions of the slab's through-thickness temperature distribution.

FEEDBACK CORRECTIONS - Information from the processing steps which follow the extraction of the slab from the furnace can be used to nondestructively infer the average temperature of the slabs. Literature sources which discuss hot rolling theories[6-8] all demonstrate the temperature sensitivity of the forces produced during the reduction of slabs and sheets. Experimental measurements made at Armco[9] have verified the relationship between average temperature and roll forces generated during the reduction of ingots. Experimental investigations at Bethlehem Steel Corporation[10] have demonstrated that a good correlation exists between surface temperature and the powers required to reduce a thin slab at the last two roughing stands in a hot strip mill. Inland Steel Company[11] has found that surface temperature measurements of bars at the exit of the final roughing stand can be used as feedback information to correct for errors in their model's[12] predictions of a slab's average temperature. These surface temperatures are useful only if they are obtained from slabs which are processed under "normal" conditions, i.e., without delays.

Since their may be delays during these post reheat processes, the inferred temperatures must either be corrected for thermal energy gains and losses by some model[13], or the information which is used must be restricted to those cases for which no delays occurred. Information modified by a model could be used to directly correct the reheat process model's temperature predictions and/or to alter reheat strategies. Uncorrected, but restricted, temperature information could be used to check the consistency of the model's predictions.

SUMMARY

This paper has described the work-in-progress toward the development of an on-line, computerized control system which will use feedforward and feedback techniques for controlling the steel slab reheating process. An on-line mathematical model, developed from an experimentally tuned, off-line, mathematical model of the reheat process, will predict the internal temperature distribution of any slab at any time while it is being reheated. The on-line control model will require an initial temperature profile for the slab, zone temperature readings, fuel flow rates, and the slab thickness in order to update its predictions of the slab's temperature profile. Optical pyrometers will measure the slab's upper and lower surface temperatures nondestructively at specific locations during the reheat process. These surface temperature measurements will be used to correct the on-line model's temperature predictions. Nondestructive techniques such as roll force measurements will be used to infer the slab's average temperature during subsequent processing steps. Optical pyrometers will also be used to measure surface temperatures during these same processing steps. After suitable restrictions have been applied or after appropriate modelling, these average temperatures and surface temperatures will be used as feedback parameters for correcting the model. The application of this control system will result in considerable energy savings during the reheating process and will reduce the incidences of overheating and underheating of the slabs.

REFERENCES

1. Veslocki, T. A., "Development and Verification of a Slab Reheating Furnace Mathematical Model," Iron and Steel Engineer, 59, 46-51 (April, 1982).
2. Kamata, M., Kuraishi, T., Murakami, S., Yabuuchi, K., Shiraishi, K., and K. Kusumoto, Nippon Kokan Technical Report No. 30, 34-41 (1980).
3. Purdy, D. J. and T. J. Williams, "Temperature Control of a Continuous, Push Type, Five Zone, Slab Reheating Furnace for Minimum Fuel Usage," Purdue Laboratory for Applied Industrial Control Report No. 91, Vol. II (1981).

4. Fontana, P., Boggiano, A., Furinghetti, A., Cabras, G., and C. A. Simmoncini, "An Advanced Computer Control System for Reheat Systems," Iron and Steel Engineer, 60, 55-62, (August, 1983).
5. Cook, J. R., and H. Wadley, "Rapid Measurement of Temperature Distribution within a Solid or Solidifying Body of Hot Steel," Presented at AISI Technical Symposium No. 12, (November 2 and 3, 1983).
6. Z. Wusatowski, "Fundamentals of Rolling," 203-386, Pergamon Press, New York, New York (1969).
7. Helmi, A. and J. M. Alexander, "Geometric Factors Affecting Roll Force and Torque During the Hot Flat Rolling of Steel," Journal of The Iron and Steel Institute, 207, 1219-1231 (September, 1969).
8. Denton, B. K. and F. A. A. Crane, "Roll Load and Torque in the Hot Rolling of Steel Strip," Journal of The Iron and Steel Institute, 210, 606-617 (August, 1972).
9. Cook, J. R. and B. Frock, "Integrated Real-Time Control Schemes for Ingot Processing Including Soaking and Slabbing Mill Dynamic Drafting," Presented at AISE Subcommittee No. 28 Meeting (April 9 and 10, 1984).
10. Hansen, S. S. and B. L. Bramfitt, "Effect of Temperature and Composition on the Power Requirements of Hot Strip Mill Roughing Stands," Mechanical Working and Steel Processing XVII, 237-259 (1979).
11. Personal Conversation with Mr. T. A. Veslocki, Inland Steel Company.
12. Veslocki, T. A. and C. C. Smith, "Applications of a Dynamic Mathematical Model of a Slab Reheating Furnace," Applications of Mathematical and Physical Models in the Iron & Steel Industry, 134-141, Third Process Technology Conference, Pittsburgh, Pennsylvania (1982).
13. Misaka, Y., Takahashi, R., Shinjo, A., Nariai, Y., and M. Kooriki, "Computer Control of a Reheat Furnace at Kashima Steel Works' Hot Strip Mill," Iron and Steel Engineer, 59, 51-55, (May, 1982).

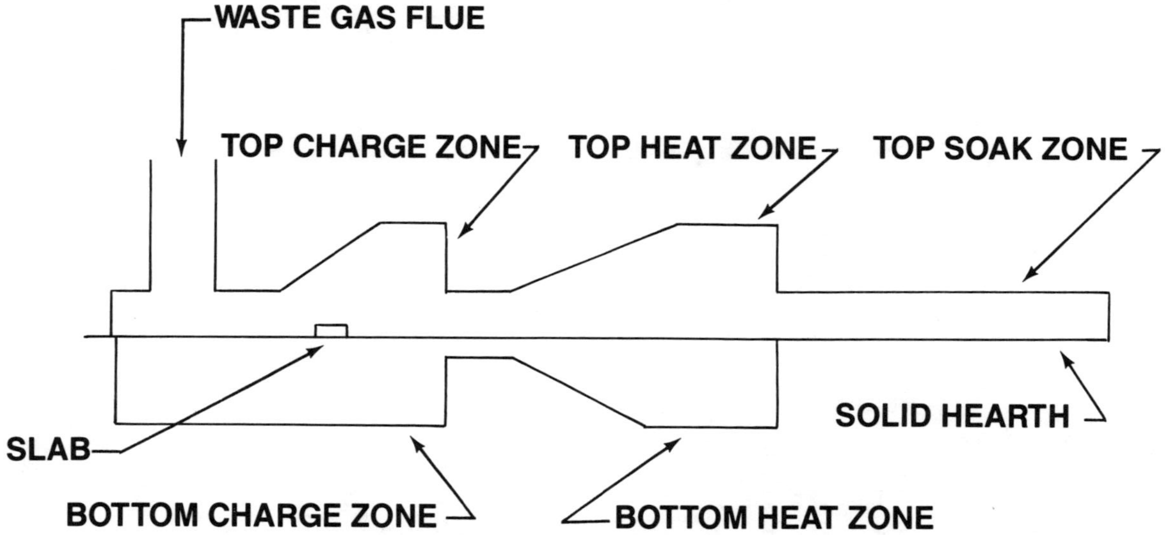

Fig. 1 – Five zone, pusher reheat furnace

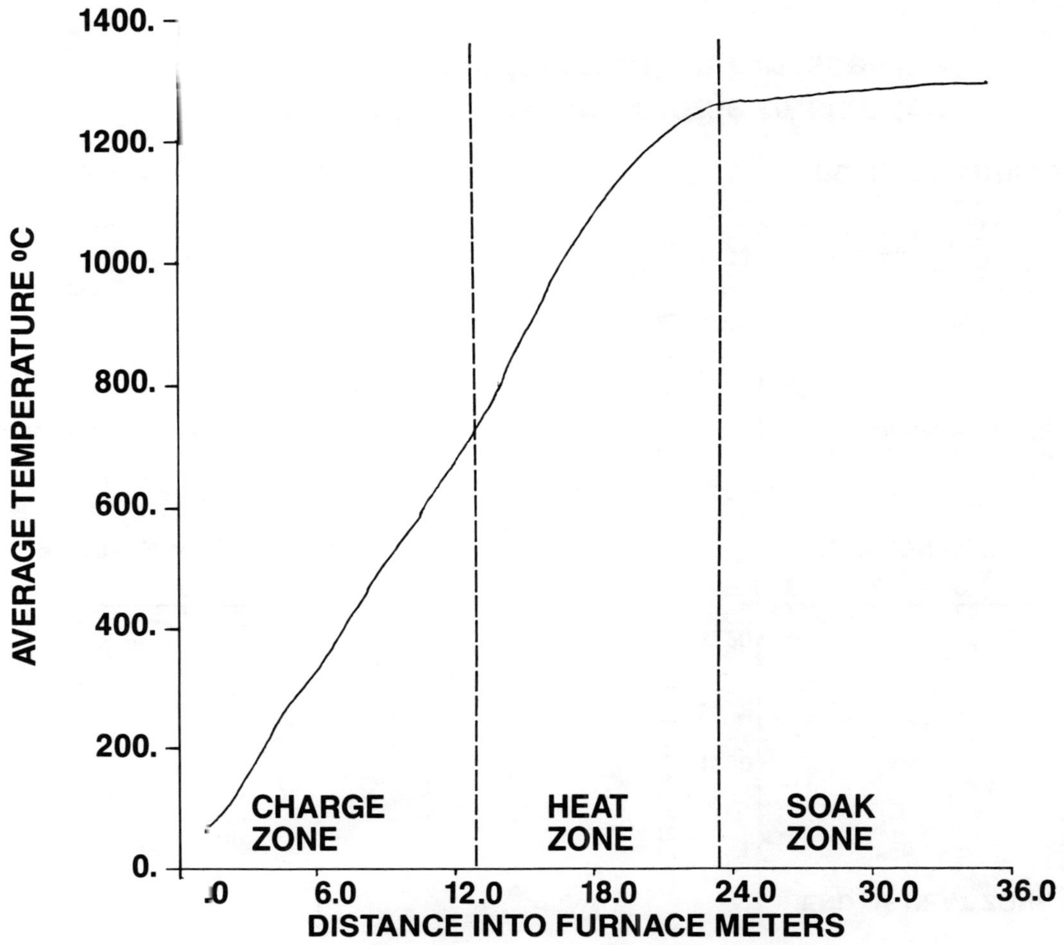

Fig. 2 – Average temperature of slab as a function of position in the furnace

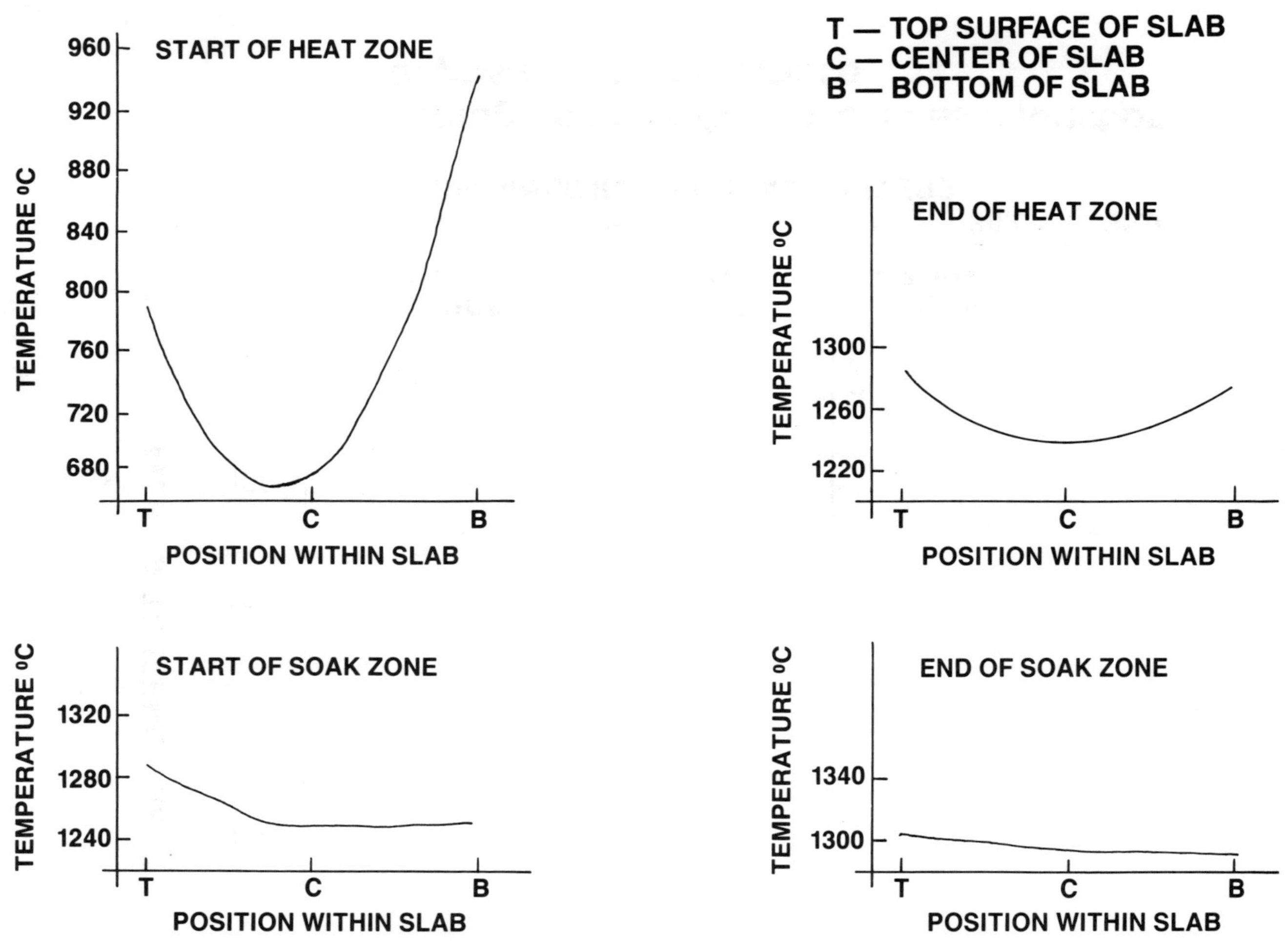

Fig 3 – Typical thermal profile of slab for selected locations in furnace

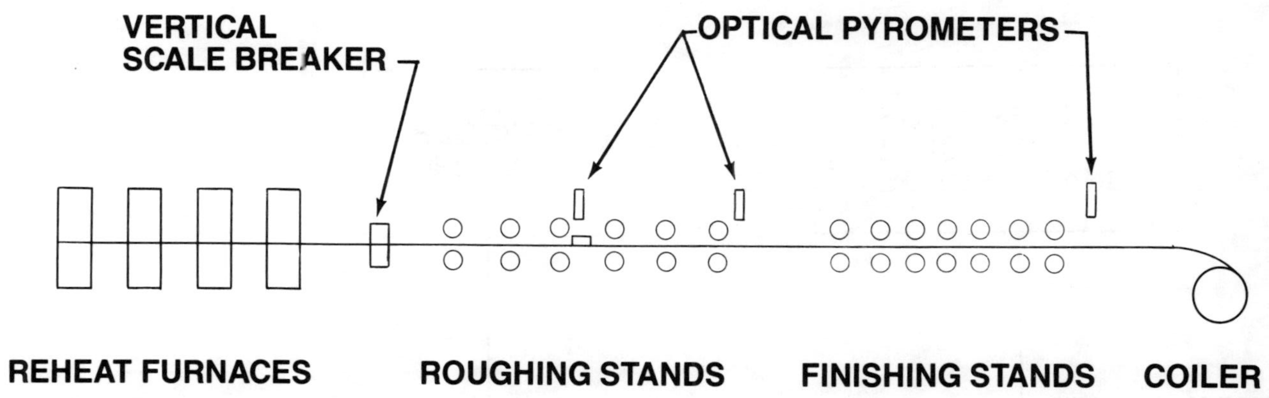

Fig. 4 – Hot strip mill

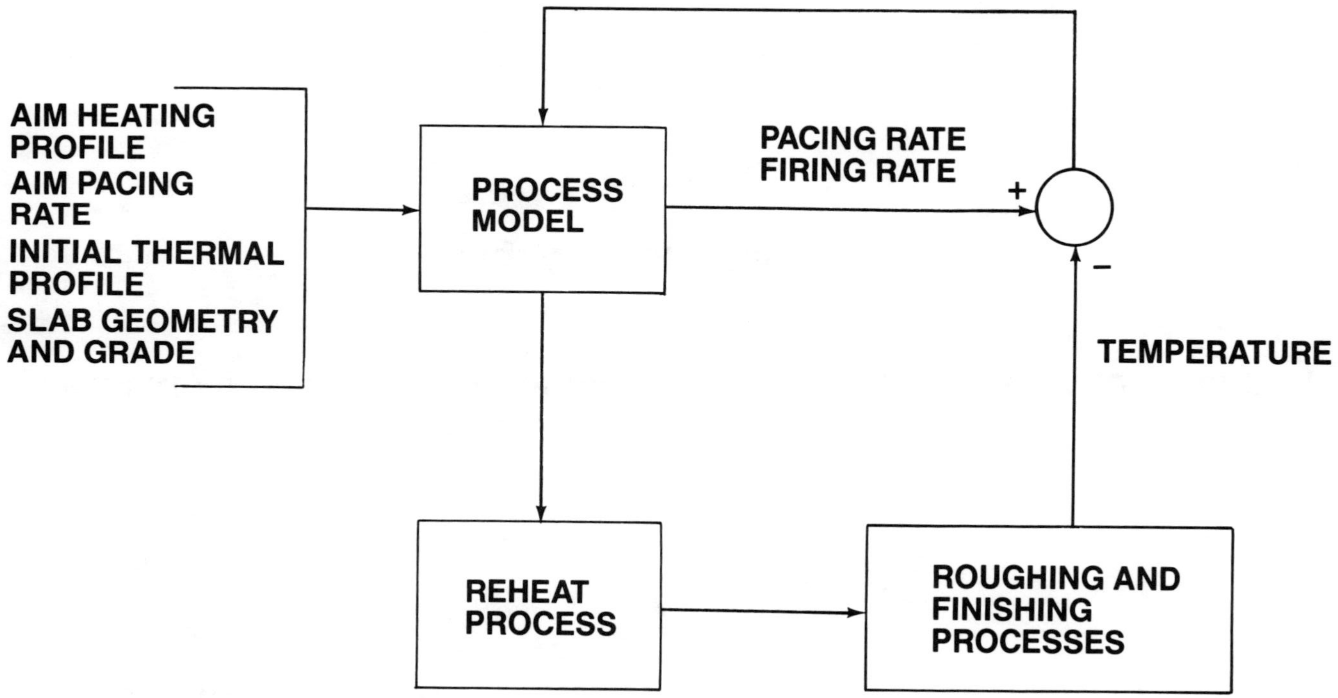

Fig. 5 – Reheat process control system

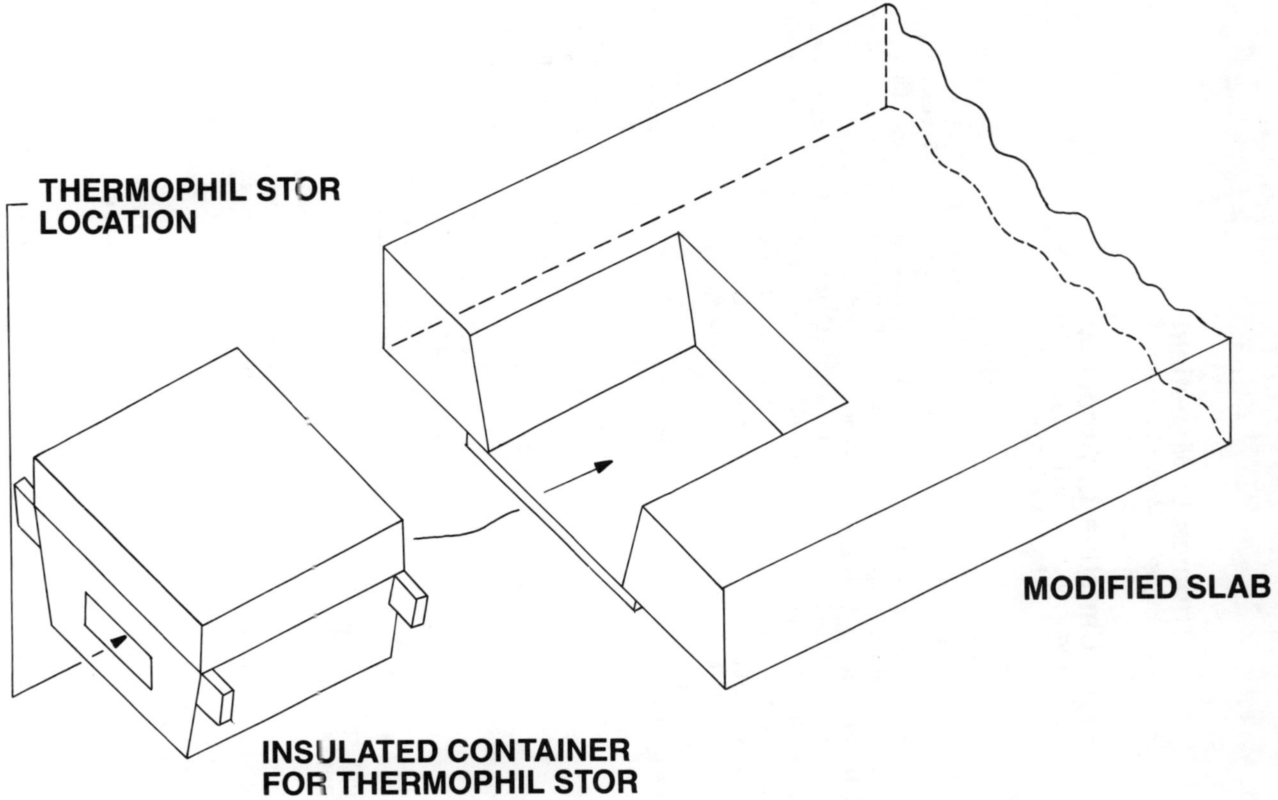

Fig. 6 – Modified slab and insulated container for Thermophil STOR

ULTRASONIC INSPECTION OF MOLTEN ALUMINUM

Thomas L. Mansfield
Reynolds Metals Company
Richmond, Virginia

Christine L. Bradshaw
Formerly of General Research Corp.
McLean, Virginia

ABSTRACT

Improved techniques for quality control of molten metal is a continuing need. Until the development of the Reynolds 4M™ System, there was no continuous real-time method available for determining the particulate/inhomogeneity content of molten aluminum. This ultrasonic system permits continuous monitoring of the quality of molten metal during casting and the evaluation of molten metal processing and/or cleaning operations. The hardware and software implementation of the Reynolds 4M™ System, and examples of results obtained through its use, will be discussed.

DURING THE PRODUCTION OF ALUMINUM, the quality control engineers and plant managers must be assured of the quality of their product. Whether the end product is specialty metals for critical applications or large-volume bulk pieces, the plant personnel must have some means of evaluating product quality and determining that the product meets the specifications and requirements of their customers.

This quality evaluation may be performed in a number of ways at various stages of the aluminum-making process:

(1) After final fabrication, the end product may be inspected by visual methods or, for example, by standard x-ray, eddy current, and ultrasonic nondestructive testing techniques.

(2) During fabrication, sophisticated inspection systems with front-end sensor units may be an integral part of rolling or extrusion processes to detect defective items as they are produced.

These are both feasible methods for evaluating product quality and screening end products for defects. However, in the first case, the cost of final processing of the product has already been expended when the defective pieces are identified, resulting in unnecessary expense in processing and/or inspection. The second testing stage is more desirable; however, there is a large initial investment to develop such sophisticated systems. In addition, unique sensor configurations and signal processing techniques would be required for each different product inspected due to variations in the dimensions, shapes, and properties of the manufactured item (i.e., rolled sheets vs. seamless pipe).

Ideally, the quality evaluation should be made prior to final fabrication and should apply in general to the production of any aluminum product. Until the development of the Mansfield Molten Metal Monitor (Reynolds 4M™ System), ultrasonic inspection within Reynolds operations had been confined to the finished or semi-finished product, e.g., extrusion, plate, sheet, and ingot, and fell into the categories mentioned above. A third possibility now exists:

(3) During the casting process, the molten metal quality can be determined to provide an immediate evaluation of the ingot before it is further processed.

When metal quality is evaluated during casting, poor quality metal is identified immediately and the expense of further processing may be avoided. Monitoring metal quality during casting also provides immediate feedback of various cast-house processes such as filtering, fluxing or holding time. For example, monitoring the aluminum quality both upstream and downstream of a filter provides valuable information about the filter operation.

An existing technique for molten metal examination involves taking "grab" samples during casting. These samples are often evaluated under a microscope at a laboratory -- a process that is subjective and time-consuming. Also, these samples may only provide an evaluation at one point in time and give no information about the rest of the drop. This can be quite misleading since distribution of impurities tend to be nonuniform throughout a casting cycle.

The Reynolds Production 4M™ System offers a means for determining the quality of molten aluminum before the casting operation on a routine basis. It is, clearly, a valuable addition to existing quality control methods. This pulse-echo ultrasonic system allows for continuous monitoring of the quality of molten metal during casting.

In the cited literature, Pitcher and Young [1-2] and others[3] show that ultrasonic techniques can be used successfully to determine the presence of nometallics, precipitation effects, and velocity of sound in molten aluminum.

For practical applications, it was found that a departure from the conventional means of ultrasonic inspection was necessary. In recent papers[4,5], a description of the prototype equipment and its capabilities was given. Reynolds involvement with the prototype system soon led to the development of a production system.

DEVELOPMENT OF THE 4M™ PROBE

Scientists at Reynolds took a new approach by developing a titanium probe[6] assembly which could be inserted into a flowing stream of molten aluminum during the casting operation. This design and construction greatly simplifies the procedures required for molten metal inspection.

The probe was designed such that it has a built-in cavity with top and bottom surfaces machined flat and parallel. The bottom surface (reflector) provides a means for continuously monitoring sound in the melt. This design allows molten metal to fill up and/or flow through the cavity. The design of the Reynolds 4M™ probe is shown in Figure 1. The patented 4M™ probe is shown in Figure 2.

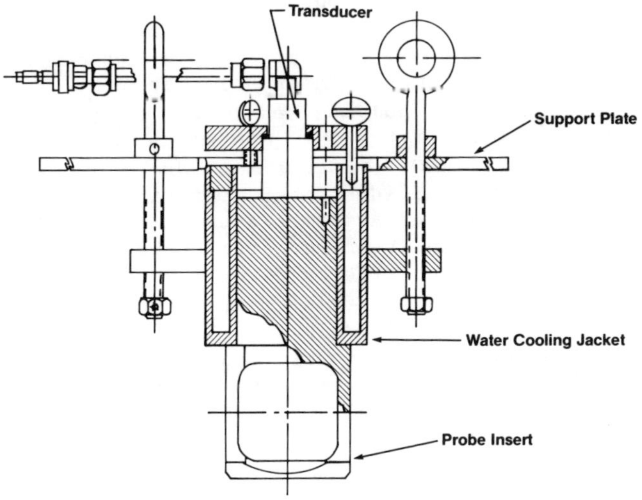

Figure 1
4-M™ Probe Assembly

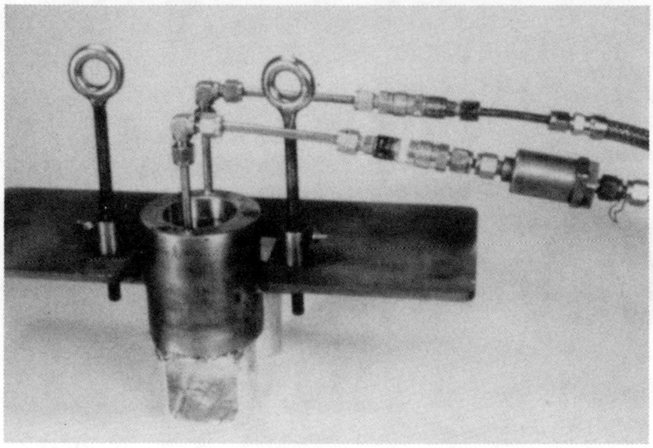

Figure 2
Reynolds 4M™ Probe

The probe can be moved about, held stationary, raised, and lowered without the occurrence of any misalignment. It is a very stable probe--thermally, ultrasonically, and physically. Probe "life" is dependent mainly on the velocity of molten aluminum flowing through the probe. Average probe "life" for most of our applications extends over 1 to 3 million lb. of aluminum inspected before replacement of the probe insert is required. Replacement of the spent probe insert is easily achieved and saves the need for replacement of the water cooling coil.

Two simultaneous ultrasonic measurements can be made with the production 4M™ System without making any changes in the probe setup. They are measurements that do not require any special preparation prior to inspection. Usually, a 15 second waiting period is all that is needed after the probe is immersed in the melt.

Figure 3 shows the probe being immersed into a flowing stream of molten metal. The ultrasonic signal is transmitted through the titanium insert into the molten metal. Two measurements from the ultrasonic signal are of interest:

(1) attenuation,

(2) discontinuity count of indications

These two measurements extracted from the ultrasonic signal provide an evaluation of metal quality. In general, an increased attenuation and/or an increased discontinuity count indicate a lesser quality of metal.[4,5]

Figure 3
4M™ Probe Assembly
Being Immersed into Flowing
Stream of Molten Aluminum

THE 4M™ SYSTEM DESCRIPTION

Reynolds developed the prototype 4M™ System in their laboratory as part of their research program, then contracted to develop the production version. Among the specification requirements were the following:

(1) Portability - could be rolled onto a rough cast house floor and negotiate small steps;

(2) Reliability - must operate in adverse environment with temperatures up to 120°F;

(3) Maintainability - must be modular in design with state of the art digital components for troubleshooting and maintenance;

(4) User-Friendly - plant personnel should not require special skills to operate the equipment;

(5) Self-Calibrating - necessary to eliminate operator-dependent calibrations and assure consistent results from unit to unit; and

(6) Flexibility - could easily be modified in the future to enhance system features or add analysis capability in software.

The project resulted in the development of the production version of the 4M™ System, shown in Figure 4 with the 4M™ probe inserted in molten aluminum.

At the top of the system is the temperature monitor. This unit attaches to a thermocouple which is inserted into the brass water jacket of the probe. The temperature of the probe assembly is constantly monitored. When this temperature exceeds the threshold set by the operator, the analysis results are printed immediately and the operator is notified.

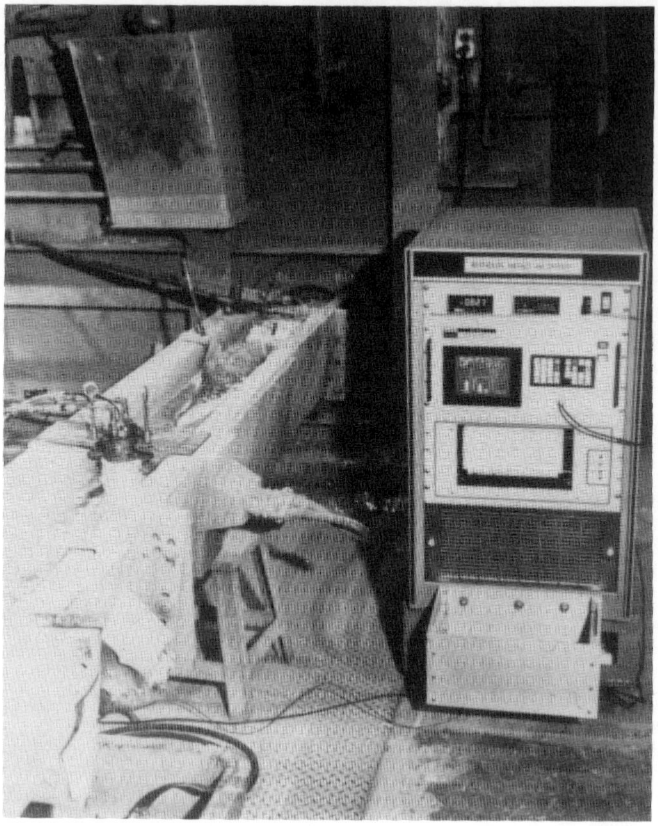

Figure 4
Reynolds Production
4M™ System

Similarly, this unit monitors the inside rack temperature to detect unacceptable operating levels.

A printer/plotter was integrated into the system to provide continuous real-time plots of attenuation and discontinuity results. A summary of these results is printed, along with identification parameter information at the conclusion of each analysis session.

At the heart of the 4M™ System is the Adaptronics ALN 5000™ Process Monitor, a microprocessor-based instrument featuring state-of-the-art high-speed arithmetic and A/D circuitry, and a built-in, digital, tone-burst pulser and receiver. The operator interfaces with the unit through a front-panel keypad and CRT. External devices for printing, plotting and temperature monitoring are controlled by ALN 5000™. The interfaces to the microprocessor or CPU (Central Processing Unit) are shown in Figure 5.

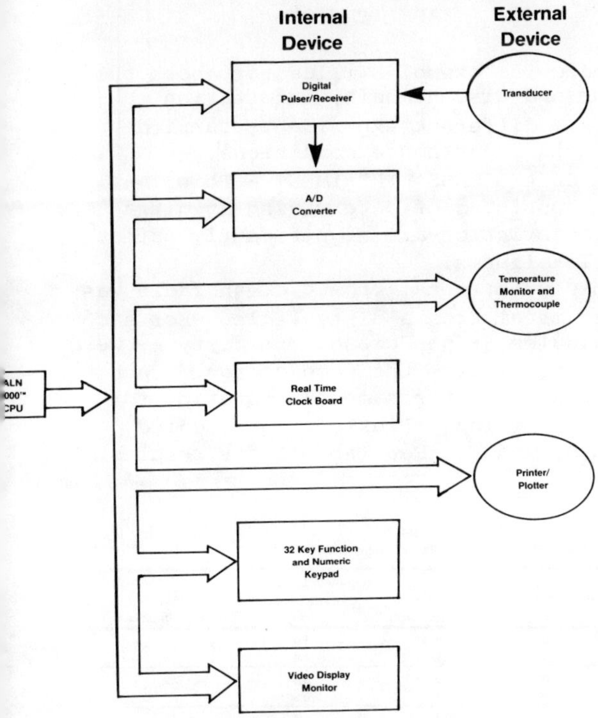

Figure 5 4M™ System Interfaces

OPERATION OF THE 4M™ SYSTEM

The system is easy to use; in many cases, the operator is prompted at each step. The front-panel keypads, shown in Figure 6, allow the operator to enter parameter values and select one of five modes of operation.

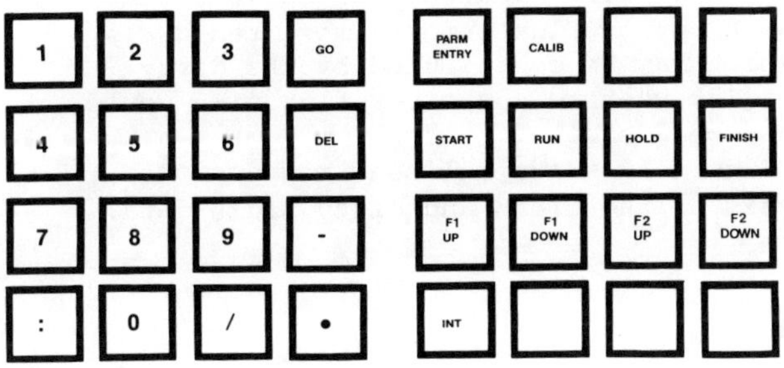

Figure 6
ALN 5000™ Front Panel

The interaction of these five modes is shown in Figure 7.

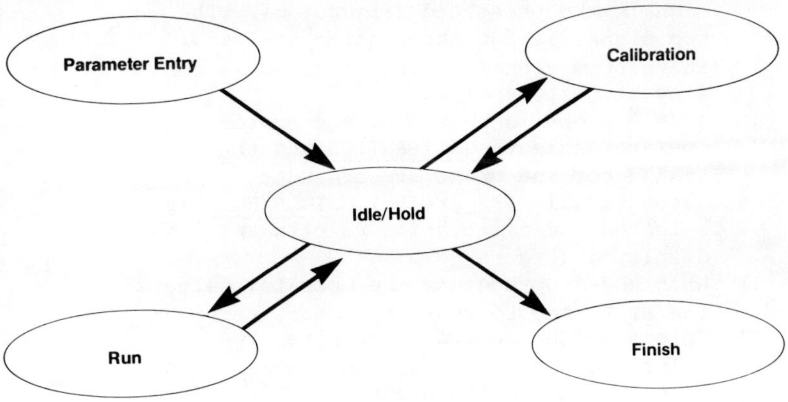

Figure 7
Five Modes of Interaction

As the operator selects one of these modes, the video display is updated appropriately.

The operating procedure is described below with references to the display screens of each mode in Figure 8.

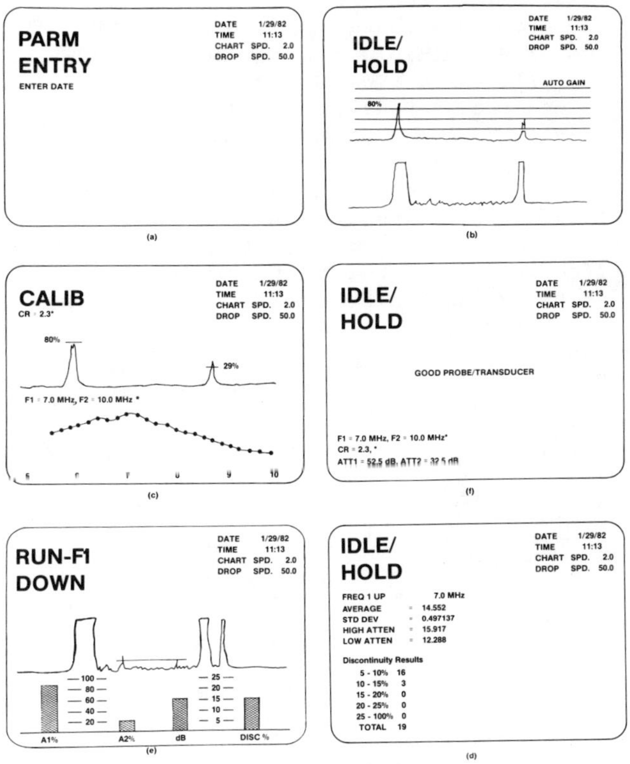

Figure 8 4M™ System Display Screens

(1) The operator powers on the system and, in the PARAMETER ENTRY mode, enters information parameters such as date, time, speed of the drop and speed of the chart recorder (Figure 8a).

(2) IDLE/HOLD is selected and two digitized signals are presented (Figure 8b). The top signal is for attenuation analysis; the bottom signal is the UT response for discontinuity analysis.

(3) A 4M™ probe is hooked to the system; the operator selects CALIBRATION and the transducer and probe are evaluated automatically (Figure 8c). IDLE/HOLD is selected and calibration results are displayed (Figure 8d).

(4) When the drop begins, the operator selects the START key to start the chart recorder (plots on the recorder are directly related to drop levels of the melt). The system is now calibrated and ready to run.

(5) The probe is positioned in the molten metal, such as upstream or downstream of a filter, and the operator presses RUN to enter the RUN mode and begin data analysis.

(6) Attenuation and discontinuity data are computed, displayed for the operator (Figure 8e) and plotted continuously on the printer/plotter.

(7) To summarize the run, the operator presses the HOLD key to print and display summary results (Figure 8f).

(8) The probe may be repositioned, and any number of multiple runs may be made at two selectable frequencies.

(9) To end the inspection, the operator presses FINISH to print final analysis totals for all runs.

This simple procedure gives the operator hard-copy output of data which can be used to determine the overall quality of the metal and the effectiveness of the cleaning processes such as filtering and fluxing. See Figure 9.

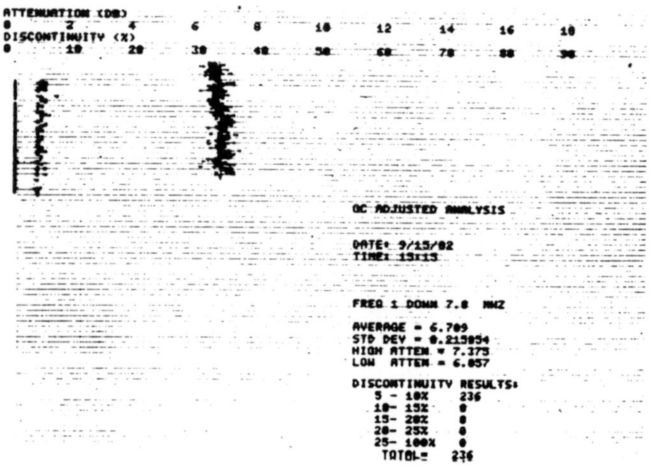

Figure 9
Hard Copy Output of 4M™ System Data

DISCUSSION: SOME EXAMPLES OF CASTHOUSE APPLICATIONS

The following examples illustrate both the attenuation and discontinuity information recorded from different experiments carried out under actual casthouse conditions.

LADLE CLEANLINESS EXPERIMENT - This experiment consisted of determining if ultrasonic characteristics are significantly affected by ladle cleanliness.

For this work, metal from a clean ladle was compared to metal from a dirty ladle. For each drop, two ladles -- one clean, one dirty -- were filled with metal from the same charge. Each ladle was fluxed at the same flux station for the same time period. Flux periods studied were A,B,C,D, and E. See Table I for results.

Table I
4M™ System Results for Clean and Dirty Ladles

Cast Series #	Ladle Condition	Flux Time (min.)	Ladle Attenuation (dB)	Standard Deviation of Attenuation (dB)	Number of Discontinuities
8	Clean	A	7.5	1.0	46
	Dirty	A	8.5	1.6	5
9	Clean	B	4.2	0.6	48
	Dirty	B	5.1	0.9	71
10	Clean	C	3.0	1.1	99
	Dirty	C	5.1	0.7	307
11	Clean	D	4.5	0.6	9
	Dirty	D	6.6	1.0	17
12	Clean	E	2.9	0.6	2
	Dirty	E	2.4	0.5	4

Two important facts emerged from this work:

A. Metal processed in clean ladle produced better ultrasonic quality metal (lower attenuation) than metal processed in dirty ladles for flux periods ranging from A to D.

B. For the E flux time, there was no real difference between the metal poured from clean or dirty ladles. We believe that an E period flux removes most residuals which may be contributed by the dirty ladle, whereas a shorter flux time would not be adequate.

Figure 10 shows a linear regression analysis of the clean and dirty ladle data.

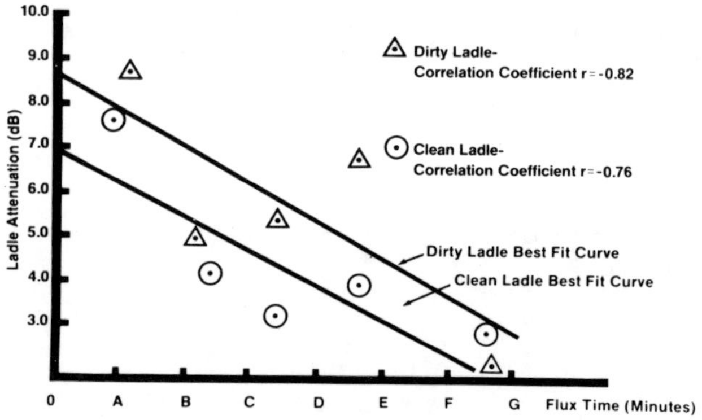

Figure 10
Ladle Attenuation (dB) Versus Flux Time
For Clean and Dirty Ladles

Statistical analysis[7] for this data shows:

A. Good correlation between attenuation and flux time for clean ladles (r= -0.76).

B. Good correlation between attenuation and flux time for dirty ladles (r= -0.82).

Using this information, an additional statistical analysis was conducted to determine if the difference between metal produced from clean ladles fluxed from A to D minutes was really different from metal produced from dirty ladles fluxed from A to D minutes. The clean and dirty ladle data were compared using a "t" test.[7] This statistical test yielded a "t" value of -4.78. To say, with a 98% confidence level, that the clean ladle metal is really different from the dirty ladle metal requires that the "t" test yield $|t| > 4.541$. Therefore, we can say with a high degree of confidence (98%) that clean ladles should be used when the metal is fluxed for a period of time D or less. Metal poured from these ladles would be of better ultrasonic quality than metal poured from dirty ladles.

FLUXING TIME EXPERIMENT - This experiment consisted in determining if ultrasonic characteristics are significantly affected by increasing the time a ladle is fluxed. Both clean and dirty ladles were used in this experiment.

For this work, several drops were monitored in which flux times were varied. The 4M™ System results indicated a trend of lower attenuation and lower standard deviation of attenuation as flux times are increased. The ultrasonic quality is not only better but also more consistent.

If we examine the 56 ladle attenuation results in terms of "periods" of flux time we see in Figure 11 that for short flux periods the mean attenuation and the standard deviation are greatest. For increased flux periods, the mean attenuation and standard deviation decline significantly. Low attenuation and low standard deviation indicate better ultrasonic quality.

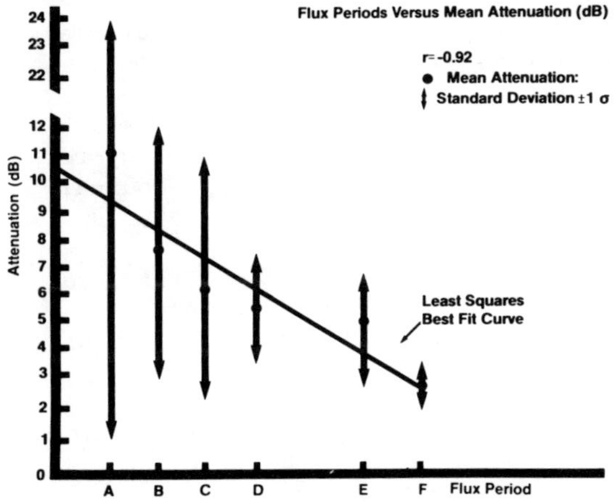

Figure 11
Results for Various Flux Periods

We conclude, therefore, that longer flux time is very important in improving the ultrasonic quality of the molten metal. For the flux times we analyzed, those times between Flux Period E&F were preferable -- ladle attenuations were low and consistent.

HOLDING TIME EXPERIMENT - This experiment consisted in determining if ultrasonic characteristics are significantly affected by increased holding time for ladles.

Holding time is the period between when a fluxed ladle is skimmed and when the ladle is first poured. In this study fourteen (14) ladles were monitored.

Figure 12 is a graphical representation of holding time versus ladle attenuation. It is apparent that ladle holding time shows little correlation with overall ultrasonic metal quality. The flux time for each ladle involved in this work was held constant.

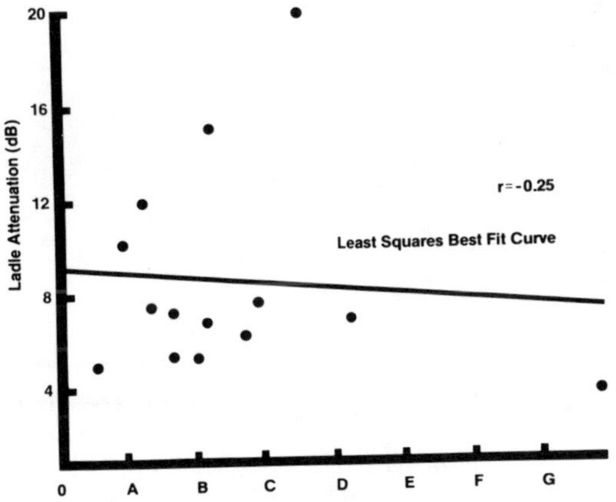

Figure 12
Ladle Holding Time Versus Ladle Attenuation

It is worth mentioning at this point that near the end of a ladle, and on occasion during the tilting of the ladle, ultrasonic metal quality often decreases. For example, attenuation and discontinuity counts increased due to residual insolubles being released when the ladles were being completely drained. Figure 13 is a recording of such a case and depicts the changes which occurred when the last remaining metal was poured from a ladle. This, we feel, in all probability, is the reason for no apparent correlation between holding time and the ultrasonic data.

If we examine the ladle situation closer, we find that three conditions can exist for exogenous material within a ladle of molten aluminum. The first being that some exogenous material will "float" to the metal surface; the second, some will stay in suspension; and third some will "sink" to the bottom. For all practical purposes, we feel that regardless of holding time in a ladle, sooner or later most of the molten aluminum and exogenous material will be poured from the ladle into the transfer trough. For critical product quality, it may be advisable to have extended holding times and partial pouring from ladles so that residual insolubles which may have collected at the ladle bottom will be retained in the ladle.

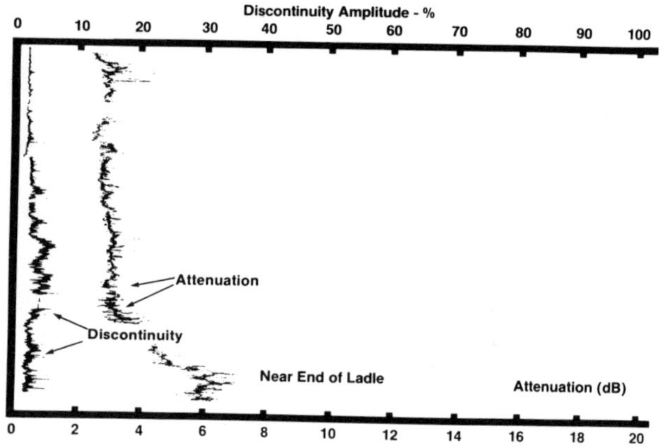

Figure 13
4M Characteristics of Metal When Ladle is Completely Drained

SUMMARY

The development of the Reynolds 4M™ Production System is a valuable addition to existing molten aluminum quality control methods. This pulse-echo ultrasonic system allows for continuous monitoring during casting and the evaluation of molten aluminum processing and/or cleaning operations. The molten metal monitor is a microprocessor based instrument featuring state-of-the-art high speed arithmetic and analog to digital circuitry, and a self-contained tone-burst pulser/receiver.

The primary benefits of the 4M™ System can be summarized as follows:

- The quality of metal is determined during the casting process. This saves time and expense by identifying poor quality metal before submitting the material to the final processing stages.
- The presence of inclusions and non-metallic particles is determined, and these defects can be directly related to locations in the solidified ingots. Defective areas can be removed or delegated to less critical applications.
- For the first time, a method is available for online evaluation of cast house processes such as filtering, fluxing and holding time.
- The automatic calibration and probe/transducer evaluation features are very important. The system can be operated by plant personnel without special ultrasonic training and the inspection results are independent of operator variations and, therefore, consistent with each 4M™ System.

The 4M™ System is software-intensive; system functions and computations are implemented in software. This provides for flexibility and adaptability since new functions or controls can be supported by modifications to the existing software without redesign of hardware components.

ACKNOWLEDGMENT

Many people have contributed to this work. The contributions of Messrs. C. E. Burley and H. A. Lecky, Reynolds Metals Co., and Messrs. J. R. Gouge and A. N. Mucciardi of General Research Corporation are particularly acknowledged.

REFERENCES

[1] Pitcher, D. E., and R. S. Young, "Improvements in or Relating to Methods and Apparatus for Testing Molten Metal," Great Britain Patent, Patent 1,127,231 (Sept. 18, 1968).

[2] Pitcher, D. E., and R. S. Young, "Methods of and Apparatus for Testing Molten Metal," U. S. Patent 3,444,726 (May 20, 1969).

[3] Stremousov, V. I., and V. V. Tekuchev, "Velocity of Ultrasound in Molten Binary Alloys of Aluminum with Tin and Zinc," Zhurnal Fizicheskoi Khimii, Vol. 50, 1976, pp. 2373-2375.

[4] Mansfield, T. L., "Ultrasonic Technology for Measuring Molten Aluminum Quality," Materials Evaluation, Vol. 41, Number 6, May, 1983, pp. 743-747.

[5] Mansfield, T. L., "Ultrasonic Technology for Measuring Molten Aluminum Quality," Journal of Metals, September, 1982, pp. 54-57.

[6] Mansfield, T. L., "Probe for the Ultrasonic Inspection of Aluminum," U. S. Patent 4,261,197 (April 14, 1981) and U. S. Patent 4,287,755 (September 8, 1981).

[7] Bowker and Lieberman, Engineering Statistics, 2nd edition, 1972, Prentice-Hall, Inc., Englewood Cliffs, New Jersey.

A STUDY OF CONTINUOUS/DISCONTINUOUS CHIP FORMATION USING ACOUSTIC EMISSION

David A. Dornfeld, Associate Professor
Chung-Shih Pan, Research Assistant
Dept. of Mechanical Engineering
University of California
Berkeley, CA 94720

ABSTRACT - Chip formation control is an important problem in unmanned machining operations. Short, discontinuous chips are often most desireable to avoid entanglement with tooling and to aid with mechanized removal systems. Chip formation conditions can change during machining, especially with single-point turning. Experiments were conducted on a machining center. Conditions of feed-rate-induced segmented chips correlate well with the count rate of acoustic emission (AE). The sensitivity of AE signals to chip congestion or entangling due to continuous chip formation is illustrated.

INTRODUCTION - In automated factories unmanned machine tools are required to operate for long periods of time with little or no operator supervision. Under these circumstances it is crucial that chip control, specifically producing discontinuous chips, be maintained. Continuous chips can cause catastrophic failures of the tooling, entangle the workpiece causing damage, as well as render mechanized chip disposal systems less efficient. Because of this problem, called one of the most serious problems in machine tool automation by Nakayama [1], many sensing devices have been proposed to detect chip form. These sensors primarily utilize variations in cutting forces [2] or heat radiation from chips [3] to detect chip breaking. This paper describes the results of recent studies using acoustic emission signal analysis to study chip formation during machining as well as provide information on chip congestion or entangling due to continuous chip formation. A comparison to variations in forces generated during machining due to chip formation is made. The work reported here is the result of continuation of a study of chip form detection during the machining of low carbon steel [4]. Acoustic emission (AE), i.e. stress waves spontaneously released in materials undergoing deformation, fracture, or both, are usually of two distinct types, continuous type signals which arrive at the transducer in large numbers (and for which distinct events cannot be detected) and burst type signals in which distinct emission events are seen. During machining, continuous-type AE signals are generated due to plastic deformation in the shear zone, at the chip-tool interface and at the tool flank-workpiece interface [5,6], while burst-type signals are generated due to chip breakage during or after chip formation [4]. Thus, the analysis of the acoustic emission signals can provide information on the chip formation during machining. Under relatively stable cutting conditions the chip form is still rather unstable and can easily vary due to [1]:

- variation of the force on the chip (due to gravity, increased length, or chip breaker influence)
- changes in temperature of the process (due to lubrication removal or tool wear)
- changes in tool geometry (due to tooling modification or wear)
- variation in work material property (as a result of through thickness hardness changes or prior surface cold work)

All of these influences are present in the machining operation. In the case of the unmanned operation where other conditions are carefully controlled, the changes due to tool wear may be most prominent. As the work material passes through the plastic region (known as the shear zone) during machining to form a chip the material is increasingly strain hardened. The chip breaks if a fracture occurs in the plastic zone and propagates through the deformed material. As the cutting tool wears the temperature of the chip-tool and tool-work interface increases due primarily to the increase in energy dissipated as friction as areas of contact grow. The ultimate strain of chips increases significantly for various metals above 400°C [3], thus, with wear induced temperature increases in the tool and chip, continuous chips can result where discontinuous chips had been originally produced. It is necessary, then, to be able to detect whether or not discontinuous chips are being formed.

ACOUSTIC EMISSION AND CHIP FORM - Chips are commonly described as either continuous, discontinuous, or segmented (i.e. partially fractured) [7,8], see Figure 1. The fracture effective in producing discontinuous or segmented chips occurs under conditions of relatively high stress and results in the creation of surfaces and the accompanying generation of acoustic emission. This AE is seen as burst emission superimposed on a background of continuous AE due to the machining process itself, Figure 2. The root mean square (RMS) energy of acoustic emission generated during a machining process producing a continuous chip is shown in Figure 3. In the event of a tangle, a random, high amplitude noisy emission is generated due to the chip impacting the tooling and is also seen in Figure 3. When discontinuous chips are formed the RMS AE signal, Figure 4, appears similar to that shown schematically in Figure 2. Using a suitable threshold, for example, the average event rate of the emission can be determined and will be shown to be related to the rate of chip breaking.

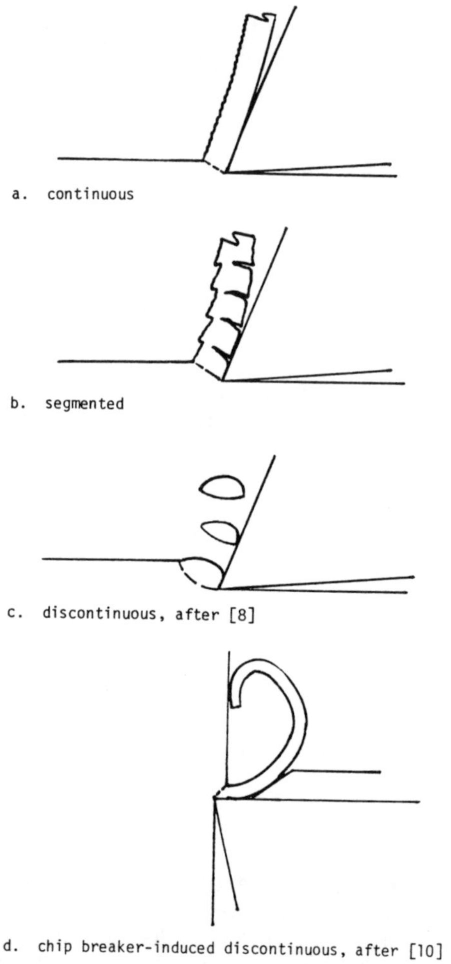

Figure 1. Chip forms in machining

EXPERIMENTAL METHOD - In this investigation the conditions for chip formation were altered by varying the feedrate of the cutting operation. This is an effective and straightforward method for chip form variation. For a chip length represented as a simple beam, the strain in the outer layer of the chip can be expressed as a function of chip thickness, t, (related to the uncut chip thickness, or feed, in orthogonal machining by the shear angle and tool rake angle) and the radius of curvature of the chip, ρ, as follows [10]:

$$e = \frac{t}{2\rho}$$

If this strain is sufficiently high the chip will fracture on exiting the cutting zone. As can be seen from this relationship, either an increase in feed or decrease in chip radius of curvature will encourage chip segmentation. Increases in feed have been shown to substantially reduce the chip curvature [11].

The tests were done on a Tree 1000 CNC lathe. The tool used is a Kennametal TPG 322 grade K68 carbide insert with CBT-12F chip breaker in a KTAR-163C tool holder (5° rake). An AE transducer (Dunegan 9201A) is mounted at the rear of the tool holder which is then clamped on a specially designed fixture with a built-in force transducer (Kistler 9251A) to measure the tangential and feed cutting forces. The raw AE signal is recorded on a modified videotape recorder and the RMS of AE and the force measurements are digitized by a LSI 11/03 computer with a sampling frequency of 1kHz.

The cutting conditions used in the investigation are as follows: 1500 sfpm; .050 in. depth of cut; .002 to .020 in/rev (ipr) feedrate; no coolant. The material cut is 6061-T651 aluminum rod for which, in general,

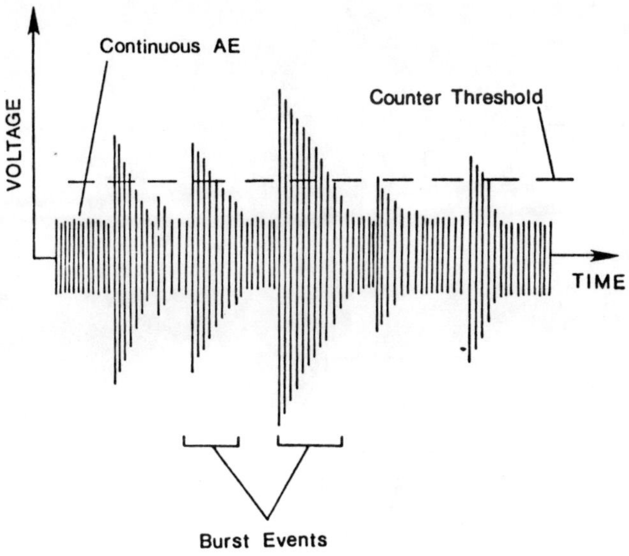

Figure 2. Acoustic Emission Signal Nomenclature

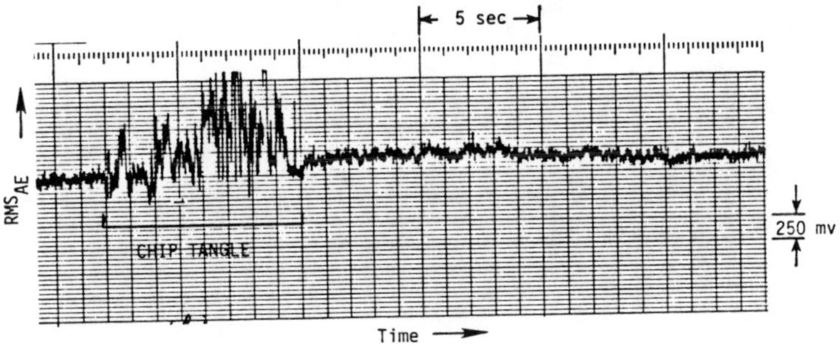

Figure 3. Acoustic Emission from Continuous Chip Formation with Chip Tangle

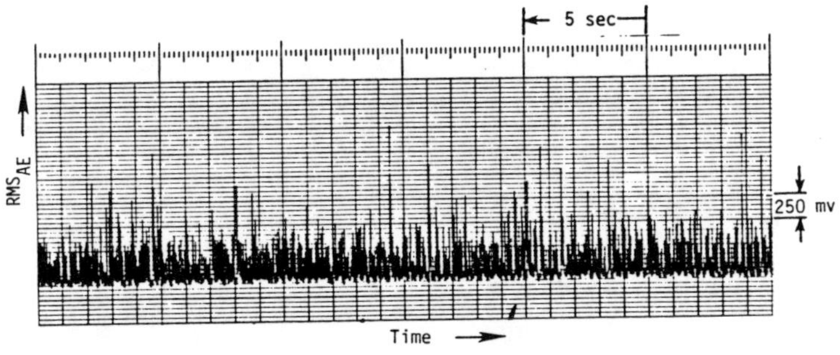

Figure 4. Acoustic Emission from Discontinuous Chip Formation

it is harder to produce discontinuous chips than for low carbon steel machining done in the previous study [4].

RESULTS AND DISCUSSION - The average chip breaking frequency is calculated from the average number of chips produced per unit of time. This can be determined by collecting the chips produced for each cutting test and counting the number of chips, n, and measuring the total weight of chips, w. The chip breaking frequency, f, can then be simply calculated by:

$$f = \frac{vdrp}{w/n}$$

where: v = cutting velocity
d = depth of cut
r = feed rate
p = material density

The table below shows the average chip breaking frequency and chip breaking conditions for each different feed rate of cut. As seen from the table, the chip breaking frequency increased as the feed rate increases. The chips also change from continuous type to discontinuous type at feed rates above .006 ipr.

The raw AE signal was analyzed using an AET 5000 acoustic emission system to measure the event rate. Each significant burst event in the AE signal can be related to the chip breaking process due to the instantaneously large energy release at chip fracture compared to AE generated during normal cutting. The count rate of burst events above a certain threshold excludes the continuous type AE signal contributed by normal cutting and is a good indicator of the chip breaking frequency. In practice it is desired to know only whether or not chips are being broken and not the frequency. For the purpose oт the research conducted here frequency is also of interest.

Figure 5 shows a typical plot of the event rate vs time for 0.014 ipr feed rate. The event rate is approximately constant in a uniform cut. After several trials of different thresholds for each cut, an appropriate threshold can be selected to give an average RMS event rate matched to the average estimated chip breaking frequency. Figure 6 shows a good correlation between averaged event rate and calculated chip breaking frequency that confirms the earlier findings of Dornfeld and Lan [4] for low carbon steel.

The RMS of AE for normal cutting is, in general, of constant level. Therefore, the spikes in the RMS of AE can be related to the burst type AE which is also related to the chip breaking process. Software was developed to count the spike rate of the RMS of AE signal above a certain threshold level which,

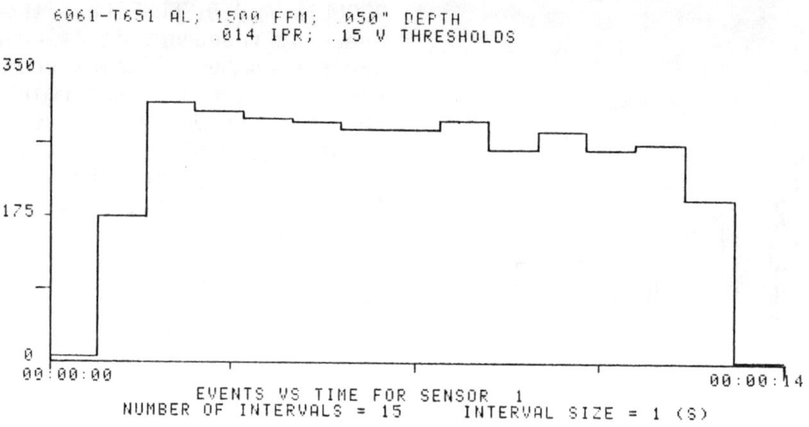

Figure 5. AE Event Count

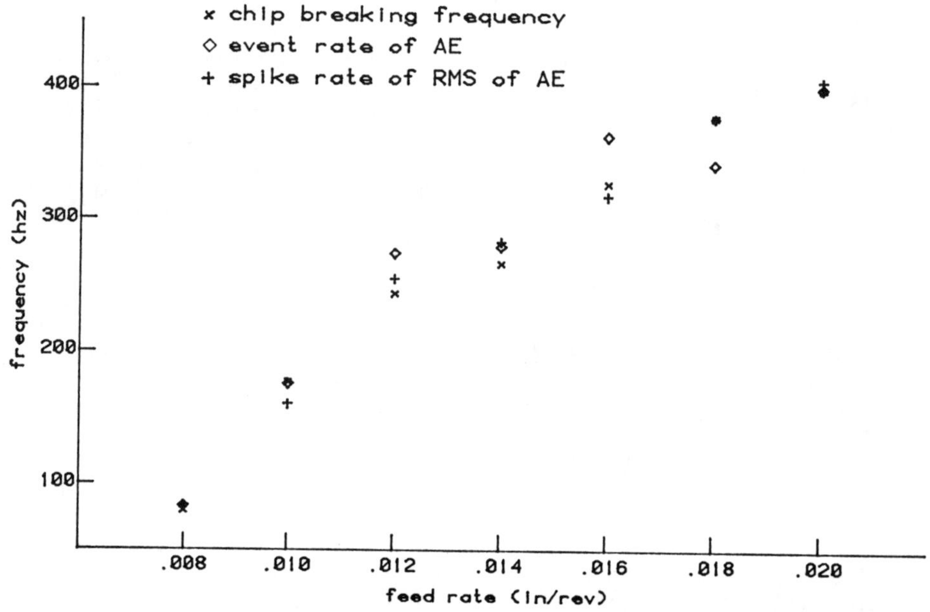

Figure 6. Average Event Rate, Average RMS Spike Rate and Estimated Chip Breaking Frequency vs. Feed Rate

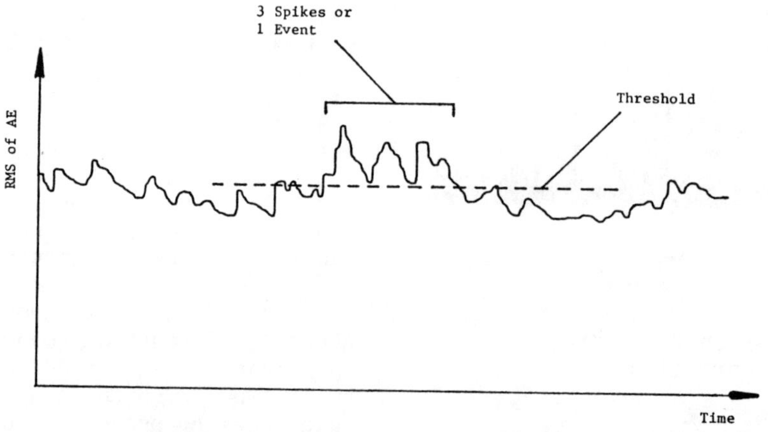

Figure 7. Schematic of Event Count & Spike Count of RMS of AE

theoretically, excludes the normal cutting contribution in energy. This is illustrated in Fig. 7 for a "well behaved" signal. Since one major goal of this investigation was determination of the chip breaking frequency from analysis of the AE signal energy, different threshold levels were selected as optimum for different cutting conditions to best achieve this goal. The ratio between average level of RMS of AE signal and the successful threshold selected for each cut is given in Fig. 8. The variation in the ratio is approximately 10% indicating there is a good possibility of selecting one average multiplier value as an initial threshold value for detecting continuous/discontinuous chip formation since the demands of the task are less stringent than measuring chip frequency.

An FFT analysis of the RMS and AE and feed force was made to determine if a dominant frequency is present that correlates to chip breaking. A relative peak can, in general, be found near the chip breaking frequency, but it is not distinguishable from the whole spectrum, Figs. 9 and 10 for AE energy and feed force, respectively. As in the earlier work [4] the FFT analysis of RMS of AE or feed force does not seem too promising at this time.

The ability to detect the presence of a chip tangle due to "stringy" or continuous chip formation is also of great interest for machining process monitoring. A typical RMS AE signal for continuous chip cutting process is shown in Fig. 11. At the beginning of cutting, the chip flows smoothly over the tool surface and flows away. The corresponding RMS of AE is nearly constant without spikes. After approximately .5 seconds the chip grows too long and begins tangling between the tool and the workpiece. Figures 12 and 13 show force data (feed and tangential, respectively) for the same test. No such dramatic change in the force level corresponding to tangle is seen. A sudden rise in the RMS of AE is seen and several spikes are generated. From this we can see that there exist three different characteristics of RMS of AE which correspond to three different chip form conditions. In brief, they are:

1. A constant and relatively uniform level RMS of AE corresponding to continuous chip formation,
2. a constant level RMS of AE with rather regular spikes corresponding to discontinuous chip cutting, and
3. a relatively noisy RMS or AE with sudden increase in level corresponding to chip congestion or other undesirable chip variations during cutting.

Although not studied in this investigation, prior work of Dornfeld [12] has shown that for segmented chip formation the RMS of AE behaves similar to that in category 2 above except that, due primarily to reduced fracture area, the event spikes in AE due to partial chip fracture are not as pronounced.

Based on this categorization, a methodology is proposed for distinguishing the three different chip conditions in cutting based on RMS of AE. The average level of RMS of AE signal for each cut is used as a reference in order to determine more quickly the threshold. Further, the number of events instead of spikes are counted per unit time for the RMS of AE above the threshold in Fig. 7. Figure 14 shows the event rate for each cut using thresholds of 1.0, 1.1 and 1.2 times (threshold multipliers) the average RMS of AE respectively. For a threshold multiplier of 1.1 a very low rate (below 10/sec) indicates continuous chip cutting. Since the RMS meter used had a 200 hz response rate, chip breaking information above 100 hz is not seen. Thus, the event rates for feed above .010 ipr are similar indicating only that the chips are discontinuous but not the approximate chip breaking frequency. The difference with respect to the continuous chip (transition from continuous to discontinuous at feed rates of .006-.008 ipr) is clearly seen. For these machining conditions a threshold multiplier of 1.1 provides the best variation of event rate with feed and most distinctly separates continuous from discontinuous chip formation.

For the threshold multiplier of 1.0 most of the AE signal contributed by normal cutting is not excluded so very high event rates are seen in Fig. 14 for all tests. The multiplier is related to the material being cut, cutting tool and transducer set up. At present the multiplier can only be obtained empirically but it will provide a quick way to determine the threshold for operations in which the same set-up and material are used under a range of cutting conditions.

The AE RMS event rate response under chip tangling conditions has not yet been quantified. A substantial increase in the event rate over background levels would occur for a tangle as seen in Figure 11.

As a comparison, a similar thresholding technique was also applied to determine "event rates" from the tangential and feed force as shown in Figs. 15 and 16. It is apparent that tangential force does not serve the purpose of distinguishing different chip forming conditions very well. The event rate of feed force with a threshold multiplier of 1.2 shows good ability to distinguish between continuous and discontinuous chip conditions (transition at .006 ipr feed). The event rate also increases as the feed rate increases closely

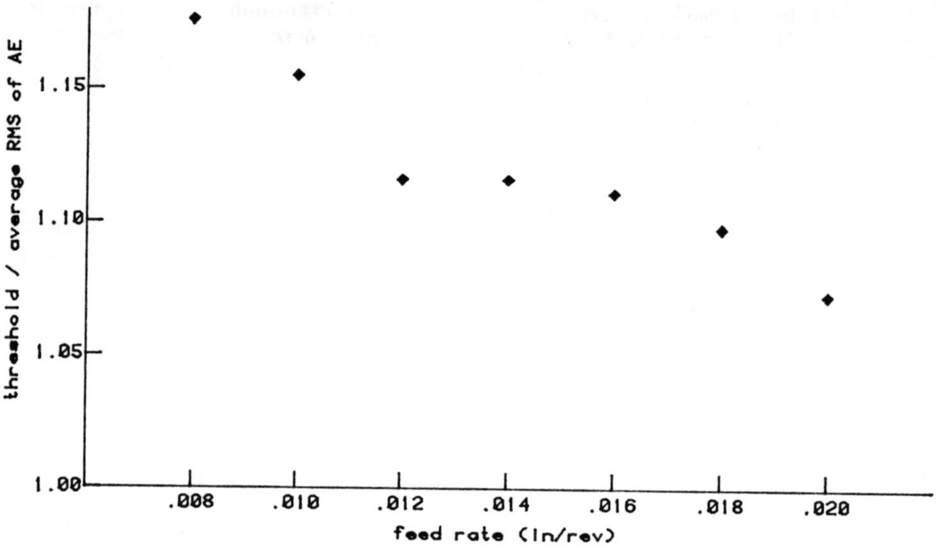

Figure 8. Ratio of Threshold to Average RMS vs. Feed Rate for Best Chip Breaking Frequency Estimation

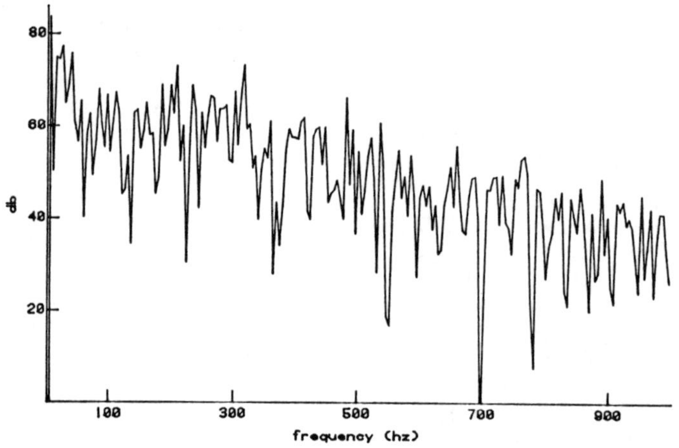

Figure 9. Power Spectrum of RMS of AE for the Machining at .014 ipr Feed Rate

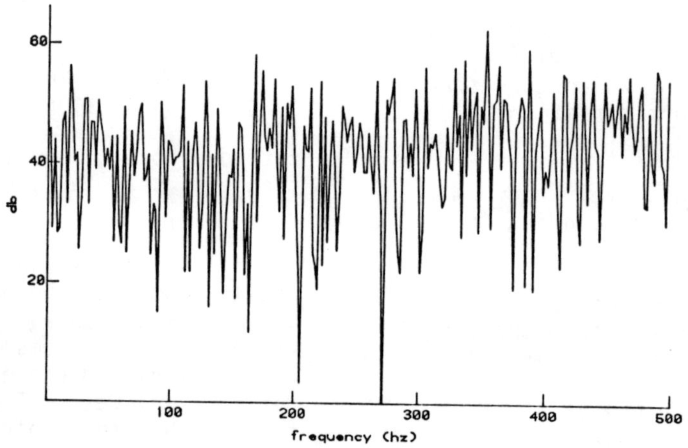

Figure 10. Power Spectrum of Feed Force for Machining at .014 ipr Feed Rate

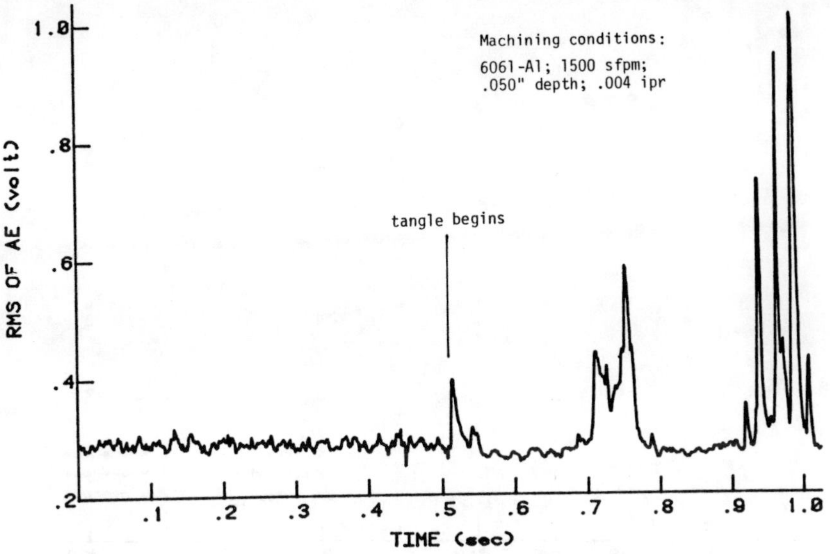

Figure 11. RMS AE vs. Time with Chip Tangle at .5 Seconds

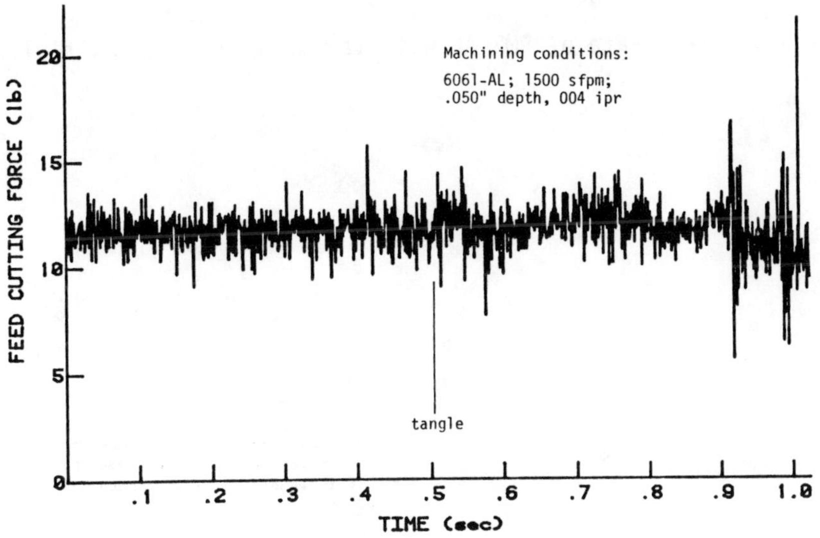

Figure 12. Feed Force with Chip Tangle

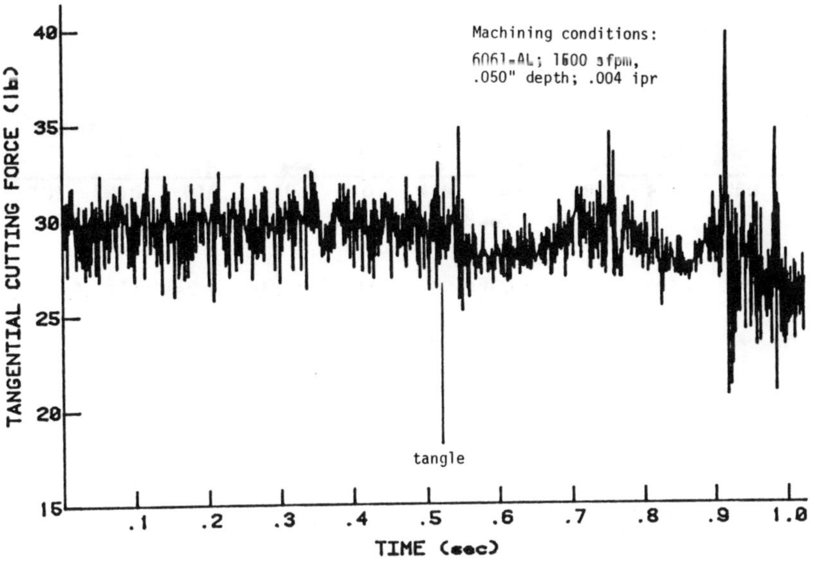

Figure 13. Tangential Force with Chip Tangle

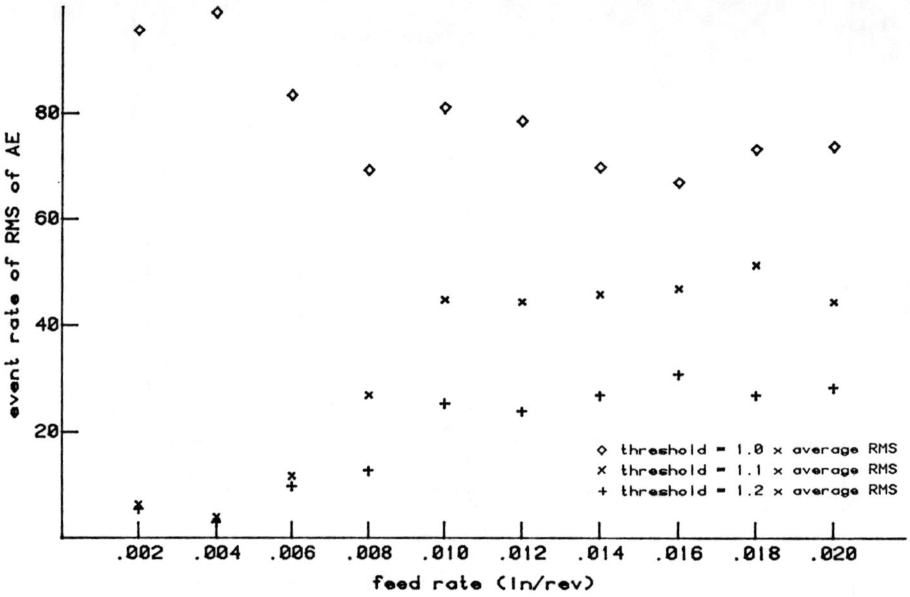

Figure 14. Event Rate of RMS AE with Various Threshold Multipliers

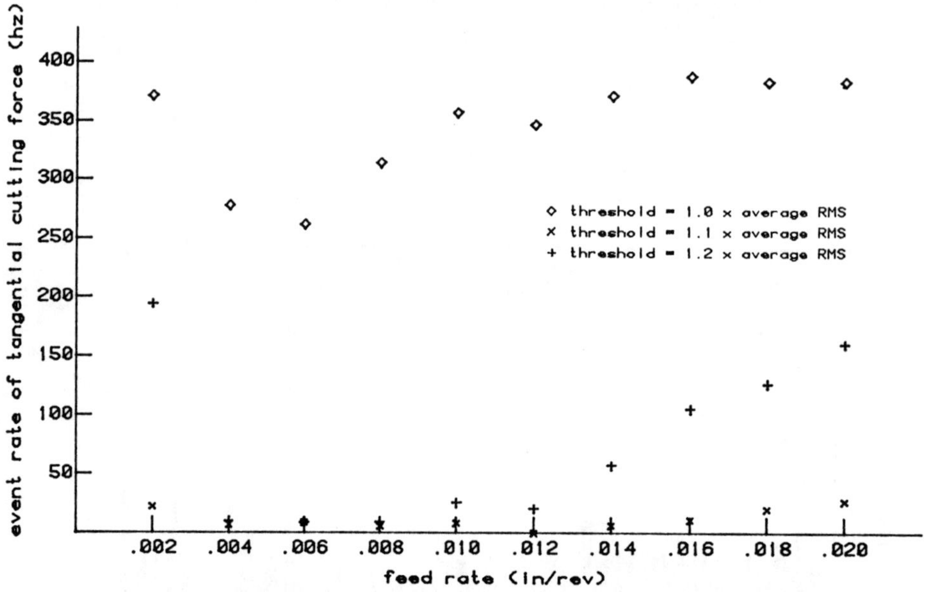

Figure 15. Event Rate of Cutting Force with Various Threshold Multipliers

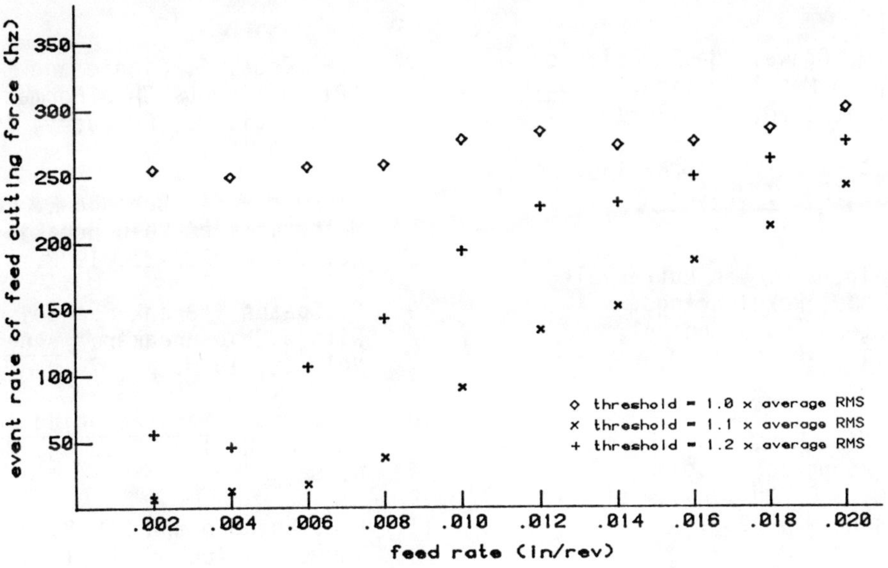

Figure 16. Event Rate of Feed Force with Various Threshold Multipliers

matching the chip breaking frequency. Here a 1000 hz sampling rate (substantially higher than the maximum chip breaking frequency) was used for digitizing force data so good correlation over the entire range is seen (compare to similar AE based data in Figure 6). The feed force may also have potential as a chip form detector although sensitivity to parameter changes (speed, depth, etc.) have not been determined and the machining conditons used here have yielded a "well behaved" force response not necessarily seen under heavier cutting conditions. The RMS of AE is not adversely affected by the more severe cutting conditions likely to reduce the effectiveness of force-based measurements.

CONCLUSIONS - The work done to date has provided the background for utilizing the analysis of acoustic emission signals in chip form detection. Research in progress at the University of California is utilizing this technique for investigation of the mechanics of discontinuous and segmented chip formation. As a result of the study reported here and earlier work [4], the following conclusions can be made:

1. The event rate of AE RMS energy shows excellent correlation to the estimated chip-breaking frequency for low carbon steel and 6061 Aluminum work materials for an appropriate threshold multiplier (1.1 for aluminum),

2. the AE RMS event rate reliably indicates the transition from continuous to discontinuous chip formation thus can indicate the presence or absence of chip breaking activity,

3. the AE RMS is sensitive to chip tangle formation during continuous chip machining and shows greater sensitivity to tangle than machining force data under similar conditions,

4. under the conditions tested here (aluminum work material) force "event rate" data correlates well with estimated chip-breaking frequency for an appropriately selected threshold multiplier, and

5. machining acoustic emission and force power spectra are not sufficiently sensitive to indicate chip breaking frequency.

ACKNOWLEDGEMENTS - Contribution of tool holders and inserts by Kennametal, Inc., Latrobe, Pennsylvania, is gratefully acknowledged. The assistance of Ms. Pat Boughtin in preparing the manuscript, Mssrs. Don Blomquist and Ken Yee of the National Bureau of Standards for helpful discussion and instrumentation development, and Dr. M.S. Lan is appreciated. This work is supported in part by a grant from the Production Research Division of the National Science Foundation, No. MEA-8112843 and a grant from the Automated Production Technology Division of the National Bureau of Standards.

REFERENCES -

1. K. Nakayama and M. Ogawa, "Basic Rules on the Form of Chip in Metal Cutting," Annals of the CIRP, 27, 1, 1978, pp. 17-21.

2. "Automatic Detection of Chip Breaking," review of article from Industrie-Anzieger, 101, 59, July 1979.

3. W. Kluft, W. Konig, C.A. van Luttervelt, K. Nakayama and A.J. Pekelharing, "Present Knowledge of Chip Control," Annals of the CIRP, 28, 2, 1979, pp. 441-455.

4. D. Dornfeld and M.S. Lan, "Chip Form Detection Using Acoustic Emission," Proc. 11th NAMRC, SME, University of Wisconsin, Madison, May 1983, pp. 386-389.

5. E. Kannatey-Asibu, Jr., and D.A. Dornfeld, "Quantitative Relationships for Acoustic Emission from Orthogonal Metal Cutting," Trans. ASME, J. Eng. Ind., 103, 3, 1981, pp. 330-340.

6. M.S. Lan and D.A. Dornfeld, "Experimental Studies of Tool Wear via Acoustic Emission Analysis," Proc. 10th NAMRC, SME, McMaster University, Hamilton, Ontario, May 1982, pp. 305-311.

7. N.H. Cook, I. Finnie and M.C. Shaw , "Discontinuous Chip Formation," Trans. ASME, Vol. 76, February, 1954, pp. 153-162.

8. S. Kaldor, A. Ber and E. Lenz, "On the Mechanisms of Chip Breaking," ASME Paper No. 78-WA/PROD-21, 1978.

9. C. Spaans, "Break Mechanism in Cutting with a Chip Breaker," Annals of the CIRP, Vol. 18, 1970, p. 87.

10. N.H. Cook, Manufacturing Analysis, Addison-Wesley, Reading, Mass., 1966, p. 58.

11. H.E. Enahoro and P.L.B. Oxley, "An Investigation of the Transition from a Continuous to a Discontinuous Chip in Orthogonal Machining," Int. J. Mech.Sci., Vol. 3, 1961, pp. 149-150.

12. D.A. Dornfeld, "Investigation of Machining and Cutting Tool Wear and Chatter Using Acoutic Emission," Review of Progress in Quantitative NDE, Plenum Press, 1981, pp. 475-483.

FEED RATE (IN/REV)	ESTIMATED CHIP BREAKING FREQUENCY (HZ)	TYPE OF CHIP
.002	0	CONTINUOUS
.004	0	"
.006	0	"
.008	79.5	DISCONTINUOUS
.010	176.4	"
.012	243.9	"
.014	266.6	"
.016	327.0	"
.018	377.6	"
.020	399.4	"

TABLE

FEED RATE AND CORRESPONDING ESTIMATED CHIP BREAKING FREQUENCY AND TYPE OF CHIP